REED
PROPERTY &
CONSTRUCTION

E
EM

P
D

&
NG

LEVEL 3

B|A
C|H
BRITISH ASSOCIATION
OF
CONSTRUCTION HEADS

OXFORD
UNIVERSITY PRESS

OXFORD
UNIVERSITY PRESS

Great Clarendon Street, Oxford, OX2 6DP, United Kingdom

Oxford University Press is a department of the University of Oxford. It furthers the University's objective of excellence in research, scholarship, and education by publishing worldwide. Oxford is a registered trade mark of Oxford University Press in the UK and in certain other countries.

© Oxford University Press 2015

The moral rights of the authors have been asserted

First published in 2015

British Library Cataloguing in Publication Data
Data available

978-1-40-852697-2

10 9 8 7 6 5 4 3 2 1

Paper used in the production of this book is a natural, recyclable product made from wood grown in sustainable forests. The manufacturing process conforms to the environmental regulations of the country of origin.

Page make-up by GreenGate Publishing Services, Tonbridge, Kent
Printed in China by 1010 International

Acknowledgements

The publishers would like to thank the following for permissions to use their photographs:

Cover image by **Lucidio Studio Inc./Getty Images**

© Adrian Sherratt/Alamy: 10.02; © Andreas von Einsiedel/Alamy: 7.012, 10.07; © Arcaid Images/Alamy: 10.04, 10.06; © Augusto Colombo - ITALIA-/Alamy: 8.034; © Avico Ltd/Alamy: 5.75; © Building Image/ Alamy: 3.14; © Charles Stirling/Alamy: 5.74; © David Lyons/Alamy: 8.030; © Ian Francis/Alamy: 5.68; © Matthew Richardson/Alamy: 5.55; © Midland Aerial Pictures/Alamy: 3.03; © paul eccleston/Alamy: 7.014; © V&A Images/Alamy: 7.004; © Zoonar GmbH/Alamy: 7.010; © ZUMA Press, Inc/Alamy: 2.01; **2013 © Energy Saving Trust**: 3.16; **A breche violette table top (marble), European School, (18th century)/ Private Collection/Photo © Christie's Images/Bridgeman Images**: 8.037; **AlexRoz/Shutterstock**: 6.39; **Anton Gvozdikov/Shutterstock**: 6.01; **architetta/iStockphoto**: 2.12; **arne thaysen/Getty Images**: 7.011; **BanksPhotos/iStockphoto**: 3.10; **bjeayes/iStockphoto**: 3.04; **BSA**: 2.05; **Calek/Shutterstock**: 3.11; **Courtesy of Anaglypta**: 7.001, 7.002; **Courtesy of Chuck Zayat, paintshaver.com**: 5.65; **Courtesy of RAL Color**: Table 6.4, page 201; **Cynthia Farmer/Shutterstock**: 4.15; **David Bishop Inc./Getty Images**: 8.058; **dem10/iStockphoto**: 5.47; **Dmitriev Lidiya/Shutterstock**: 5.50; **e-paint.co.uk [granted by Patrick Benson Sales@e-paint.co.uk]**: 6.41, 6.42; **Ella Sayce/ Oxford University Press**: 6.49, 6.50, 6.51, 6.52, 6.53, 6.54, 6.55, 7.000, 7.007, 7.008, 7.041, 7.042, 7.043, 7.044, 7.048, 7.069, 7.070, 7.071, 7.093, 7.094, 7.095, 7.096, 7.097, 7.098, 7.099, 7.100, 7.101, 7.102, 7.103, 7.104, 7.105, 7.115, 7.116, 7.117, 8.000, 8.012, 8.013, 8.014, 8.015, 8.028, 8.031, 8.032, 8.044, 8.050, 8.051, 8.052, 8.054, 8.056, 8.057, 8.063, 8.064, 8.065, 8.066, 8.067, 8.083, 8.095, 8.096, 8.097, 8.098, 8.099, 8.100, 8.101, 8.102, 8.119, 8.120, 8.121, 8.122, 8.123, 8.124, 8.125, 8.137, 9.00, 9.03, 9.14, 9.20, 9.21, 9.22, 9.23, 9.24, 9.25, 9.26, 9.27, 9.28, 9.29, 9.30, 9.31, 9.32, 9.33, 10.08, 10.09, 10.10, 10.11, 10.12, 10.13, 10.14, 10.15, 10.16, 10.17, 10.18, 10.20, 10.24, 10.25, 10.26, 10.27, 10.28, 10.29, 10.30, 10.31, 10.32, 10.33, 10.34, 10.35, 10.36, 10.37, 10.38, 10.39, 10.40, 10.41, 10.42, 10.43, 10.44, 10.45, 10.46, 10.47, 10.48, 10.49, 10.50; **Eric Strand/Shutterstock**: 5.51; **FedeCandoniPhoto/Shutterstock**: 5.67; **ffotocymru/Alamy**: 4.22; **Fotolia**: 1.01, 1.02, 1.03, 1.05, 1.06, 1.07, 1.08, 1.14, 1.15, 1.16, 3.02, 3.08; **Furtseff/Shutterstock**: 8.036; **GEORGE BERNARD/ SCIENCE PHOTO LIBRARY**: 8.029; **George Clerk/Getty Images**: 10.01; **hal pand/Shutterstock**: 10.05; **Helfen**: 2.04; **Hmelnitkaia Ana/Shutterstock**: 7.015; **ictor/iStockphoto**: 3.15; **imageBROKER/ Alamy**: 5.15; **InCommunicado/iStockphoto**: 3.13; **iStockphoto**: 5.30, 5.34; **Ivan Neru/Fotolia**: 6.02; **Joe Gough/Shutterstock**: 3.07; **kropic/iStockphoto**: 3.01; **LenaTru/Shutterstock**: 5.69; **Lindasj22/ Shutterstock**: 6.38; **Lisa S./Shutterstock**: 3.09; **marchkimoo/ Shutterstock**: 5.43; **mark higgins/Shutterstock**: 5.49; **Mark William Richardson/Shutterstock**: 3.00; **Marquestra/Shutterstock**: 5.53; **Martin Barraud/Getty Images**: 6.09; **mexrix/Shutterstock**: 5.46; **Mika Heittola/Shutterstock**: 3.06; **mlbalexander/iStockphoto**: 3.05; **MPFphotography/Shutterstock**: 7.013; **ndoeljindoel/ Shutterstock**: 2.00; **Nelson Thornes/Oxford University Press**: 1.09, 1.10, 1.12, 1.13; **Nicholas Eveleigh/Getty Images**: 6.33; **NinaMalyna/ Shutterstock**: 6.07; **ogressie/Fotolia**: 5.64; **OneSmallSquare/ Shutterstock**: 7.003; **Pablo Scapinachis/Shutterstock**: 6.40; **Permission granted by David Crowley, Scafit Ltd (www.scafit. co.nz)**: 4.27; **Peter Brett**: 2.03, 2.06, 2.07; **Peter Davey/Alamy**: 1.00; **PETER GARDINER/SCIENCE PHOTO LIBRARY**: 1.04; **photolia/ iStockphoto**: 1.11; **prill/iStockphoto**: 4.08; **Richard Wilson/Oxford University Press**: 4.00, 4.04, 4.05, 4.06, 4.07, 4.09, 4.10, 4.11, 4.12, 4.13, 4.26, 4.30, 4.31, 4.32, 4.33, 4.34, 4.35, 4.36, 4.37, 4.38, 5.00, 5.01, 5.02, 5.03, 5.04, 5.05, 5.06, 5.07, 5.08, 5.09, 5.10, 5.11, 5.12, 5.13, 5.14, 5.16, 5.17, 5.18, 5.19, 5.20, 5.21, 5.22, 5.23, 5.24, 5.25, 5.26, 5.27, 5.28, 5.29, 5.31, 5.32, 5.33, 5.35, 5.36, 5.37, 5.38, 5.39, 5.40, 5.41, 5.54, 5.61, 5.62, 5.63, 5.73, 5.76, 5.78, 5.79, 5.80, 5.81, 5.82, 5.83, 6.00, 6.03, 6.05, 6.06, 6.17, 6.18, 6.19, 6.20, 6.21, 6.22, 6.23, 6.24, 6.25, 6.26, 6.27, 6.28, 6.30, 6.48, 7.017, 7.018, 7.019, 7.020, 7.021, 7.022, 7.023, 7.024, 7.025, 7.026, 7.027, 7.028, 7.029, 7.030, 7.031, 7.032, 7.033, 7.034, 7.035, 7.036, 7.037, 7.038, 7.039, 7.040, 7.049, 7.052, 7.054, 7.060, 7.061, 7.062, 7.063, 7.064, 7.065, 7.066, 7.067, 7.068, 7.072, 7.073, 7.074, 7.075, 7.076, 7.077, 7.078, 7.079, 7.080, 7.081, 7.082, 7.083, 7.084, 7.085, 7.086, 7.087, 7.088, 7.089, 7.090, 7.091, 7.092, 7.106, 7.107, 7.108, 7.109, 7.110, 7.111, 7.112, 7.113, 7.114, 8.001, 8.002, 8.003, 8.004, 8.005, 8.006, 8.007, 8.008, 8.009, 8.011, 8.018, 8.019, 8.020, 8.021, 8.022, 8.023, 8.026, 8.027, 8.033, 8.038, 8.039, 8.040, 8.041, 8.042, 8.043, 8.045, 8.046, 8.047, 8.048, 8.049, 8.053, 8.068, 8.069, 8.070, 8.071, 8.072, 8.073, 8.076, 8.077, 8.078, 8.079, 8.080, 8.081, 8.082, 8.084, 8.085, 8.086, 8.087, 8.088, 8.089, 8.090, 8.091, 8.092, 8.093, 8.094, 8.103, 8.104, 8.105, 8.106, 8.107, 8.108, 8.109, 8.110, 8.111, 8.112, 8.113, 8.114, 8.115, 8.116, 8.117, 8.118, 8.136, 9.08, 9.10, 9.12, 9.13, 9.15, 9.16, 9.17, 9.18, 9.19, 10.00; **Ron Ellis/Shutterstock**: 5.59; **S-F/ Shutterstock**: 8.055; **Scrofula/iStock**: 5.48; **severija/Shutterstock**: 7.009; **siro46/Shutterstock**: 5.45; **SSPL via Getty Images**: 6.43; **starekase/Fotolia**: 5.66; **stuartpitkin/Getty Images**: 6.10; **Tanya Gorelova/Shutterstock**: 5.77; **tharrison/iStockphoto**: 2.13; **The Drawing Room at the Old Rectory, Quenington (photo)/© Country Life/Bridgeman Images**: 10.03; **the palms/Shutterstock**: 7.006; **thehague/iStockphoto**: 3.12; **Tom Gowanlock/Shutterstock**: 6.08; **toschro/iStockphoto**: 4.14; **Valery Kraynov/Shutterstock**: 8.035; **Vinod K Pillai/Shutterstock**: 5.42; **With kind permission of Tiranti**: 8.059, 8.060, 8.061, 8.062

Although we have made every effort to trace and contact all copyright holders before publication this has not been possible in all cases. If notified, the publisher will rectify any errors or omissions at the earliest opportunity.

Links to third party websites are provided by Oxford in good faith and for information only. Oxford disclaims any responsibility for the materials contained in any third party website referenced in this work.

Note to learners and tutors

This book clearly states that a risk assessment should be undertaken and the correct PPE worn for the particular activities before any practical activity is carried out. Risk assessments were carried out before photographs for this book were taken and the models are wearing the PPE deemed appropriate for the activity and situation. This was correct at the time of going to print. Colleges may prefer that their learners wear additional items of PPE not featured in the photographs in this book and should instruct learners to do so in the standard risk assessments they hold for activities undertaken by their learners. Learners should follow the standard risk assessments provided by their college for each activity they undertake which will determine the PPE they wear.

CONTENTS

INTRODUCTION

About this book

This book has been written for the Cskills Awards Level 3 Diploma in Painting & Decorating. It covers all the units of the qualification, so you can feel confident that your book fully covers the requirements of your course.

This book contains a number of features to help you acquire the knowledge you need. It also demonstrates the practical skills you will need to master to successfully complete your qualification. We've included additional features to show how the skills and knowledge can be applied to the workplace, as well as tips and advice on how you can improve your chances of gaining employment.

The features include:

* chapter openers which list the learning outcomes you must achieve in each unit

* key terms that provide explanations of important terminology that you will need to know and understand

* Did you know? margin notes to provide key facts that are helpful to your learning

* practical tips to explain facts or skills to remember when undertaking practical tasks

* Reed tips to offer advice about work, building your CV and how to apply the skills and knowledge you have learnt in the workplace

* case studies that are based on real tradespeople who have undertaken apprenticeships and explain why the skills and knowledge you learn with your training provider are useful in the workplace

* practical tasks that provide step-by-step directions and illustrations for a range of projects you may do during your course

* Test yourself multiple choice questions that appear at the end of each unit to give you the chance to revise what you have learnt and to practise your assessment (your tutor will give you the answers to these questions).

You can download unit CSAL3Occ122 from: https://global.oup.com/education/secondary/subjects/vocational/building/free-resources.

Further support for this book can be found at our website, www.planetvocational.com/subjects/build.

KEY TERMS

DID YOU KNOW?

PRACTICAL TIP

REED TIP

CASE STUDY

PRACTICAL TASK

TEST YOURSELF

Planet Vocational

CONTRIBUTORS TO THIS BOOK

British Association of Construction Heads

The British Association of Construction Heads is an association formed largely from those managing and delivering the construction curriculum from pre-apprenticeship to post graduate level. The Association is a voluntary organisation and was formed in 1983 and has grown to a position where it can demonstrate that BACH members now manage over 90% of the Learners studying the construction curriculum and includes membership of 80% of the Colleges offering the Construction curriculum in England, Northern Ireland, Scotland and Wales. It accepts membership applications from Colleges and other organisations who are passionate about quality and standards in construction education and training. Visit www.bach.uk.com for more information.

A huge thank you to Mike Morson at Riverside College and David Kehoe at Vision West Nottinghamshire College for their technical expertise in reviewing, advising and facilitating the photo shoot.

Weston College

Special thanks to Weston College for their technical expertise and enthusiasm in facilitating the photo shoot.

Reed Property & Construction

Reed Property & Construction specialises in placing staff at all levels, in both temporary and permanent positions, across the complete lifecycle of the construction process. Our consultants work with most major construction companies in the UK and our clients are involved with the design, build and maintenance of infrastructure projects throughout the UK.

Expert help

As a leading recruitment consultancy for mid–senior level construction staff in the UK, Reed Property & Construction is ideally placed to advise new workers entering the sector, from building a CV to providing expertise and sharing our extensive sector knowledge with you. That's why you will find helpful hints from our highly experienced consultants, designed to help you find that first step on the construction career ladder. These tips range from advice on CV writing to interview tips and techniques, and are linked with the learning material in this book.

Work-related advice

Reed Property & Construction has gained insights from some of our biggest clients to help you understand the mind-set of potential employers. This includes the traits and skills that they would like to see in their new employees, why you need the skills taught in this book and how they are used on a day to day basis within their organisations.

Getting your first job

This invaluable information is not available anywhere else and is geared to helping you gain a position with an employer once you've completed your studies. Entry level positions are not usually offered by recruitment companies, but our advice will help you to apply for jobs in construction and hopefully gain your first position as a skilled worker.

CONTRIBUTORS TO THIS BOOK

The case studies in this book feature staff from Laing O'Rourke and South Tyneside Homes.

Laing O'Rourke is an international engineering company that constructs large-scale building projects all over the world. Originally formed from two companies, John Laing (founded in 1848) and R O'Rourke and Son (founded in 1978) joined forces in 2001.

At Laing O'Rourke, there is a strong and unique apprenticeship programme. It runs a four-year 'Apprenticeship Plus' scheme in the UK, combining formal college education with on-the-job training. Apprentices receive support and advice from mentors and experienced tradespeople, and are given the option of three different career pathways upon completion: remaining on site, continuing into a further education programme, or progressing into supervision and management.

The company prides itself on its people development, supporting educational initiatives and investing in its employees. Laing O'Rourke believes in collaboration and teamwork as a path to achieving greater success, and strives to maintain exceptionally high standards in workplace health and safety.

South Tyneside Homes

South Tyneside Council's Housing Company

South Tyneside Homes was launched in 2006, and was previously part of South Tyneside Council. It now works in partnership with the council to repair and maintain 18,000 properties within the borough, including delivering parts of the Decent Homes Programme.

South Tyneside Homes believes in putting back into the community, with 90 per cent of its employees living in the borough itself. Equality and diversity, as well as health and wellbeing of staff, is a top priority, and it has achieved the Gold Status Investors in People Award.

South Tyneside Homes is committed to the development of its employees, providing opportunities for further education and training and great career paths within the company – 80 per cent of its management team started as apprentices with the company. As well as looking after its staff and their community, the company looks after the environment too, running a renewable energy scheme for council tenants in order to reduce carbon emissions and save tenants money.

The apprenticeship programme at South Tyneside Homes has been recognised nationally, having trained over 80 young people in five main trade areas over the past six years. One of the UK's Top 100 Apprenticeship Employers, it is an Ambassador on the panel of the National Apprentice Service. It has won the Large Employer of the Year Award at the National Apprenticeship Awards and several of its apprentices have been nominated for awards, including winning the Female Apprentice of the Year for the local authority.

Chapter 1

Unit CSA–L1Core01
HEALTH, SAFETY AND WELFARE IN CONSTRUCTION AND ASSOCIATED INDUSTRIES

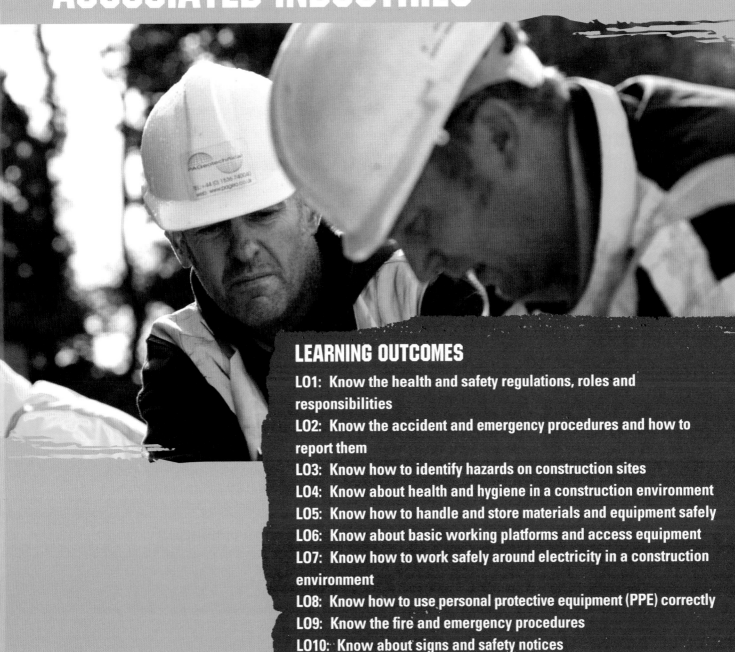

LEARNING OUTCOMES

LO1: Know the health and safety regulations, roles and responsibilities

LO2: Know the accident and emergency procedures and how to report them

LO3: Know how to identify hazards on construction sites

LO4: Know about health and hygiene in a construction environment

LO5: Know how to handle and store materials and equipment safely

LO6: Know about basic working platforms and access equipment

LO7: Know how to work safely around electricity in a construction environment

LO8: Know how to use personal protective equipment (PPE) correctly

LO9: Know the fire and emergency procedures

LO10: Know about signs and safety notices

INTRODUCTION

The aim of this chapter is to:

* help you to source relevant safety information
* help you to use the relevant safety procedures at work.

KEY TERMS

HASAWA

– the Health and Safety at Work etc. Act outlines your and your employer's health and safety responsibilities.

COSHH

– the Control of Substances Hazardous to Health Regulations are concerned with controlling exposure to hazardous materials.

DID YOU KNOW?

In 2011 to 2012, there were 49 fatal accidents in the construction industry in the UK. (*Source* HSE, www.hse.gov.uk)

KEY TERMS

HSE

– the Health and Safety Executive, which ensures that health and safety laws are followed.

Accident book

– this is required by law under the Social Security (Claims and Payments) Regulations 1979. Even minor accidents need to be recorded by the employer. For the purposes of RIDDOR, hard copy accident books or online records of incidents are equally acceptable.

HEALTH AND SAFETY REGULATIONS, ROLES AND RESPONSIBILITIES

The construction industry can be dangerous, so keeping safe and healthy at work is very important. If you are not careful, you could injure yourself in an accident or perhaps use equipment or materials that could damage your health. Keeping safe and healthy will help ensure that you have a long and injury-free career.

Although the construction industry is much safer today than in the past, more than 2,000 people are injured and around 50 are killed on site every year. Many others suffer from long-term ill-health such as deafness, spinal damage, skin conditions or breathing problems.

Key health and safety legislation

Laws have been created in the UK to try to ensure safety at work. Ignoring the rules can mean injury or damage to health. It can also mean losing your job or being taken to court.

The two main laws are the Health and Safety at Work etc. Act **(HASAWA)** and the Control of Substances Hazardous to Health Regulations **(COSHH)**.

The Health and Safety at Work etc. Act (HASAWA) (1974)

This law applies to all working environments and to all types of worker, sub-contractor, employer and all visitors to the workplace. It places a duty on everyone to follow rules in order to ensure health, safety and welfare. Businesses must manage health and safety risks, for example by providing appropriate training and facilities. The Act also covers first aid, accidents and ill health.

Reporting of Injuries, Diseases and Dangerous Occurrences Regulations (RIDDOR) (1995)

Under RIDDOR, employers are required to report any injuries, diseases or dangerous occurrences to the **Health and Safety Executive (HSE)**. The regulations also state the need to maintain an **accident book**.

Control of Substances Hazardous to Health (COSHH) (2002)

In construction, it is common to be exposed to substances that could cause ill health. For example, you may use oil-based paints or preservatives, or work in conditions where there is dust or bacteria.

Employers need to protect their employees from the risks associated with using hazardous substances. This means assessing the risks and deciding on the necessary precautions to take.

Any control measures (things that are being done to reduce the risk of people being hurt or becoming ill) have to be introduced into the workplace and maintained; this includes monitoring an employee's exposure to harmful substances. The employer will need to carry out health checks and ensure that employees are made aware of the dangers and are supervised.

Control of Asbestos at Work Regulations (2012)

Asbestos was a popular building material in the past because it was a good insulator, had good fire protection properties and also protected metals against corrosion. Any building that was constructed before 2000 is likely to have some asbestos. It can be found in pipe insulation, boilers and ceiling tiles. There is also asbestos cement roof sheeting and there is a small amount of asbestos in decorative coatings such as Artex.

Asbestos has been linked with lung cancer, other damage to the lungs and breathing problems. The regulations require you and your employer to take care when dealing with asbestos:

* You should always assume that materials contain asbestos unless it is obvious that they do not.

* A record of the location and condition of asbestos should be kept.

* A risk assessment should be carried out if there is a chance that anyone will be exposed to asbestos.

The general advice is as follows:

* Do not remove the asbestos. It is not a hazard unless it is removed or damaged.

* Remember that not all asbestos presents the same risk. Asbestos cement is less dangerous than pipe insulation.

* Call in a specialist if you are uncertain.

Provision and Use of Work Equipment Regulations (PUWER) (1998)

PUWER concerns health and safety risks related to equipment used at work. It states that any risks arising from the use of equipment must either be prevented or controlled, and all suitable safety measures must have been taken. In addition, tools need to be:

* suitable for their intended use

* safe

* well maintained

* used only by those who have been trained to do so.

Manual Handling Operations Regulations (1992)

These regulations try to control the risk of injury when lifting or handling bulky or heavy equipment and materials. The regulations state as follows:

* Hazardous manual handling should be avoided if possible.

* An assessment of hazardous manual handling should be made to try to find alternatives.

* You should use mechanical assistance where possible.

* The main idea is to look at how manual handling is carried out and finding safer ways of doing it.

REED
TIP

Employers will want to know that you understand the importance of health and safety. Make sure you know the reasons for each safe working practice.

KEY TERMS

PPE

– personal protective equipment can include gloves, goggles and hard hats.

Competent

– to be competent an organisation or individual must have:

- sufficient knowledge of the tasks to be undertaken and the risks involved

- the experience and ability to carry out their duties in relation to the project, to recognise their limitations and take appropriate action to prevent harm to those carrying out construction work, or those affected by the work.

(*Source* HSE)

Personal Protection at Work Regulations (PPE) (1992)

This law states that employers must provide employees with personal protective equipment **(PPE)** at work whenever there is a risk to health and safety. PPE needs to be:

- suitable for the work being done
- well maintained and replaced if damaged
- properly stored
- correctly used (which means employees need to be trained in how to use the PPE properly).

Work at Height Regulations (2005)

Whenever a person works at any height there is a risk that they could fall and injure themselves. The regulations place a duty on employers or anyone who controls the work of others. This means that they need to:

- plan and organise the work
- make sure those working at height are **competent**
- assess the risks and provide appropriate equipment
- manage work near or on fragile surfaces
- ensure equipment is inspected and maintained.

In all cases the regulations suggest that, if it is possible, work at height should be avoided. Perhaps the job could be done from ground level? If it is not possible, then equipment and other measures are needed to prevent the risk of falling. When working at height measures also need to be put in place to minimise the distance someone might fall.

Figure 1.1 Examples of personal protective equipment

Employer responsibilities under HASAWA

HASAWA states that employers with five or more staff need their own health and safety policy. Employers must assess any risks that may be involved in their workplace and then introduce controls to reduce these risks. These risk assessments need to be reviewed regularly.

Employers also need to supply personal protective equipment (PPE) to all employees when it is needed and to ensure that it is worn when required.

Specific employer responsibilities are outlined in Table 1.1.

Employee responsibilities under HASAWA

HASAWA states that all those operating in the workplace must aim to work in a safe way. For example, they must wear any PPE provided and look after their equipment. Employees should not be charged for PPE or any actions that the employer needs to take to ensure safety.

Specific employer responsibilities are outlined in Table 1.1. Table 1.2 identifies the key employee responsibilities.

Employer responsibility	Explanation
Safe working environment	Where possible all potential **risks** and **hazards** should be eliminated.
Adequate staff training	When new employees begin a job their induction should cover health and safety. There should be ongoing training for existing employees on risks and control measures.
Health and safety information	Relevant information related to health and safety should be available for employees to read and have their own copies.
Risk assessment	Each task or job should be investigated and potential risks identified so that measures can be put in place. A risk assessment and method statement should be produced. The method statement will tell you how to carry out the task, what PPE to wear, equipment to use and the sequence of its use.
Supervision	A competent and experienced individual should always be available to help ensure that health and safety problems are avoided.

Table 1.1 Employer responsibilities under HASAWA

Employee responsibility	Explanation
Working safely	Employees should take care of themselves, only do work that they are competent to carry out and remove obvious hazards if they are seen.
Working in partnership with the employer	Co-operation is important and you should never interfere with or misuse any health and safety signs or equipment. You should always follow the site rules.
Reporting hazards, **near misses** and accidents correctly	Any health and safety problems should be reported and discussed, particularly a near miss or an actual accident.

Table 1.2 Employee responsibilities under HASAWA

Health and Safety Executive

The Health and Safety Executive (HSE) is responsible for health, safety and welfare. It carries out spot checks on different workplaces to make sure that the law is being followed.

HSE inspectors have access to all areas of a construction site and can also bring in the police. If they find a problem then they can issue an **improvement notice**. This gives the employer a limited amount of time to put things right.

In serious cases, the HSE can issue a **prohibition notice**. This means all work has to stop until the problem is dealt with. An employer, the employees or **sub-contractors** could be taken to court.

The roles and responsibilities of the HSE are outlined in Table 1.3.

KEY TERMS

Risk
– the likelihood that a person may be harmed if they are exposed to a hazard.

Hazard
– a potential source of harm, injury or ill-health.

Near miss
– any incident, accident or emergency that did not result in an injury but could have done so.

Improvement notice
– this is issued by the HSE if a health or safety issue is found and gives the employer a time limit to make changes to improve health and safety.

Prohibition notice
– this is issued by the HSE if a health or safety issue involving the risk of serious personal injury is found and stops all work until the improvements to health and safety have been made.

Sub-contractor
– an individual or group of workers who are directly employed by the main contractor to undertake specific parts of the work.

Responsibility	Explanation
Enforcement	It is the HSE's responsibility to reduce work-related death, injury and ill health. It will use the law against those who put others at risk.
Legislation and advice	The HSE will use health and safety legislation to serve improvement or prohibition notices or even to prosecute those who break health and safety rules. Inspectors will provide advice either face-to-face or in writing on health and safety matters.
Inspection	The HSE will look at site conditions, standards and practices and inspect documents to make sure that businesses and individuals are complying with health and safety law.

Table 1.3 HSE roles and responsibilities

Sources of health and safety information

There is a wide variety of health and safety information. Most of it is available free of charge, while other organisations may make a charge to provide information and advice. Table 1.4 outlines the key sources of health and safety information.

Source	Types of information	Website
Health and Safety Executive (HSE)	The HSE is the primary source of work-related health and safety information. It covers all possible topics and industries.	www.hse.gov.uk
Construction Industry Training Board (CITB)	The national training organisation provides key information on legislation and site safety.	www.citb.co.uk
British Standards Institute (BSI)	Provides guidelines for risk management, PPE, fire hazards and many other health and safety-related areas.	www.bsigroup.com
Royal Society for the Prevention of Accidents (RoSPA)	Provides training, consultancy and advice on a wide range of health and safety issues that are aimed to reduce work related accidents and ill health.	www.rospa.com
Royal Society for Public Health (RSPH)	Has a range of qualifications and training programmes focusing on health and safety.	www.rsph.org.uk

Table 1.4 Health and safety information

Informing the HSE

The HSE requires the reporting of:

* deaths and injuries – any **major injury**, **over 7-day injury** or death
* occupational disease
* dangerous occurrence – a collapse, explosion, fire or collision
* gas accidents – any accidental leaks or other incident related to gas.

Enforcing guidance

Work-related injuries and illnesses affect huge numbers of people. According to the HSE, 1.1 million working people in the UK suffered from a work-related illness in 2011 to 2012. Across all industries, 173 workers were killed, 111,000 other injuries were reported and 27 million working days were lost.

The construction industry is a high risk one and, although only around 5 per cent of the working population is in construction, it accounts for 10 per cent of all major injuries and 22 per cent of fatal injuries.

KEY TERMS

Major injury

– any fractures, amputations, dislocations, loss of sight or other severe injury.

Over 7-day injury

– an injury that has kept someone off work for more than seven days.

The good news is that enforcing guidance on health and safety has driven down the numbers of injuries and deaths in the industry. Only 20 years ago over 120 construction workers died in workplace accidents each year. This is now reduced to fewer than 60 a year.

However, there is still more work to be done and it is vital that organisations such as the HSE continue to enforce health and safety and continue to reduce risks in the industry.

On-site safety inductions and toolbox talks

The HSE suggests that all new workers arriving on site should attend a short induction session on health and safety. It should:

* show the commitment of the company to health and safety
* explain the health and safety policy
* explain the roles individuals play in the policy
* state that each individual has a legal duty to contribute to safe working
* cover issues like excavations, work at height, electricity and fire risk
* provide a layout of the site and show evacuation routes
* identify where fire fighting equipment is located
* ensure that all employees have evidence of their skills
* stress the importance of signing in and out of the site.

Behaviour and actions that could affect others

It is the responsibility of everyone on site not only to look after their own health and safety, but also to ensure that their actions do not put anyone else at risk.

Trying to carry out work that you are not competent to do is not only dangerous to yourself but could compromise the safety of others.

Simple actions, such as ensuring that all of your rubbish and waste is properly disposed of, will go a long way to removing hazards on site that could affect others.

Just as you should not create a hazard, ignoring an obvious one is just as dangerous. You should always obey site rules and particularly the health and safety rules. You should follow any instructions you are given.

ACCIDENT AND EMERGENCY PROCEDURES

All sites will have specific procedures for dealing with accidents and emergencies. An emergency will often mean that the site needs to be evacuated, so you should know in advance where to assemble and who to report to. The site should never be re-entered without authorisation from an individual in charge or the emergency services.

DID YOU KNOW?

Workplace injuries cost the UK £13.4bn in 2010 to 2011.

DID YOU KNOW?

Toolbox talks are normally given by a supervisor and often take place on site, either during the course of a normal working day or when someone has been seen working in an unsafe way. CITB produces a book called *GT700 Toolbox Talks* which covers a range of health and safety topics, from trying a new process and using new equipment to particular hazards or work practices.

PRACTICAL TIP

If you come across any health and safety problems you should report them so that they can be controlled.

Figure 1.2 It's important that you know where your company's fire-fighting equipment is located

Types of emergencies

Emergencies are incidents that require immediate action. They can include:

* fires
* spillages or leaks of chemicals or other hazardous substances, such as gas
* failure of a scaffold
* collapse of a wall or trench
* a health problem
* an injury
* bombs and security alerts.

Legislation and reporting accidents

RIDDOR (1995) puts a duty on employers, anyone who is self-employed, or an individual in control of the work, to report any serious workplace accidents, occupational diseases or dangerous occurrences (also known as near misses).

The report has to be made by these individuals and, if it is serious enough, the responsible person may have to fill out a RIDDOR report.

Injuries, diseases and dangerous occurrences

Construction sites can be dangerous places, as we have seen. The HSE maintains a list of all possible injuries, diseases and dangerous occurrences, particularly those that need to be reported.

Injuries

There are two main classifications of injuries: minor and major. A minor injury can usually be handled by a competent first aider, although it is often a good idea to refer the individual to their doctor or to the hospital. Typical minor injuries can include:

* minor cuts
* minor burns
* exposure to fumes.

Major injuries are more dangerous and will usually require the presence of an ambulance with paramedics. Major injuries can include:

* bone fracture
* unconsciousness
* concussion
* electric shock.

Diseases

There are several different diseases and health issues that have to be reported, particularly if a doctor notifies that a disease has been diagnosed. These include:

* poisoning
* skin diseases
* lung diseases
* infections
* occupational cancer
* hand/arm vibration syndrome.

Dangerous occurrences

Even if something happens that does not result in an injury, but could easily have done so, it is classed as a dangerous occurrence. It needs to be reported immediately and then followed up by an accident report form. Dangerous occurrences can include:

* accidental release of a substance that could damage health
* anything coming into contact with overhead power lines

* an electrical problem that caused a fire or explosion
* collapse or partial collapse of scaffolding over 5m high.

Recording accidents and emergencies

The Reporting of Injuries, Diseases and Dangerous Occurrences Regulations (RIDDOR) (1995) requires employers to:

* report any relevant injuries, diseases or dangerous occurrences to the Health and Safety Executive (HSE)
* keep records of incidents in a formal and organised manner (for example, in an accident book or online database).

After an accident, you may need to complete an accident report form – either in writing or online. This form may be completed by the person who was injured or the first aider.

On the accident report form you need to note down:

* the casualty's personal details, e.g. name, address, occupation
* the name of the person filling in the report form
* the details of the accident.

In addition, the person reporting the accident will need to sign the form.

On site a trained first aider will be the first individual to try and deal with the situation. In addition to trying to save life, stop the condition from getting worse and getting help, they will also record the occurrence.

On larger sites there will be a health and safety officer, who would keep records and documentation detailing any accidents and emergencies that have taken place on site. All companies should keep such records; it may be a legal requirement for them to do so under RIDDOR and it is good practice to do so in case the HSE asks to see it.

Importance of reporting accidents and near misses

Reporting incidents is not just about complying with the law or providing information for statistics. Each time an accident or near miss takes place it means lessons can be learned and future problems avoided.

The accident or near miss can alert the business or organisation to a potential problem. They can then take steps to ensure that it does not occur in the future.

Major and minor injuries and near misses

RIDDOR defines a major injury as:

* a fracture (but not to a finger, thumb or toes)
* a dislocation
* an amputation
* a loss of sight in an eye
* a chemical or hot metal burn to the eye
* a penetrating injury to the eye
* an electric shock or electric burn leading to unconsciousness and/or requiring resuscitation
* hyperthermia, heat-induced illness or unconsciousness

REED TIP

On some construction sites, you may get a Health and Safety Inspector come to look round without any notice – one more reason to always be thinking about working safely.

* asphyxia

* exposure to a harmful substance

* inhalation of a substance

* acute illness after exposure to toxins or infected materials.

A minor injury could be considered as any occurrence that does not fall into any of the above categories.

A near miss is any incident that did not actually result in an injury but which could have caused a major injury if it had done so. Non-reportable near misses are useful to record as they can help to identify potential problems. Looking at a list of near misses might show patterns for potential risk.

Accident trends

We have already seen that the HSE maintains statistics on the number and types of construction accidents. The following are among the 2011/2012 construction statistics:

* There were 49 fatalities.

* There were 5,000 occupational cancer patients.

* There were 74,000 cases of work-related ill health.

* The most common types of injury were caused by falls, although many injuries were caused by falling objects, collapses and electricity. A number of construction workers were also hurt when they slipped or tripped, or were injured while lifting heavy objects.

Accidents, emergencies and the employer

Even less serious accidents and injuries can cost a business a great deal of money. But there are other costs too:

* Poor company image – if a business does not have health and safety controls in place then it may get a reputation for not caring about its employees. The number of accidents and injuries may be far higher than average.

* Loss of production – the injured individual might have to be treated and then may need a period of time off work to recover. The loss of production can include those who have to take time out from working to help the injured person and the time of a manager or supervisor who has to deal with all the paperwork and problems.

* Insurance – each time there is an accident or injury claim against the company's insurance the premiums will go up. If there are many accidents and injuries the business may find it impossible to get insurance. It is a legal requirement for a business to have insurance so in the end that company might have to close down.

* Closure of site – if there is a serious accident or injury then the site may have to be closed while investigations take place to discover the reason, or who was responsible. This could cause serious delays and loss of income for workers and the business.

Accident and emergency authorised personnel

Several different groups of people could be involved in dealing with accident and emergency situations. These are listed in Table 1.5.

Authorised personnel	Role
First aiders and emergency responders	These are employees on site and in the workforce who have been trained to be the first to respond to accidents and injuries. The minimum provision of an appointed person would be someone who has had basic first aid training. The appointment of a first aider is someone who has attained a higher or specific level of training. A construction site with fewer than 5 employees needs an appointed first aider. A construction site with up to 50 employees requires a trained first aider, and for bigger sites at least one trained first aider is required for every 50 people.
Supervisors and managers	These have the responsibility of managing the site and would have to organise the response and contact emergency services if necessary. They would also ensure that records of any accidents are completed and up to date and notify the HSE if required.
Health and Safety Executive	The HSE requires businesses to investigate all accidents and emergencies. The HSE may send an inspector, or even a team, to investigate and take action if the law has been broken.
Emergency services	Calling the emergency services depends on the seriousness of the accident. Paramedics will take charge of the situation if there is a serious injury and if they feel it necessary will take the individual to hospital.

Table 1.5 People who deal with accident and emergency situations

The basic first aid kit

BS 8599 relates to first aid kits, but it is not legally binding. The contents of a first aid box will depend on an employer's assessment of their likely needs. The HSE does not have to approve the contents of a first aid box but it states that where the work involves low level hazards the minimum contents of a first aid box should be:

* a copy of its leaflet on first aid – *HSE Basic advice on first aid at work*
* 20 sterile plasters of assorted size
* 2 sterile eye pads
* 4 sterile triangular bandages
* 6 safety pins
* 2 large sterile, unmedicated wound dressings
* 6 medium-sized sterile unmedicated wound dressings
* 1 pair of disposable gloves.

The HSE also recommends that no tablets or medicines are kept in the first aid box.

> **DID YOU KNOW?**
>
> The three main emergency services in the UK are: the Fire Service (for fire and rescue); the Ambulance Service (for medical emergencies); the Police (for an immediate police response). Call them on 999 only if it is an emergency.

What to do if you discover an accident

When an accident happens it may not only injure the person involved directly, but it may also create a hazard that could then injure others. You need to make sure that the area is safe enough for you or someone else to help the injured person. It may be necessary to turn off the electrical supply or remove obstructions to the site of the accident.

Figure 1.3 A typical first aid box

The first thing that needs to be done if there is an accident is to raise the alarm. This could mean:

* calling for the first aider

* phoning for the emergency services

* dealing with the problem yourself.

How you respond will depend on the severity of the injury.

You should follow this procedure if you need to contact the emergency services:

* Find a telephone away from the emergency.

* Dial 999.

* You may have to go through a switchboard. Carefully listen to what the operator is saying to you and try to stay calm.

* When asked, give the operator your name and location, and the name of the emergency service or services you require.

* You will then be transferred to the appropriate emergency service, who will ask you questions about the accident and its location. Answer the questions in a clear and calm way.

* Once the call is over, make sure someone is available to help direct the emergency services to the location of the accident.

IDENTIFYING HAZARDS

As we have already seen, construction sites are potentially dangerous places. The most effective way of handling health and safety on a construction site is to spot the hazards and deal with them before they can cause an accident or an injury. This begins with basic housekeeping and carrying out risk assessments. It also means having a procedure in place to report hazards so that they can be dealt with.

Good housekeeping

Work areas should always be clean and tidy. Sites that are messy, strewn with materials, equipment, wires and other hazards can prove to be very dangerous. You should:

* always work in a tidy way

* never block fire exits or emergency escape routes

* never leave nails and screws scattered around

* ensure you clean and sweep up at the end of each working day

* not block walkways

* never overfill skips or bins

* never leave food waste on site.

Risk assessments and method statements

It is a legal requirement for employers to carry out risk assessments. This covers not only those who are actually working on a particular job, but other workers in the immediate area, and others who might be affected by the work.

It is important to remember that when you are carrying out work your actions may affect the safety of other people. It is important, therefore, to know whether there are any potential hazards. Once you know what these hazards are you can do something to either prevent or reduce them as a risk. Every job has potential hazards.

There are five simple steps to carrying out a risk assessment, which are shown in Table 1.6, using the example of repointing brickwork on the front face of a dwelling.

Step	Action	Example
1	Identify hazards	The property is on a street with a narrow pavement. The damaged brickwork and loose mortar need to be removed and placed in a skip below. Scaffolding has been erected. The road is not closed to traffic.
2	Identify who is at risk	The workers repointing are at risk as they are working at height. Pedestrians and vehicles passing are at risk from the positioning of the skip and the chance that debris could fall from height.
3	What is the risk from the hazard that may cause an accident?	The risk to the workers is relatively low as they have PPE and the scaffolding has been correctly erected. The risk to those passing by is higher, as they are unaware of the work being carried out above them.
4	Measures to be taken to reduce the risk	Station someone near the skip to direct pedestrians and vehicles away from the skip while the work is being carried out. Fix a secure barrier to the edge of the scaffolding to reduce the chance of debris falling down. Lower the bricks and mortar debris using a bucket or bag into the skip and not throwing them from the scaffolding. Consider carrying out the work when there are fewer pedestrians and less traffic on the road.
5	Monitor the risk	If there are problems with the first stages of the job, you need to take steps to solve them. If necessary consider taking the debris by hand through the building after removal.

Table 1.6 A five-step risk assessment for repointing brickwork

Your employer should follow these working practices, which can help to prevent accidents or dangerous situations occurring in the workplace:

* *Risk assessments* look carefully at what could cause an individual harm and how to prevent this. This is to ensure that no one should be injured or become ill as a result of their work. Risk assessments identify how likely it is that an accident might happen and the consequences of it happening. A risk factor is worked out and control measures created to try to offset them.

* *Method statements,* however brief, should be available for every risk assessment. They summarise risk assessments and other findings to provide guidance on how the work should be carried out.

* *Permit to work systems* are used for very high risk or even potentially fatal activities. They are checklists that need to be completed before the work begins. They must be signed by a supervisor.

* *A hazard book* lists standard tasks and identifies common hazards. These are useful tools to help quickly identify hazards related to particular tasks.

Types of hazards

Typical construction accidents can include:

* fires and explosions
* slips, trips and falls
* burns, including those from chemicals
* falls from scaffolding, ladders and roofs
* electrocution
* injury from faulty machinery
* power tool accidents
* being hit by construction debris
* falling through holes in flooring.

We will look at some of the more common hazards in a little more detail.

Fires

Fires need oxygen, heat and fuel to burn. Even a spark can provide enough heat needed to start a fire, and anything flammable, such as petrol, paper or wood, provides the fuel. It may help to remember the 'triangle of fire' – heat, oxygen and fuel are all needed to make fire so remove one or more to help prevent or stop the fire.

Tripping

Leaving equipment and materials lying around can cause accidents, as can trailing cables and spilt water or oil. Some of these materials are also potential fire hazards.

Chemical spills

If the chemicals are not hazardous then they just need to be mopped up. But sometimes they do involve hazardous materials and there will be an existing plan on how to deal with them. A risk assessment will have been carried out.

Falls from height

A fall even from a low height can cause serious injuries. Precautions need to be taken when working at height to avoid permanent injury. You should also consider falls into open excavations as falls from height. All the same precautions need to be in place to prevent a fall.

Burns

Burns can be caused not only by fires and heat, but also from chemicals and solvents. Electricity and wet concrete and cement can also burn skin. PPE is often the best way to avoid these dangers. Sunburn is a common and uncomfortable form of burning and sunscreen should be made available. For example, keeping skin covered up will help to prevent sunburn. You might think a tan looks good, but it could lead to skin cancer.

Electrical

Electricity is hazardous and electric shocks can cause burns and muscle damage, and can kill.

Exposure to hazardous substances

We look at hazardous substances in more detail on pages 18–19. COSHH regulations identify hazardous substances and require them to be labelled. You should always follow the instructions when using them.

Plant and vehicles

On busy sites there is always a danger from moving vehicles and heavy plant. Although many are fitted with reversing alarms, it may not be easy to hear them over other machinery and equipment. You should always ensure you are not blocking routes or exits. Designated walkways separate site traffic and pedestrians – this includes workers who are walking around the site. Crossing points should be in place for ease of movement on site.

Reporting hazards

We have already seen that hazards have the potential to cause serious accidents and injuries. It is therefore important to report hazards and there are different methods of doing this.

The first major reason to report hazards is to prevent danger to others, whether they are other employees or visitors to the site. It is vital to prevent accidents from taking place and to quickly correct any dangerous situations.

Injuries, diseases and actual accidents all need to be reported and so do dangerous occurrences. These are incidents that do not result in an actual injury, but could easily have hurt someone.

Accidents need to be recorded in an accident book, computer database or other secure recording system, as do near misses. Again it is a legal requirement to keep appropriate records of accidents and every company will have a procedure for this which they should tell you about. Everyone should know where the book is kept or how the records are made. Anyone that has been hurt or has taken part in dealing with an occurrence should complete the details of what has happened. Typically this will require you to fill in:

* the date, time and place of the incident

* how it happened

* what was the cause

* how it was dealt with

* who was involved

* signature and date.

The details in the book have to be transferred onto an official HSE report form.

As far as is possible, the site, company or workplace will have set procedures in place for reporting hazards and accidents. These procedures will usually be found in the place where the accident book or records are stored. The location tends to be posted on the site notice board.

How hazards are created

Construction sites are busy places. There are constantly new stages in development. As each stage is begun a whole new set of potential hazards need to be considered.

At the same time, new workers will always be joining the site. It is mandatory for them to be given health and safety instruction during induction. But sometimes this is impossible due to pressure of work or availability of trainers.

Construction sites can become even more hazardous in times of extreme weather:

* Flooding – long periods of rain can cause trenches to fill with water, cellars to be flooded and smooth surfaces to become extremely wet and slippery.

* Wind – strong winds may prevent all work at height. Scaffolding may have become unstable, unsecured roofing materials may come loose, dry-stored materials such as sand and cement may have been blown across the site.

* Heat – this can change the behaviour of materials: setting quicker, failing to cure and melting. It can also seriously affect the health of the workforce through dehydration and heat exhaustion.

* Snow – this can add enormous weight to roofs and other structures and could cause collapse. Snow can also prevent access or block exits and can mean that simple and routine work becomes impossible due to frozen conditions.

Storing combustibles and chemicals

A combustible substance can be both flammable and explosive. There are some basic suggestions from the HSE about storing these:

* Ventilation – the area should be well ventilated to disperse any vapours that could trigger off an explosion.

* Ignition – an ignition is any spark or flame that could trigger off the vapours, so materials should be stored away from any area that uses electrical equipment or any tool that heats up.

* Containment – the materials should always be kept in proper containers with lids and there should be spillage trays to prevent any leak seeping into other parts of the site.

* Exchange – in many cases it can be possible to find an alternative material that is less dangerous. This option should be taken if possible.

* Separation – always keep flammable substances away from general work areas. If possible they should be partitioned off.

Combustible materials can include a large number of commonly used substances, such as cleaning agents, paints and adhesives.

HEALTH AND HYGIENE

Just as hazards can be a major problem on site, other less obvious problems relating to health and hygiene can also be an issue. It is both your responsibility and that of your employer to make sure that you stay healthy.

The employer will need to provide basic welfare facilities, no matter where you are working and these must have minimum standards.

Welfare facilities

Welfare facilities can include a wide range of different considerations, as can be seen in Table 1.7.

DID YOU KNOW?

You do not have to be involved in specialist work to come into contact with combustibles.

Facilities	Purpose and minimum standards
Toilets	If there is a lock on the door there is no need to have separate male and female toilets. There should be enough for the site workforce. If there is no flushing water on site they must be chemical toilets.
Washing facilities	There should be a wash basin large enough to be able to wash up to the elbow. There should be soap, hot and cold water and, if you are working with dangerous substances, then showers are needed.
Drinking water	Clean drinking water should be available; either directly connected to the mains or bottled water. Employers must ensure that there is no **contamination.**
Dry room	This can operate also as a store room, which needs to be secure so that workers can leave their belongings there and also use it as a place to dry out if they have been working in wet weather, in which case a heater needs to be provided.
Work break area	This is a shelter out of the wind and rain, with a kettle, a microwave, tables and chairs. It should also have heating.

Table 1.7 Welfare facilities in the workplace

KEY TERMS

Contamination

– this is when a substance has been polluted by some harmful substance or chemical.

CASE STUDY

South Tyneside Homes

South Tyneside Council's Housing Company

Staying safe on site

Johnny McErlane finished his apprenticeship at South Tyneside Homes a year ago.

'I've been working on sheltered accommodation for the last year, so there are a lot of vulnerable and elderly people around. All the things I learnt at college from doing the health and safety exams comes into practice really, like taking care when using extension leads, wearing high-vis and correct footwear. It's not just about your health and safety, but looking out for others as well.

On the shelters, you can get a health and safety inspector who just comes around randomly, so you have to always be ready. It just becomes a habit once it's been drilled into you. You're health and safety conscious all the time.

The shelters also have a fire alarm drill every second Monday, so you've got to know the procedure involved there. When it comes to the more specialised skills, such as mouth-to-mouth and CPR, you might have a designated first aider on site who will have their skills refreshed regularly. Having a full first aid certificate would be valuable if you're working in construction.

You cover quite a bit of the first aid skills in college and you really have to know them because you're not always working on large sites. For example, you might be on the repairs team, working in people's houses where you wouldn't have a first aider, so you've got to have the basic knowledge yourself, just in case. All our vans have a basic first aid kit that's kept fully stocked.

The company keeps our knowledge current with these "toolbox talks", which are like refresher courses. They give you any new information that needs to be passed on to all the trades. It's a good way of keeping everyone up to date.'

Noise

Ear defenders are the best precaution to protect the ears from loud noises on site. Ear defenders are either basic ear plugs or ear muffs, which can be seen in Fig 1.13 on page 29.

The long-term impact of noise depends on the intensity and duration of the noise. Basically, the louder and longer the noise exposure, the more damage is caused. There are ways of dealing with this:

* Remove the source of the noise.

* Move the equipment away from those not directly working with it.

* Put the source of the noise into a soundproof area or cover it with soundproof material.

* Ask a supervisor if they can move all other employees away from that part of the site until the noise stops.

Substances hazardous to health

COSHH Regulations (see pages 2–3) identify a wide variety of substances and materials that must be labelled in different ways.

Controlling the use of these substances is always difficult. Ideally, their use should be eliminated (stopped) or they should be replaced with something less harmful. Failing this, they should only be used in controlled or restricted areas. If none of this is possible then they should only be used in controlled situations.

If a hazardous situation occurs at work, then you should:

* ensure the area is made safe

* inform the supervisor, site manager, safety officer or other nominated person.

You will also need to report any potential hazards or near misses.

Personal hygiene

Construction sites can be dirty places to work. Some jobs will expose you to dust, chemicals or substances that can make contact with your skin or may stain your work clothing. It is good practice to wear suitable PPE as a first line of defence as chemicals can penetrate your skin. Whenever you have finished a job you should always wash your hands. This is certainly true before eating lunch or travelling home. It can be good practice to have dedicated work clothing, which should be washed regularly.

Always ensure you wash your hands and face and scrub your nails. This will prevent dirt, chemicals and other substances from contaminating your food and your home.

Make sure that you regularly wash your work clothing and either repair it or replace it if it becomes too worn or stained.

Health risks

The construction industry uses a wide variety of substances that could harm your health. You will also be carrying out work that could be a health risk to you, and you should always be aware that certain activities could cause long-term damage or even kill you if things go wrong. Unfortunately not all health risks are immediately obvious. It is important to make sure that from time to time you have health checks, particularly if you have been using hazardous substances. Table 1.8 outlines some potential health risks in a typical construction site.

Health risk	Potential future problems
Dust	The most dangerous potential dust is, of course, asbestos, which **should only be handled by specialists under controlled conditions**. But even brick dust and other fine particles can cause eye injuries, problems with breathing and even cancer.
Chemicals	Inhaling or swallowing dangerous chemicals could cause immediate, long-term damage to lungs and other internal organs. Skin problems include burns or skin can become very inflamed and sore. This is known as **dermatitis**.
Bacteria	Contact with waste water or soil could lead to a bacterial infection. The germs in the water or dirt could cause infection which will require treatment if they enter the body. The most extreme version is **leptospirosis**.
Heavy objects	Lifting heavy, bulky or awkward objects can lead to permanent back injuries that could require surgery. Heavy objects can also damage the muscles in all areas of the body.
Noise	Failure to wear ear defenders when you are exposed to loud noises can permanently affect your hearing. This could lead to deafness in the future.
Vibrating tools	Using machines that vibrate can cause a condition known as hand/arm vibration syndrome (HAVS) or vibration white finger, which is caused by injury to nerves and blood vessels. You will feel tingling that could lead to permanent numbness in the fingers and hands, as well as muscle weakness.
Cuts	Any open wound, no matter how small, leaves your body exposed to potential infections. Cuts should always be cleaned and covered, preferably with a waterproof dressing. The blood loss from deep cuts could make you feel faint and weak, which may be dangerous if you are working at height or operating machinery.
Sunlight	Most construction work involves working outside. There is a temptation to take advantage of hot weather and get a tan. But long-term exposure to sunshine means risking skin cancer so you should cover up and apply sun cream.
Head injuries	You should seek medical attention after any bump to the head. Severe head injuries could cause epilepsy, hearing problems, brain damage or death.

Table 1.8 Health risks in construction

KEY TERMS

Dermatitis

– this is an inflammation of the skin. The skin will become red and sore, particularly if you scratch the area. A GP should be consulted.

Leptospirosis

– this is also known as Weil's disease. It is spread by touching soil or water contaminated with the urine of wild animals infected with the leptospira bacteria. Symptoms are usually flu-like but in extreme cases it can cause organ failure.

HANDLING AND STORING MATERIALS AND EQUIPMENT

On a busy construction site it is often tempting not to even think about the potential dangers of handling equipment and materials. If something needs to be moved or collected you will just pick it up without any thought. It is also tempting just to drop your tools and other equipment when you have finished with them to deal with later. But abandoned equipment and tools can cause hazards both for you and for other people.

Safe lifting

Lifting or handling heavy or bulky items is a major cause of injuries on construction sites. So whenever you are dealing with a heavy load, it is important to carry out a basic risk assessment.

The first thing you need to do is to think about the job to be done and ask:

* Do I need to lift it manually or is there another way of getting the object to where I need it?

Consider any mechanical methods of transporting loads or picking up materials. If there really is no alternative, then ask yourself:

1. Do I need to bend or twist?

2. Does the object need to be lifted or put down from high up?

3. Does the object need to be carried a long way?

4. Does the object need to be pushed or pulled for a long distance?

5. Is the object likely to shift around while it is being moved?

If the answer to any of these questions is 'yes', you may need to adjust the way the task is done to make it safer.

Think about the object itself. Ask:

1. Is it just heavy or is it also bulky and an awkward shape?

2. How easy is it to get a good hand-hold on the object?

3. Is the object a single item or are there parts that might move around and shift the weight?

4. Is the object hot or does it have sharp edges?

Again, if you have answered 'yes' to any of these questions, then you need to take steps to address these issues.

It is also important to think about the working environment and where the lifting and carrying is taking place. Ask yourself:

1. Are the floors stable?

2. Are the surfaces slippery?

3. Will a lack of space restrict my movement?

4. Are there any steps or slopes?

5. What is the lighting like?

DID YOU KNOW?

Although many people regard the weight limit for lifting and/or moving heavy or awkward objects to be 20 kg, the HSE does not recommend safe weights. There are many things that will affect the ability of an individual to lift and carry particular objects and the risk that this creates, so manual handling should be avoided altogether where possible.

Before lifting and moving an object, think about the following:

* Check that your pathway is clear to where the load needs to be taken.
* Look at the product data sheet and assess the weight. If you think the object is too heavy or difficult to move then ask someone to help you. Alternatively, you may need to use a mechanical lifting device.

When you are ready to lift, gently raise the load. Take care to ensure the correct posture – you should have a straight back, with your elbows tucked in, your knees bent and your feet slightly apart.

Once you have picked up the load, move slowly towards your destination. When you get there, make sure that you do not drop the load but carefully place it down.

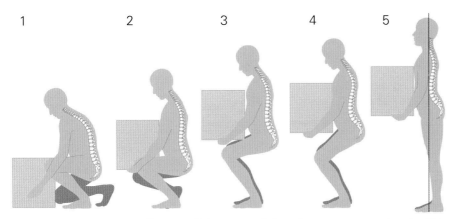

Figure 1.4 Take care to follow the correct procedure for lifting

Figure 1.5 Pallet truck

Sack trolleys are useful for moving heavy and bulky items around. Gently slide the bottom of the sack trolley under the object and then raise the trolley to an angle of 45° before moving off. Make sure that the object is properly balanced and is not too big for the trolley.

Trailers and forklift trucks are often used on large construction sites, as are dump trucks. Never use these without proper training.

Site safety equipment

You should always read the construction site safety rules and when required wear your PPE. Simple things, such as wearing the right footwear for the right job, are important.

Figure 1.6 Sack trolley

Safety equipment falls into two main categories:

* PPE – including hard hats, footwear, gloves, glasses and safety vests
* perimeter safety – this includes screens, netting and guards or clamps to prevent materials from falling or spreading.

Construction safety is also directed by signs, which will highlight potential hazards.

Safe handling of materials and equipment

All tools and equipment are potentially dangerous. It is up to you to make sure that they do not cause harm to yourself or others. You should always know how to use tools and equipment. This means either instruction from someone else who is experienced, or at least reading the manufacturer's instructions.

You should always make sure that you:

* use the right tool – don't be tempted to use a tool that is close to hand instead of the one that is right for the job

* wear your PPE – the one time you decide not to bother could be the time that you injure yourself

* never try to use a tool or a piece of equipment that you have not been trained to use.

You should always remember that if you are working on a building that was constructed before 2000 it may contain asbestos.

Correct storage

We have already seen that tools and equipment need to be treated with respect. Damaged tools and equipment are not only less effective at doing their job, they could also cause you to injure yourself.

Table 1.9 provides some pointers on how to store and handle different types of materials and equipment.

Materials and equipment	Safe storage and handling
Hand tools	Store hand tools with sharp edges either in a cover or a roll. They should be stored in bags or boxes. They should always be dried before putting them away as they will rust.
Power tools	Never carry them by the cable. Store them in their original carrying case. Always follow the manufacturer's instructions.
Wheelbarrows	Check the tyres and metal stays regularly. Always clean out after use and never overload.
Bricks and blocks	Never store more than two packs high. When cutting open a pack, be careful as the bricks could collapse.
Slabs and curbs	Store slabs flat on their edges on level ground, preferably with wood underneath to prevent damage. Store curbs the same way. To prevent weather damage, cover them with a sheet.
Tiles	Always cover them and protect them from damage as they are relatively fragile. Ideally store them in a hut or container.
Aggregates	Never store aggregates under trees as leaves will drop on them and contaminate them. Cover them with plastic sheets.
Plaster and plasterboard	Plaster needs to be kept dry, so even if stored inside you should take the precaution of putting the bags on pallets. To prevent moisture do not store against walls and do not pile higher than five bags. Plasterboard can be awkward to manage and move around. It also needs to be stored in a waterproof area. It should be stored flat and off the ground but should not be stored against walls as it may bend. Use a rotation system so that the materials are not stored in the same place for long periods.
Wood	Always keep wood in dry, well-ventilated conditions. If it needs to be stored outside it should be stored on bearers that may be on concrete. If wood gets wet and bends it is virtually useless. Always be careful when moving large cuts of wood or sheets of ply or MDF as they can easily become damaged.
Adhesives and paint	Always read the manufacturer's instructions. Ideally they should always be stored on clearly marked shelves. Make sure you rotate the stock using the older stock first. Always make sure that containers are tightly sealed. Storage areas must comply with fire regulations and display signs to advise of their contents.

Table 1.9 Safe storing and handling of materials and equipment

Waste control

The expectation within the building services industry is increasingly that working practices conserve energy and protect the environment. Everyone can play a part in this. For example, you can contribute by turning off hose pipes when you have finished using water, or not running electrical items when you don't need to.

Simple things, such as keeping construction sites neat and orderly, can go a long way to conserving energy and protecting the environment. A good way to remember this is Sort, Set, Shine, Standardise:

* Sort – sort and store items in your work area, eliminate clutter and manage deliveries.

* Set – everything should have its own place and be clearly marked and easy to access. In other words, be neat!

* Shine – clean your work area and you will be able to see potential problems far more easily.

* Standardise – by using standardised working practices you can keep organised, clean and safe.

Reducing waste is all about good working practice. By reducing wastage disposal, and recycling materials on site, you will benefit from savings on raw materials and lower transportation costs.

Planning ahead, and accurately measuring and cutting materials, means that you will be able to reduce wastage.

Figure 1.7 It's important to create as little waste as possible on the construction site

BASIC WORKING PLATFORMS AND ACCESS EQUIPMENT

Working at height should be eliminated or the work carried out using other methods where possible. However, there may be situations where you may need to work at height. These situations can include:

* roofing

* repair and maintenance above ground level

* working on high ceilings.

Any work at height must be carefully planned. Access equipment includes all types of ladder, scaffold and platform. You must always use a working platform that is safe. Sometimes a simple step ladder will be sufficient, but at other times you may have to use a tower scaffold.

Generally, ladders are fine for small, quick jobs of less than 30 minutes. However, for larger, longer jobs a more permanent piece of access equipment will be necessary.

Working platforms and access equipment: good practice and dangers of working at height

Table 1.10 outlines the common types of equipment used to allow you to work at heights, along with the basic safety checks necessary.

Equipment	Main features	Safety checks
Step ladder	Ideal for confined spaces. Four legs give stability	• Knee should remain below top of steps • Check hinges, cords or ropes • Position only to face work
Ladder	Ideal for basic access, short-term work. Made from aluminium, fibreglass or wood	• Check rungs, tie rods, repairs, and ropes and cords on stepladders • Ensure it is placed on firm, level ground • Angle should be no greater than 75° or 1 in 4
Mobile mini towers or scaffolds	These are usually aluminium and foldable, with lockable wheels	• Ensure the ground is even and the wheels are locked • Never move the platform while it has tools, equipment or people on it
Roof ladders and crawling boards	The roof ladder allows access while crawling boards provide a safe passage over tiles	• The ladder needs to be long enough and supported • Check boards are in good condition • Check the welds are intact • Ensure all clips function correctly
Mobile tower scaffolds	These larger versions of mini towers usually have edge protection	• Ensure the ground is even and the wheels are locked • Never move the platform while it has tools, equipment or people on it • Base width to height ratio should be no greater than 1:3
Fixed scaffolds and edge protection	Scaffolds fitted and sized to the specific job, with edge protection and guard rails	• There needs to be sufficient braces, guard rails and scaffold boards • The tubes should be level • There should be proper access using a ladder
Mobile elevated work platforms	Known as scissor lifts or cherry pickers	• Specialist training is required before use • Use guard rails and toe boards • Care needs to be taken to avoid overhead hazards such as cables

Table 1.10 Equipment for working at height and safety checks

Figure 1.8 A tower scaffold

You must be trained in the use of certain types of access equipment, like mobile scaffolds. Care needs to be taken when assembling and using access equipment. These are all examples of good practice:

* Step ladders should always rest firmly on the ground. Only use the top step if the ladder is part of a platform.

* Do not rest ladders against fragile surfaces, and always use both hands to climb. It is best if the ladder is steadied (footed) by someone at the foot of the ladder. Always maintain three points of contact – two feet and one hand.

* A roof ladder is positioned by turning it on its wheels and pushing it up the roof. It then hooks over the ridge tiles. Ensure that the access ladder to the roof is directly beside the roof ladder.

* A mobile scaffold is put together by slotting sections until the required height is reached. The working platform needs to have a suitable edge protection such as guard-rails and toe-boards. Always push from the bottom of the base and not from the top to move it, otherwise it may lean or topple over.

WORKING SAFELY WITH ELECTRICITY

It is essential whenever you work with electricity that you are competent and that you understand the common dangers. Electrical tools must be used in a safe manner on site. There are precautions that you can take to prevent possible injury, or even death.

Precautions

Whether you are using electrical tools or equipment on site, you should always remember the following:

* Use the right tool for the job.
* Use a transformer with equipment that runs on 110V.
* Keep the two voltages separate from each other. You should avoid using 230V where possible but, if you must, use a residual current device (RCD) if you have to use 230V.
* When using 110V, ensure that leads are yellow in colour.
* Check the plug is in good order.
* Confirm that the fuse is the correct rating for the equipment.
* Check the cable (including making sure that it does not present a tripping hazard).
* Find out where the mains switch is, in case you need to turn off the power in the event of an emergency.
* Never attempt to repair electrical equipment yourself.
* Disconnect from the mains power before making adjustments, such as changing a drill bit.
* Make sure that the electrical equipment has a sticker that displays a recent test date.

Visual inspection and testing is a three-stage process:

1. The user should check for potential danger signs, such as a frayed cable or cracked plug.

2. A formal visual inspection should then take place. If this is done correctly then most faults can be detected.

3. Combined inspections and **PAT** should take place at regular intervals by a competent person.

Watch out for the following causes of accidents – they would also fail a safety check:

* damage to the power cable or plug
* taped joints on the cable
* wet or rusty tools and equipment
* weak external casing
* loose parts or screws

* signs of overheating
* the incorrect fuse
* lack of cord grip
* electrical wires attached to incorrect terminals
* bare wires.

KEY TERMS

PAT

– Portable Appliance Testing – regular testing is a health and safety requirement under the Electricity at Work Regulations (1989).

When preparing to work on an electrical circuit, do not start until a permit to work has been issued by a supervisor or manager to a competent person.

Make sure the circuit is broken before you begin. A 'dead' circuit will not cause you, or anybody else, harm. These steps must be followed:

* Switch off – ensure the supply to the circuit is switched off by disconnecting the supply cables or using an isolating switch.

* Isolate – disconnect the power cables or use an isolating switch.

* Warn others – to avoid someone reconnecting the circuit, place warning signs at the isolation point.

* Lock off – this step physically prevents others from reconnecting the circuit.

* Testing – is carried out by electricians but you should be aware that it involves three parts:

    1. testing a voltmeter on a known good source (a live circuit) so you know it is working properly

    2. checking that the circuit to be worked on is dead

    3. rechecking your voltmeter on the known live source, to prove that it is still working properly.

It is important to make sure that the correct point of isolation is identified. Isolation can be next to a local isolation device, such as a plug or socket, or a circuit breaker or fuse.

The isolation should be locked off using a unique key or combination. This will prevent access to a main isolator until the work has been completed. Alternatively, the handle can be made detachable in the OFF position so that it can be physically removed once the circuit is switched off.

Dangers

You are likely to encounter a number of potential dangers when working with electricity on construction sites or in private houses. Table 1.11 outlines the most common dangers.

Danger	Identifying the danger
Faulty electrical equipment	Visually inspect for signs of damage. Equipment should be double insulated or incorporate an earth cable.
Damaged or worn cables	Check for signs of wear or damage regularly. This includes checking power tools and any wiring in the property.
Trailing cables	Cables lying on the ground, or worse, stretched too far, can present a tripping hazard. They could also be cut or damaged easily.
Cables and pipe work	Always treat services you find as though they are live. This is very important as services can be mistaken for one another. You may have been trained to use a cable and pipe locator that finds cables and metal pipes.
Buried or hidden cables	Make sure you have plans. Alternatively, use a cable and pipe locator, mark the positions, look out for signs of service connection cables or pipes and hand-dig trial holes to confirm positions.
Inadequate over-current protection	Check circuit breakers and fuses are the correct size current rating for the circuit. A qualified electrician may have to identify and label these.

Table 1.11 Common dangers when working with electricity

Each year there are around 1,000 accidents at work involving electric shocks or burns from electricity. If you are working in a construction site you are part of a group that is most at risk. Electrical accidents happen when you are working close to equipment that you think is disconnected but which is, in fact, live.

Another major danger is when electrical equipment is either misused or is faulty. Electricity can cause fires and contact with the live parts can give you an electric shock or burn you.

Different voltages

The two most common voltages that are used in the UK are 230V and 110V:

* 230V: this is the standard domestic voltage. But on construction sites it is considered to be unsafe and therefore 110V is commonly used.

* 110V: these plugs are marked with a yellow casement and they have a different shaped plug. A transformer is required to convert 230V to 110V.

Some larger homes, as well as industrial and commercial buildings, may have 415V supplies. This is the same voltage that is found on overhead electricity cables. In most houses and other buildings the voltage from these cables is reduced to 230V. This is what most electrical equipment works from. Some larger machinery actually needs 415V.

In these buildings the 415V comes into the building and then can either be used directly or it is reduced so that normal 230V appliances can be used.

Colour coded cables

Normally you will come across three differently coloured wires: Live, Neutral and Earth. These have standard colours that comply with European safety standards and to ensure that they are easily identifiable. However, in some older buildings the colours are different.

Wire type	Modern colour	Older colour
Live	Brown	Red
Neutral	Blue	Black
Earth	Yellow and Green	Yellow and Green

Table 1.12 Colour coding of cables

Working with equipment with different electrical voltages

You should always check that the electrical equipment that you are going to use is suitable for the available electrical supply. The equipment's power requirements are shown on its rating plate. The voltage from the supply needs to match the voltage that is required by the equipment.

Storing electrical equipment

Electrical equipment should be stored in dry and secure conditions. Electrical equipment should never get wet but – if it does happen – it should be dried before storage. You should always clean and adjust the equipment before connecting it to the electricity supply.

PERSONAL PROTECTIVE EQUIPMENT (PPE)

Personal protective equipment, or PPE, is a general term that is used to describe a variety of different types of clothing and equipment that aim to help protect against injuries or accidents. Some PPE you will use on a daily basis and others you may use from time to time. The type of PPE you wear depends on what you are doing and where you are. For example, the practical exercises in this book were photographed at a college, which has rules and requirements for PPE that are different to those on large construction sites. Follow your tutor's or employer's instructions at all times.

Types of PPE

PPE literally covers from head to foot. Here are the main PPE types.

Protective clothing

Clothing protection such as overalls:

Figure 1.9 A hi-vis jacket

* provides some protection from spills, dust and irritants
* can help protect you from minor cuts and abrasions
* reduces wear to work clothing underneath.

Sometimes you may need waterproof or chemical-resistant overalls.

High visibility (hi-vis) clothing stands out against any background or in any weather conditions. It is important to wear high visibility clothing on a construction site to ensure that people can see you easily. In addition, workers should always try to wear light-coloured clothing underneath, as it is easier to see.

You need to keep your high visibility and protective clothing clean and in good condition.

Employers need to make sure that employees understand the reasons for wearing high visibility clothing and the consequences of not doing so.

Eye protection

Figure 1.10 Safety glasses and goggles

For many jobs, it is essential to wear goggles or safety glasses to prevent small objects, such as dust, wood or metal, from getting into the eyes. As goggles tend to steam up, particularly if they are being worn with a mask, safety glasses can often be a good alternative.

Hand protection

Figure 1.11 Hand protection

Wearing gloves will help to prevent damage or injury to the hands or fingers. For example, general purpose gloves can prevent cuts, and rubber gloves can prevent skin irritation and inflammation, such as contact dermatitis caused by handling hazardous substances. There are many different types of gloves available, including specialist gloves for working with chemicals.

Head protection

Hard hats or safety helmets are compulsory on building sites. They can protect you from falling objects or banging your head. They need to fit well and they should be regularly inspected and checked for cracks. Worn straps mean that the helmet should be replaced, as a blow to the head can be fatal. Hard hats bear a date of manufacture and should be replaced after about 3 years.

Hearing protection

Ear defenders, such as ear protectors or plugs, aim to prevent damage to your hearing or hearing loss when you are working with loud tools or are involved in a very noisy job.

Figure 1.12 Head protection

Respiratory protection

Breathing in fibre, dust or some gases could damage the lungs. Dust is a very common danger, so a dust mask, face mask or respirator may be necessary.

Make sure you have the right mask for the job. It needs to fit properly otherwise it will not give you sufficient protection.

Foot protection

Foot protection is compulsory on site, particularly if you are undertaking heavy work. Footwear should include steel toecaps (or equivalent) to protect feet against dropped objects, midsole protection (usually a steel plate) to protect against puncture or penetration from things like nails on the floor and soles with good grip to help prevent slips on wet surfaces.

Figure 1.13 Hearing protection

Legislation covering PPE

The most important piece of legislation is the Personal Protective Equipment at Work Regulations (1992). It covers all sorts of PPE and sets out your responsibilities and those of the employer. Linked to this are the Control of Substances Hazardous to Health (2002) and the Provision and Use of Work Equipment Regulations (1992 and 1998).

Figure 1.14 Respiratory protection

Storing and maintaining PPE

All forms of PPE will be less effective if they are not properly maintained. This may mean examining the PPE and either replacing or cleaning it, or if relevant testing or repairing it. PPE needs to be stored properly so that it is not damaged, contaminated or lost. Each type of PPE should have a CE mark. This shows that it has met the necessary safety requirements.

Importance of PPE

PPE needs to be suitable for its intended use and it needs to be used in the correct way. As a worker or an employee you need to:

* make sure you are trained to use PPE

* follow your employer's instructions when using the PPE and always wear it when you are told to do so

* look after the PPE and if there is a problem with it report it.

Your employer will:

* know the risks that the PPE will either reduce or avoid

* know how the PPE should be maintained

* know its limitations.

Consequences of not using PPE

The consequences of not using PPE can be immediate or long-term. Immediate problems are more obvious, as you may injure yourself. The longer-term consequences could be ill health in the future. If your employer has provided PPE, you have a legal responsibility to wear it.

FIRE AND EMERGENCY PROCEDURES

If there is a fire or an emergency, it is vital that you raise the alarm quickly. You should leave the building or site and then head for the **assembly point.**

When there is an emergency a general alarm should sound. If you are working on a larger and more complex construction site, evacuation may begin by evacuating the area closest to the emergency. Areas will then be evacuated one-by-one to avoid congestion of the escape routes.

Three elements essential to creating a fire

Three ingredients are needed to make something combust (burn):

* oxygen   * heat   * fuel.

The fuel can be anything which burns, such as wood, paper or flammable liquids or gases, and oxygen is in the air around us, so all that is needed is sufficient heat to start a fire.

The fire triangle represents these three elements visually. By removing one of the three elements the fire can be prevented or extinguished.

Figure 1.15 Assembly point sign   Figure 1.16 The fire triangle

How fire is spread

Fire can easily move from one area to another by finding more fuel. You need to consider this when you are storing or using materials on site, and be aware that untidiness can be a fire risk. For example, if there are wood shavings on the ground the fire can move across them, burning up the shavings.

Heat can also transfer from one source of fuel to another. If a piece of wood is on fire and is against or close to another piece of wood, that too will catch fire and the fire will have spread.

On site, fires are classified according to the type of material that is on fire. This will determine the type of fire-fighting equipment you will need to use. The five different types of fire are shown in Table 1.13.

Class of fire	Fuel or material on fire
A	Wood, paper and textiles
B	Petrol, oil and other flammable liquids
C	LPG, propane and other flammable gases
D	Metals and metal powder
E	Electrical equipment

Table 1.13 Different classes of fire

There is also F, cooking oil, but this is less likely to be found on site, except in a kitchen.

Taking action if you discover a fire and fire evacuation procedures

During induction, you will have been shown what to do in the event of a fire and told about assembly points. These are marked by signs and somewhere on the site there will be a map showing their location.

If you discover a fire you should:

* sound the alarm

* not attempt to fight the fire unless you have had fire marshal training

* otherwise stop work, do not collect your belongings, do not run, and do not re-enter the site until the all clear has been given.

Different types of fire extinguishers

Extinguishers can be effective when tackling small localised fires. However, you must use the correct type of extinguisher. For example, putting water on an oil fire could make it explode. For this reason, you should not attempt to use a fire extinguisher unless you have had proper training.

When using an extinguisher it is important to remember the following safety points:

* Only use an extinguisher at the early stages of a fire, when it is small.

* The instructions for use appear on the extinguisher.

* If you do choose to fight the fire because it is small enough, and you are sure you know what is burning, position yourself between the fire and the exit, so that if it doesn't work you can still get out.

Type of fire risk	Fire class Symbol	White label	Cream label	Black label	Blue label	Yellow label
		Water	Foam	Carbon dioxide	Dry powder	Wet chemical
A – Solid (e.g. wood or paper)	A	✓	✓	✗	✓	✓
B – Liquid (e.g. petrol)	B	✗	✓	✓	✓	✗
C – Gas (e.g. propane)	C	✗	✗	✓	✓	✗
D – Metal (e.g. aluminium)	D METAL	✗	✗	✗	✓	✗
E – Electrical (i.e. any electrical equipment)	E	✗	✗	✓	✓	✗
F – Cooking oil (e.g. a chip pan)	F	✗	✗	✗	✗	✓

Table 1.14 Types of fire extinguishers

There are some differences you should be aware of when using different types of extinguisher:

* *CO$_2$ extinguishers* – do not touch the nozzle; simply operate by holding the handle. This is because the nozzle gets extremely cold when ejecting the CO$_2$, as does the canister. Fires put out with a CO$_2$ extinguisher may reignite, and you will need to ventilate the room after use.

* *Powder extinguishers* – these can be used on lots of kinds of fire, but can seriously reduce visibility by throwing powder into the air as well as on the fire.

SIGNS AND SAFETY NOTICES

In a well-organised working environment safety signs will warn you of potential dangers and tell you what to do to stay safe. They are used to warn you of hazards. Their purpose is to prevent accidents. Some will tell you what to do (or not to do) in particular parts of the site and some will show you where things are, such as the location of a first aid box or a fire exit.

Types of signs and safety notices

There are five basic types of safety sign, as well as signs that are a combination of two or more of these types. These are shown in Table 1.15.

Type of safety sign	What it tells you	What it looks like	Example
Prohibition sign	Tells you what you must *not* do	Usually round, in red and white	Do not use ladder
Hazard sign	Warns you about hazards	Triangular, in yellow and black	Caution Slippery floor
Mandatory sign	Tells you what you *must* do	Round, usually blue and white	Masks must be worn in this area
Safe condition or information sign	Gives important information, e.g. about where to find fire exits, assembly points or first aid kit, or about safe working practices	Green and white	First aid
Firefighting sign	Gives information about extinguishers, hydrants, hoses and fire alarm call points, etc.	Red with white lettering	Fire alarm call point
Combination sign	These have two or more of the elements of the other types of sign, e.g. hazard, prohibition and mandatory		DANGER Isolate before removing cover

Table 1.15 Different types of safety signs

TEST YOURSELF

1. Which of the following requires you to tell the HSE about any injuries or diseases?

   a. HASAWA

   b. COSHH

   c. RIDDOR

   d. PUWER

2. What is a prohibition notice?

   a. An instruction from the HSE to stop all work until a problem is dealt with

   b. A manufacturer's announcement to stop all work using faulty equipment

   c. A site contractor's decision not to use particular materials

   d. A local authority banning the use of a particular type of brick

3. Which of the following is considered a major injury?

   a. Bruising on the knee

   b. Cut

   c. Concussion

   d. Exposure to fumes

4. If there is an accident on a site who is likely to be the first to respond?

   a. First aider

   b. Police

   c. Paramedics

   d. HSE

5. Which of the following is a summary of risk assessments and is used for high risk activities?

   a. Site notice board

   b. Hazard book

   c. Monitoring statement

   d. Method statement

6. Some substances are combustible. Which of the following are examples of combustible materials?

   a. Adhesives

   b. Paints

   c. Cleaning agents

   d. All of these

7. What is dermatitis?

   a. Inflammation of the skin

   b. Inflammation of the ear

   c. Inflammation of the eye

   d. Inflammation of the nose

8. Screens, netting and guards on a site are all examples of which of the following?

   a. PPE

   b. Signs

   c. Perimeter safety

   d. Electrical equipment

9. Which of the following are also known as scissor lifts or cherry pickers?

   a. Bench saws

   b. Hand-held power tools

   c. Cement additives

   d. Mobile elevated work platforms

10. In older properties the neutral electricity wire is which colour?

    a. Black

    b. Red

    c. Blue

    d. Brown

Chapter 2

Unit CSA–L3Core07

ANALYSING TECHNICAL INFORMATION, QUANTITIES AND COMMUNICATION WITH OTHERS

LEARNING OUTCOMES

LO1: Know how to produce different types of drawings and information in the construction industry

LO2: Know how to estimate quantities and price work for contracts

LO3: Know how to ensure good working practices

INTRODUCTION

The aims of this chapter are to:

* help you to interpret information
* help you to estimate quantities
* help you to organise the building process and communicate the design work to colleagues and others.

PRODUCING DIFFERENT TYPES OF DRAWING AND INFORMATION

Accurate construction requires the creation of accurate drawings and matching supporting information. Supporting information can be found in a variety of different types of documents. These include:

* drawings and plans
* programmes of work
* procedures
* specifications
* policies
* schedules
* manufacturers' technical information
* organisational documentation
* training and development records
* risk and method statements
* Construction (Design and Management) (CDM) Regulations
* Building Regulations.

Each different type of construction plan has a definite look and purpose. Typical construction drawings focus in on floor plans or elevation.

Construction drawings are drawn to scale and need to be accurate, so that relative sizes are correct. The scale will be stated on the drawing, to avoid inaccuracies.

Working alongside these construction drawings are matched and linked specifications and schedules. It is these that outline all of the materials and tasks required to complete specific jobs.

In order to understand construction drawings you not only need to understand their purpose and what they are showing, but also a range of hatchings and symbols that act as shortcuts on the documents.

Electronic and traditional drawing methods

Construction drawings are only part of a long process in the design of buildings. In fact the construction drawings are the final stage or final version of these drawings. The design process begins with a basic concept, which is followed by outline drawings. By the end of the design stage working drawings, technical specifications and contract drawings have been completed.

The project is put out to tender. This is a process that involves companies bidding for the job based on the information that they have been given.

There are further changes just before the construction phase gets under way. The chosen construction company may have noted issues with the design, which means that the drawings may have to be amended. It is also at this stage that the construction company will begin the process of pricing up each phase of the job.

Electronic drawing methods

Many construction drawings are based on a system known as computer aided design (CAD). CAD basically produces two-dimensional electronic drawings using the similar lines, hatches and text that can be seen in traditional paper drawings.

Each different CAD drawing is created independently, so each design change has to be followed up on other CAD drawings.

Increasingly, however, a new electronic system is being used. It is known as building information modelling (BIM). This creates drawings in 3D. The buildings are virtually modelled from real construction elements, such as walls, windows and roofs. The big advantages are:

* it allows architects to design buildings in a similar way to the way in which the building will actually be built

* a central virtual building model stores all the data, so any changes to this are applied to individual drawings

* better coordinated designs can be created meaning that construction should be more straightforward.

Systems such as BIM provide 3D models, which can be viewed from any angle or perspective. It also includes:

* scheduling information

* labour required

* estimated costs

* a detailed breakdown of the construction phases.

Figure 2.1 BIM generated model

Traditional drawing methods

The development of the computer, laptops and hand-held tablets such as iPads is gradually making manual drafting of construction drawings obsolete. The majority of drawings are now created using CAD or BIM software.

Traditionally, drawings were limited to the available paper size and what would be convenient to transport.

As each of the traditional construction drawings were hand drawn, there was always a danger that the information on one drawing would not match the information on another. The only way to check that both were accurate was to cross-reference every detail.

One advantage is that paper plans are easier to carry around site, as computers can be broken or stolen. However, damaging or losing a paper plan can cause delays while it is replaced.

Types of supporting information

Drawings and plans

Drawings are an important part of construction work. You will need to understand how they provide you with the information you need to carry out the work. The drawings show what the building will look like and how it will be constructed. This means that there are several different drawings of the building from different viewpoints. In practice most of the drawings are shown on the same sheet.

Block plans

Block plans show the construction site and the surrounding area. Normally block plans are at a ratio of 1:1250 and 1:2500. This means that 1 millimetre on a block plan is equal to 1,250mm (12.5m) or 2,500mm (25m) or on the ground.

Site plan

Often location drawings are also known as block plans or site plans. The site plan drawing shows what is planned for the site. It is often an important drawing because it has been created in order to get approval for the project from planning committees or funding sources. In most cases the site plan is actually an architectural plan, showing the basic arrangement of buildings and any landscaping.

The site plan will usually show:

* directional orientation (i.e. the north point)

* location and size of the building or buildings

* existing structures

* clear measurements.

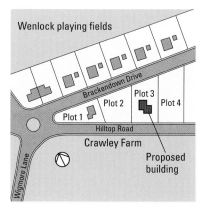

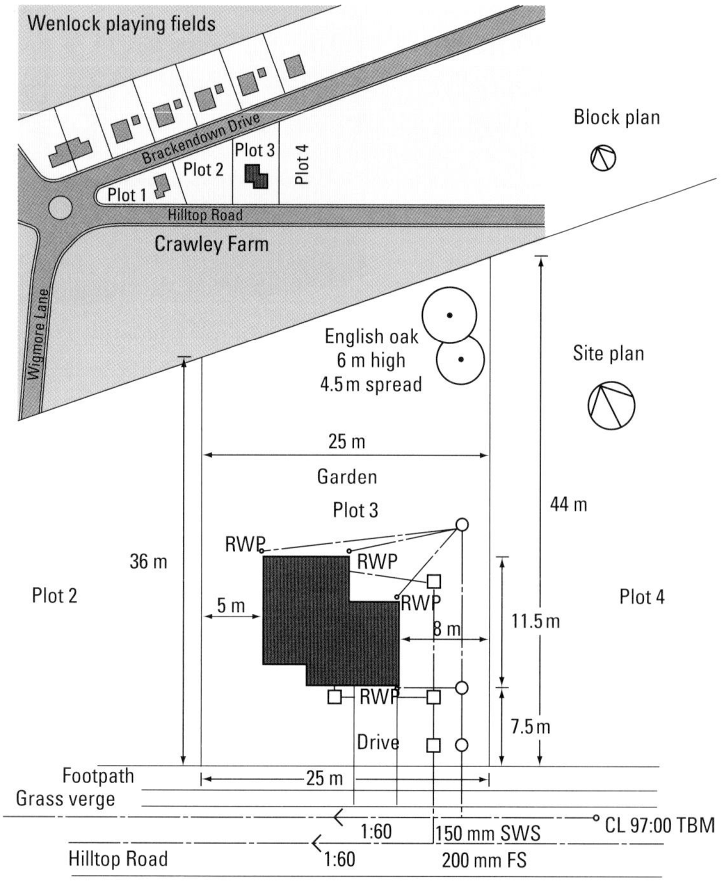

Figure 2.2 Block plan

General location

Location drawings show the site or building in relation to its surroundings. It will therefore show details such as boundaries, other buildings and roads. It will also contain other vital information, including:

* access

* drainage

* sewers

* the north point.

As with all plan drawings, the scale will be shown and the drawing will be given a title. It will be given a job or project number to help identify it easily, as well as an address, the date of the drawing and the name of the client. A version number will also be on the drawing with an amendment date if there have been any changes. You'll need to make sure you have the latest drawing.

Normally location drawings are either 1:200 or 1:500 (that is, 1 mm of the drawing represents 200mm (2m) or 500mm (5m) on the ground).

Figure 2.3 Location plan

Assembly

These are detailed drawings that illustrate the different elements and components of the construction. They tend to be 1:5, 1:10 or 1:20 (1 mm of the drawing represents 5, 10 or 20 mm on the ground). This larger scale allows more detail to be shown, to ensure accurate construction.

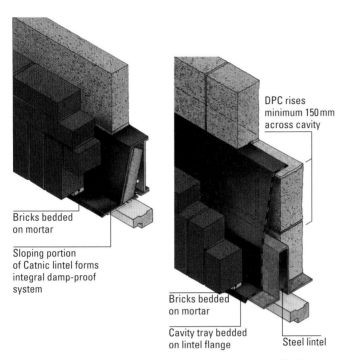

Bricks bedded on mortar

Sloping portion of Catnic lintel forms integral damp-proof system

DPC rises minimum 150 mm across cavity

Bricks bedded on mortar

Cavity tray bedded on lintel flange

Steel lintel

Figure 2.4 Assembly drawing

Sectional

These drawings aim to provide:

● vertical and horizontal measurements and details

● constructional details.

They can be used to show the height of ground levels, damp-proof courses, foundations and other aspects of the construction.

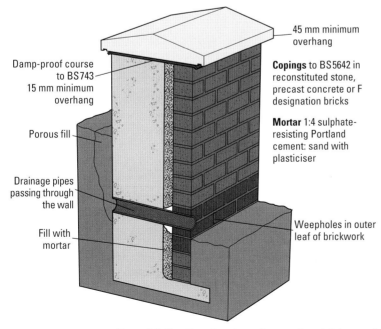

Damp-proof course to BS 743 15 mm minimum overhang

Porous fill

Drainage pipes passing through the wall

Fill with mortar

45 mm minimum overhang

Copings to BS 5642 in reconstituted stone, precast concrete or F designation bricks

Mortar 1:4 sulphate-resisting Portland cement: sand with plasticiser

Weepholes in outer leaf of brickwork

Figure 2.5 Section drawing of an earth retaining wall

serving hatch   vertical section

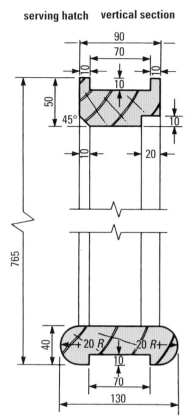

Figure 2.7 Detail drawing

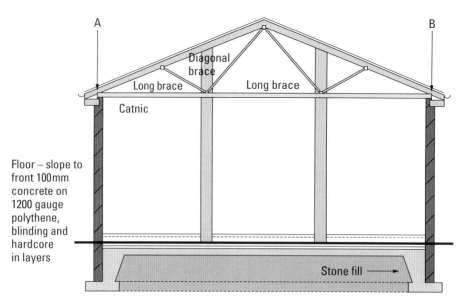

Floor – slope to front 100mm concrete on 1200 gauge polythene, blinding and hardcore in layers

Figure 2.6 Section drawing of a garage

Details

These drawings show how a component needs to be manufactured. They can be shown in various scales, but mainly 1:10, 1:5 and 1:1 (the same size as the actual component if it is small).

Programmes of work

Programmes of work show the actual sequence of any work activities on a construction project. Part of the work programme plan is to show target times. They are usually shown in the form of a Gantt chart (a special type of bar chart), as can be seen in Fig 2.8.

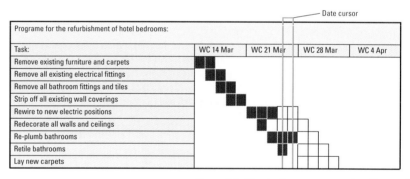

Figure 2.8 Single line contract plan Gantt chart

In this figure:

* on the left-hand side all of the tasks are listed – note this is in logical order

* on the right the blocks show the target start and end date for each of the individual tasks

* the timescale can be days, weeks or months.

Far more complex forms of work programmes can also be created. Fig 2.9 shows the planning for the construction of a house.

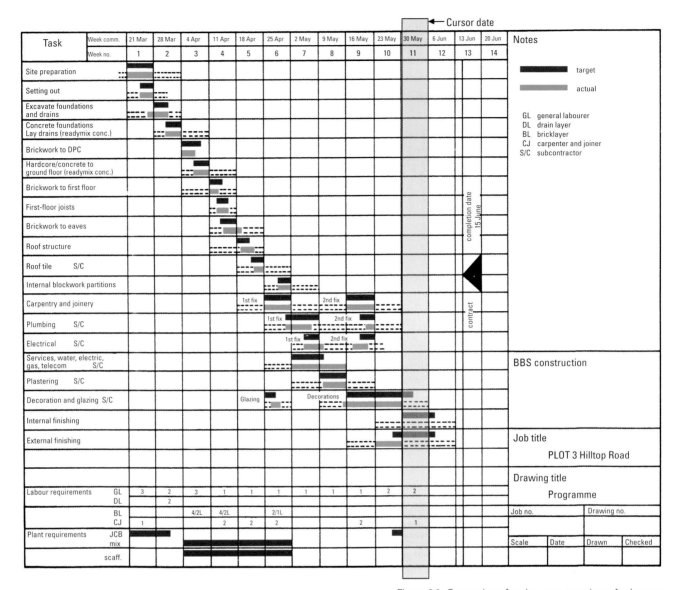

Figure 2.9 Gantt chart for the construction of a house

This more complex example shows the following:

* There are two lines – they show the target dates and actual dates. The actual dates are shaded, showing when the work actually began and how long it took.

* If this bar chart is kept up to date an accurate picture of progress and estimated completion time can be seen.

Procedures

When you work for a construction company they will have a series of procedures they will expect you to follow. A good example is the emergency procedure. This will explain precisely what is required in the case of an emergency on site and who will have responsibility to carry out particular duties. Procedures are there to show you the right way of doing something.

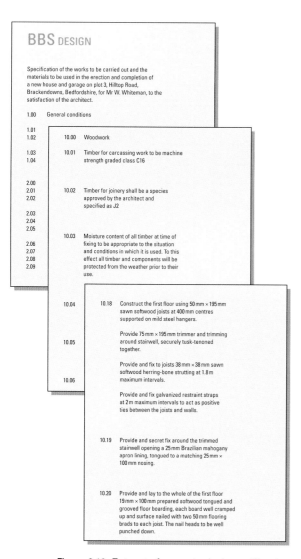

Figure 2.10 Extracts from a typical specification

Another good example of a procedure is the procurement or buying procedure. This will outline:

* who is authorised to buy what, and how much individuals are allowed to spend

* any forms or documents that have to be completed when buying.

Specifications

In addition to drawings it is usually necessary to have documents known as specifications. These provide much more information, as can be seen in Fig 2.10.

The specifications give you a precise description. They will include:

* the address and description of the site

* on-site services (e.g. water and electricity)

* materials description, outlining the size, finish, quality and tolerances

* specific requirements, such as the individual that will authorise or approve work carried out

* any restrictions on site, such as working hours.

Policies

Policies are sets of principles or a programme of actions. The following are two good examples:

* environmental policy – how the business goes about protecting the environment

* safety policy – how the business deals with health and safety matters and who is responsible for monitoring and maintaining it.

You will normally find both policies and procedures in site rules. These are usually explained to each new employee when they first join the company. Sometimes there may be additional site rules, depending on the job and the location of the work.

Schedules

Schedules are cross-referenced to drawings that have been prepared by an architect. They will show specific design information. Usually they are prepared for jobs that will crop up regularly on site, such as:

* working on windows, doors, floors, walls or ceilings

* working on drainage, lintels or sanitary ware.

A schedule can be seen in Fig 2.11.

The schedule is very useful for a number of reasons:

* working out the quantities of materials needed

* ordering materials and components and then checking them against deliveries

* locating where specific materials will be used.

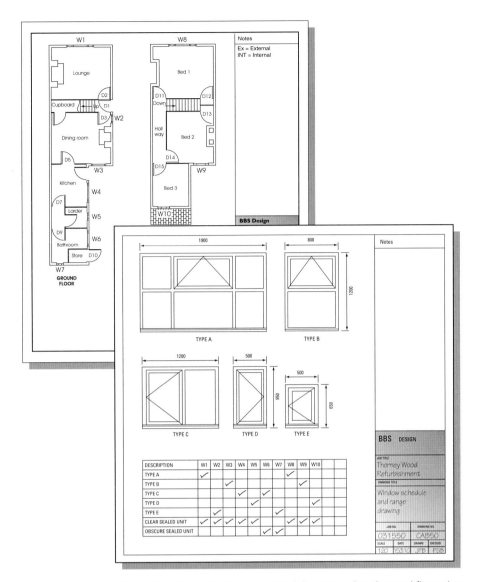

Figure 2.11 Typical windows schedule, range drawing and floor plans

Manufacturers' technical information

Almost everything that is bought to be used on site will come with a variety of information. The basic technical information provided will show what the equipment or material is intended to be used for, how it should be stored and any particular requirements it may have, such as handling or maintenance.

Technical information from the manufacturer can come from a variety of different sources:

* printed or downloadable data sheets

* printed or downloadable user instructions

* manufacturers' catalogues or brochures

* manufacturers' websites.

Organisational documentation

The potential list of organisational documentation and paperwork is massive. Examples are outlined in the following table.

Document	Purpose
Timesheet	Record of hours that you have worked and the jobs that you have carried out. They are used to help work out your wages and the total cost of the job.
Day worksheet	These detail work that has been carried out without providing an estimate beforehand. They usually include repairs or extra work and alterations.
Variation order	These are provided by the architect and given to the builder, showing any alterations, additions or omissions to the original job.
Confirmation notice	Provided by the architect to confirm any verbal instructions.
Daily report or site diary	Include things that might affect the project like detailed weather conditions, late deliveries or site visitors.
Orders and requisitions	These are order forms, requesting the delivery of materials.
Delivery notes	These are provided by the supplier of materials as a list of all materials being delivered. These need to be checked against materials actually delivered.
Delivery record	These are lists of all materials that have been delivered on site.
Memorandum	These are used for internal communications and are usually brief.
Letters	These are used for external communications, usually to customers or suppliers.
Fax	Even though email is commonly used, the industry still likes faxes, because they provide an exact copy of an original document.

Table 2.1

Training and development records

Training and development is an important part of any job, as it ensures that employees have all the skills and knowledge that they need to do their work. Most medium to large employers will have training policies that set out how they intend to do this.

To make sure that they are on track and to keep records they will have a range of different documents. These will record all the training that an employee has undertaken.

Training can take place in a number of different ways:

* induction
* toolbox talks
* in-house training
* specialist training
* training or education leading to formal qualifications.

Details required for floor plans

The floor plans shows the arrangement of the building, rather like a map. It is a cut through of the building, which shows openings, walls and other features usually at around 1 m above floor level.

The floor plan also includes elements of the building that can be seen below the 1 m level, such as the floor or part of the stairs. The drawing will show elements above the 1 m level as dotted lines. The floor plan is a vertical orthographic projection onto a horizontal plane. In effect the horizontal plane cuts through the building.

The floor plan will detail the following:

* Vertical and horizontal sections – these show the building cut along an axis to show the interior structure.

* Datum levels – these are taken from a nearby and convenient datum point. They show the building's levels in relation to the datum point.

* Wall constructions – this is revealed through the section or cross sections shown in the diagram. It details the wall construction methods and materials.

* Material codes – these will contain notes and links to specific materials and may also note particular parts of the Building Regulations that these construction materials comply with.

* Depth and height dimensions – these are drawn between the walls to show the room sizes and wall lengths. They are noted as width × depth.

* Schedules – these note repeated design information, such as types of door, windows and other features.

* Specifications – these outline the type, size and quality of materials, methods of fixing and quality of work and finish expected.

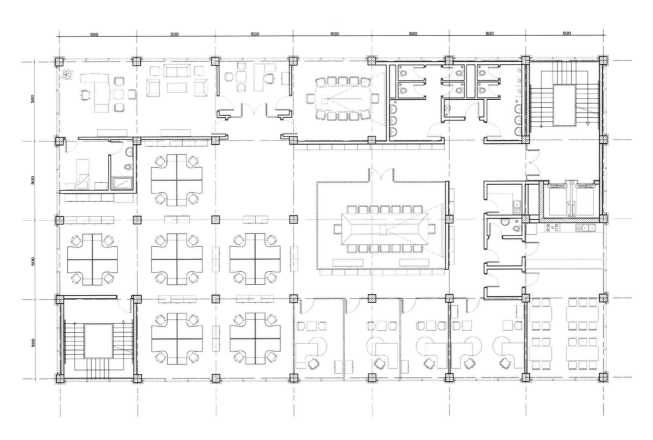

Figure 2.12 Example of a traditional floor plan

Details required for elevations

The details required for elevations in construction drawings are the same as those required for floor plans. An elevation is the view of the building as seen from one side. It can be used to show what the exterior of the building will look like. The elevation is labelled in relation to the compass direction. The elevation is a horizontal orthographic projection of the building onto a vertical plane. Usually the vertical plane is parallel to one side of the building (orthographic drawing).

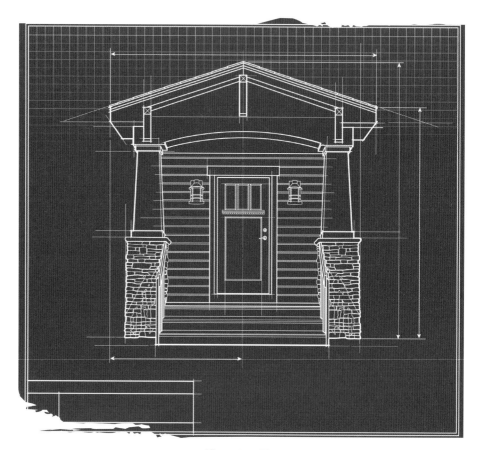

Figure 2.13 The details of the elevation of a building

Linking schedules to drawings

The schedule of work and the drawings create a single set of information. These documents need to be clear and comprehensive.

Before construction gets under way the specification schedule is the most important set of documents. It is used by the construction company to price up the job, work out how to tackle it, and then put in a bid for the work.

The construction company can look at each task in detail and see what materials are needed. This, along with all the construction information documents, will help them to make an estimate as to how long the task will take to complete and to what standard it should be completed.

During the construction period the most important documents are the drawings. Each piece of work is linked to those drawings and a schedule of work is set up. This might incorporate a Gantt chart or critical path analysis, showing expected dates and duration of on-site and off-site activities. This might need a good deal of cross-referencing. Obviously you cannot fit windows until the relevant cavity wall has been built and the opening formed.

It is important that the drawings and the specification schedules are closely linked. Reference numbers and headings that are on the drawings need to appear with exactly the same numbers and words on the schedule. This will avoid any confusion. It should be possible to look at the drawings, find a reference number or heading and then look through the schedule to find the details of that particular task. It also allows the drawings to be slightly clearer, as they won't need to have detailed information on them that can be found in the schedule.

Reasons for different projections in construction drawings

Designers will use a range of drawings in order to get across their requirements. Each is a 2D image. They show what the building will look like, along with the components or layout.

Orthographic projections

Orthographic projections are used to show the different elevations or views of an object. Each of the views is at right angles to the face.

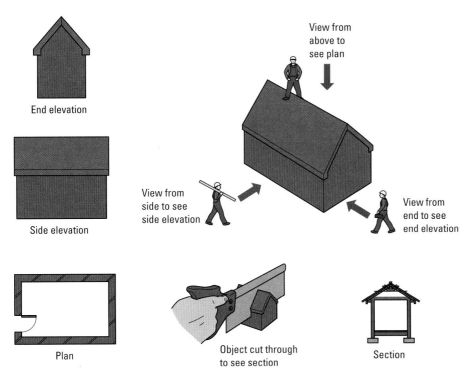

End elevation

Side elevation

Plan

View from above to see plan

View from side to see side elevation

View from end to see end elevation

Object cut through to see section

Section

Figure 2.14 Plans, elevations and sections

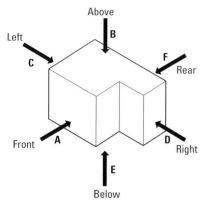

Figure 2.15 Isometric diagram showing the various views that can be portrayed in orthographic projection

Orthographic projection can be seen either as a first angle European projection or a third angle American projection. The following table shows the difference between these two views and there are examples in Figs 2.16 and 2.17, which relate to the shape shown in Fig 2.15.

Projection	Description
First angle	Everything is drawn in relation to the front view. The view from above is drawn below and the view from below is drawn from above. The view from the left is to the right and the right to the left. So all views, in effect, are reversed.
Third angle	This is often referred to as being an American projection. Everything again is in relation to the front elevation. The views from above and below are drawn in their correct position. Anything on the left is drawn to the left and the right to the right.

Table 2.2

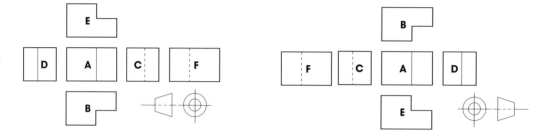

Figure 2.16 First angle projection

Figure 2.17 Third angle projection

Pictorial projections

Pictorial projections show objects in a 3D form. There are different ways of showing the view by varying the angles of the base line and the scale of any side projections. The most common is isometric. Vertical lines are drawn vertically, and horizontal lines are drawn at an angle of 30° to the horizontal. All of the other measurements are drawn to the same scale. This type of pictorial projection can be seen in Fig 2.18.

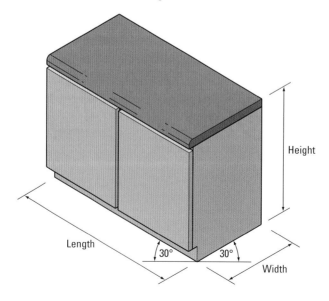

Figure 2.18 Isometric projection

There are four other different types of pictorial projection. These are not used as commonly as isometric projections.

Pictorial projection	Description
Planometric	Vertical lines are drawn vertically and horizontal lines on the front elevation of the object are drawn at 30°. The horizontal lines on the side elevation are drawn at 60° to horizontal.
Axonometric	The horizontal lines on all elevations are drawn at 45° to the horizontal. Otherwise the look is very similar to planometric.
Oblique	All of the vertical lines are drawn vertically. The horizontal lines on the front elevation are drawn horizontally but all the other horizontal lines are drawn at 45° to the horizontal.
Perspective	Horizontal lines are drawn so that they disappear into an imaginary horizon, known as a vanishing point. A one-point perspective drawing has all the sides disappearing to one vanishing point. An angular perspective, or two point perspective, has the elevations disappearing to two vanishing points.

Table 2.3

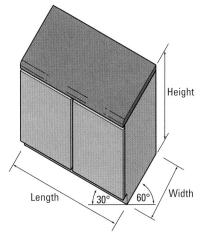

Figure 2.19 Planometric projection

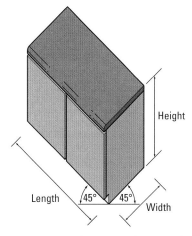

Figure 2.20 Axonometric projection

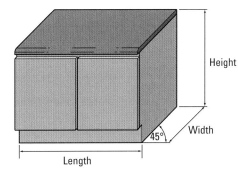

Figure 2.21 Oblique projection

VP = viewpoint

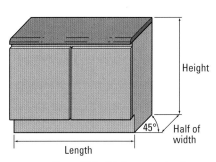

Figure 2.22 Parallel (one point) perspective projection

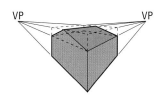

Figure 2.23 Angular (two point) perspective projection

Hatchings and symbols

Different materials and components are shown using symbols and hatchings. Abbreviations are also used. This makes the working drawings far less cluttered and easier to read.

Examples of symbols and abbreviations can be seen in Figs 2.24 and 2.25.

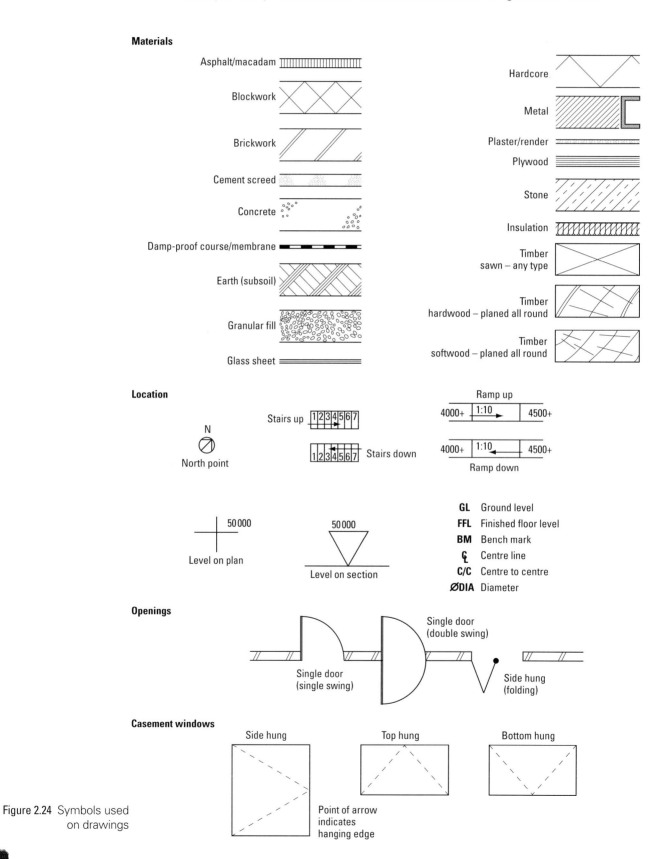

Figure 2.24 Symbols used on drawings

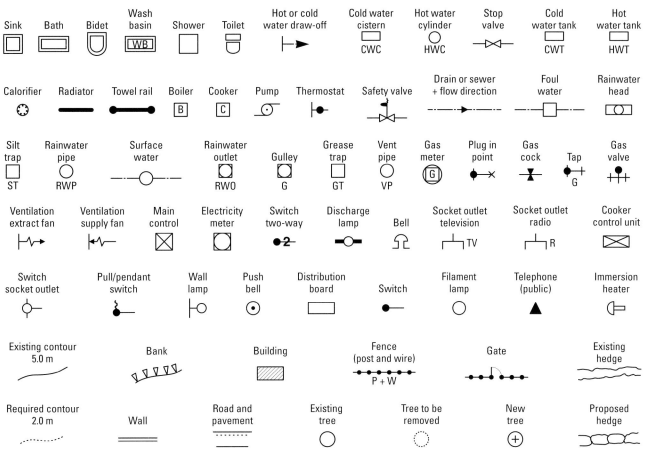

Figure 2.24 Symbols used on drawings *continued*

Aggregate	agg	BS tee	BST	Foundation	fdn	Polyvinyl acetate	PVA
Air brick	AB	Building	bldg	Fresh air inlet	FAI	Polyvinylchloride	PVC
Aluminium	al	Cast iron	CI	Glazed pipe	GP	Rainwater head	RWH
Asbestos	abs	Cement	ct	Granolithic	grano	Rainwater pipe	RWP
Asbestos cement	absct	Cleaning eye	CE	Hardcore	hc	Reinforced concrete	RC
Asphalt	asph	Column	col	Hardboard	hdbd	Rodding eye	RE
Bitumen	bit	Concrete	conc	Hardwood	hwd	Foul water sewer	FWS
Boarding	bdg	Copper	Copp cu	Inspection chamber	IC	Surface water sewer	SWS
Brickwork	bwk	Cupboard	cpd	Insulation	insul	Softwood	swd
BS* Beam	BSB	Damp-proof course	DPC	Invert	inv	Tongued and grooved	T&G
BS Universal beam	BSUB	Damp-proof membrane	DPM	Joist	jst	Unglazed pipe	UGP
BS Channel	BSC	Discharge pipe	DP	Mild steel	MS	Vent pipe	VP
BS equal angle	BSEA	Drawing	dwg	Pitch fibre	PF	Wrought iron	WI
BS unequal angle	BSUA	Expanding metal lathing	EML	Plasterboard	pbd		

Figure 2.25 Abbreviations commonly used on drawings

ESTIMATING QUANTITIES AND PRICING WORK FOR CONTRACTS

Working out the quantity and cost of resources that are needed to do a particular job can be difficult. In most cases you or the company you work for will be asked to provide a price for the work. It is generally accepted that there are three ways of doing this:

* Estimate – which is an approximate price, though estimation is a skill based on many factors.

* Quotation – which is a fixed price.

* Tender – which is a competitive quotation against other companies for a prescribed amount of work to a certain standard.

As we will see a little later in this section, these three ways of costing are very different and each of them has its own issues.

Resource requirements

As you become more experienced you will be able to estimate the amount of materials that will be needed on particular construction projects though this depends on the size and complexity of the job. This is also true of working out the best place to buy materials and how much the labour costs will be to get the job finished.

In order to work out how much a job will cost, you will need to know some basic information:

* What type of contract is agreed?

* What materials will be used?

* What are the costs of the materials?

Much of this information can be gained from the drawings, specification and other construction information for the proposed building.

To help work out the price of a job, many businesses use the *UK Building Blackbook,* which provides a construction cost guide. It breaks down all types of work and shows an average cost for each of them.

Computerised estimating packages are available, which will give a comprehensive detailed estimate that looks very professional. This will also help to estimate quantities and timescales.

Measurement

The standard unit for measurement is based on the metre (m). There are 100 centimetres (cm) and 1,000 millimetres (mm) in a metre. It is important to remember that drawings and plans have different scales, so these need to be converted to work out quantities of materials.

The most basic thing to work out is length, from which you can calculate perimeter, area and then volume, capacity, mass and weight, as can be seen in the following table.

Measurement	Explanation
Length	This is the distance from one end to the other. For most jobs metres will be sufficient, although for smaller work such as brick length or lengths of screws, millimetres are used.
Perimeter	This is the distance around a shape, such as the size of a room or a garden. It will help you estimate the length of a wall, for example. You just need to measure each side and then add them together.
Area	You can work out the area of a room, for example, by measuring its length and its width. Then you multiply the width by the length to give the number of square metres (m^2).
Volume and capacity	Volume shows how much space is taken up by an object, such as a room. Again this is simply worked out by multiplying the width of the room by its length and then by its height. This gives you the number of cubic metres (m^3).
	Capacity works in exactly the same way but instead of showing the figure as cubic metres you show it as litres. This is ideal if you are trying to work out the capacity of the water tank or a garden pond.
Mass or weight	Mass is measured usually in kilograms or in grams. Mass is the actual weight of a particular object, such as a brick.

Table 2.4

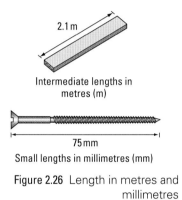

Intermediate lengths in metres (m)

75mm
Small lengths in millimetres (mm)

Figure 2.26 Length in metres and millimetres

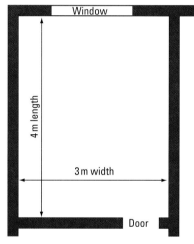

Figure 2.27 Measuring area and perimeter

$1m^3 = 10 \times 10 \times 10$
$= 1,000$ litres

Figure 2.28 Relationship between volume and capacity

Formulae

These can appear to be complicated, but using formulae is essential for working out quantities of materials. Each formula is related to different shapes. In construction you will often have to work out quantities of materials needed for odd shaped areas.

Area

To work out the area of a triangular shape, you use the following formula:

Area (A) = Base (B) × Height (H) ÷ 2

So if a triangle has a base of 4.5 and a height of 3.5 the calculation is:

4.5 × 3.5 ÷ 2

Or 4.5 × 3.5 = 15.75 ÷ 2 = 7.875 m^2

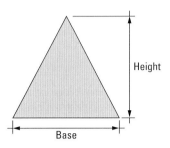

Figure 2.29 Triangle

Height

If you want to work out the height of a triangle you switch the formulae around. To give us height = 2 × Area ÷ Base

Perimeter

To work out the perimeter of a rectangle use the formula:

$$\text{Perimeter} = 2 \times (\text{Length} + \text{Width})$$

It is important to remember this because you need to count the length and the width twice to ensure you have calculated the total distance around the object.

Circles

To work out the circumference or perimeter of a circle you use the formula:

$$\text{Circumference} = \pi \text{ (pi)} \times \text{diameter}$$

π (pi) is always the same for all circles and is 3.142.

Diameter is the length of the widest part.

If you know the circumference and need to work out the diameter of the circle the formula is:

$$\text{Diameter} = \text{circumference} \div \pi \text{ (pi)}$$

For example if a circle has a circumference of 15.39 m then to work out the diameter:

$$15.39 \div 3.142 = 4.89 \text{ m}$$

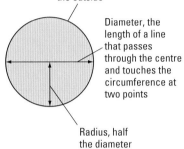

Figure 2.30 Parts of a circle

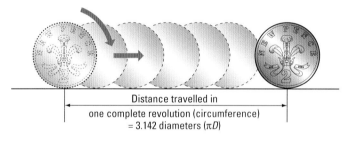

Distance travelled in
one complete revolution (circumference)
= 3.142 diameters (πD)

Figure 2.31 Relationship between circumference and diameter

Complex areas

Land, for example, is rarely square or rectangular. It is made up of odd shapes. Never be overwhelmed by complex areas, as all you need to do is to break them down into regular shapes.

By accurately measuring the perimeter you can then break down the shape into a series of triangles or rectangles. All you need to do then is to work out the area of each of the shapes within the overall shape and then add them together.

Shape	Area equals	Perimeter equals
Square 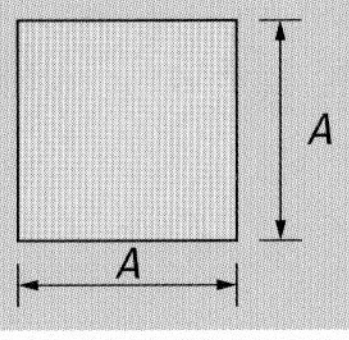	AA (or A multiplied by A)	4A (or A multiplied by 4)
Rectangle 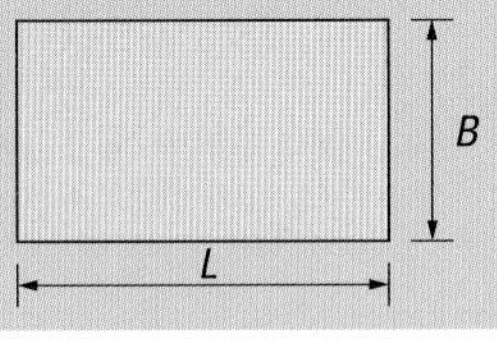	LB (or L multiplied by B)	2(L + B) (or L plus B multiplied by 2)
Trapezium 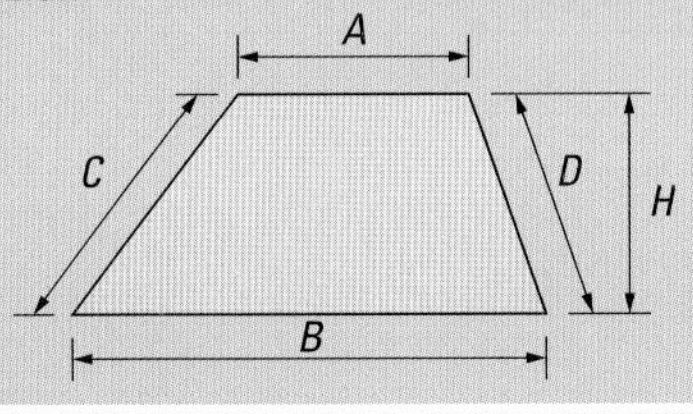	$\dfrac{(A + B)H}{2}$ (or A plus B multiplied by H and then divided by 2)	A + B + C + D
Triangle 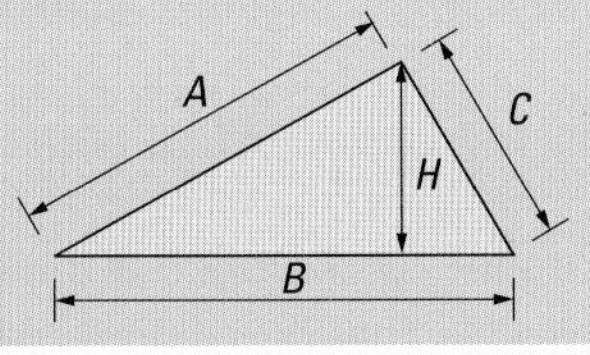	$\dfrac{BH}{2}$ (or B multiplied by H and then divided by 2)	A + B + C
Circle 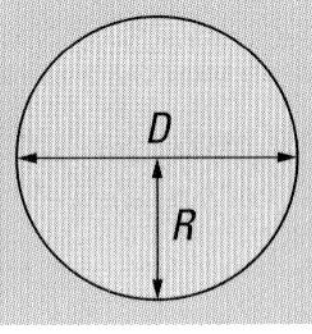	πR^2 (or R multiplied by itself and then multiplied by pi (3.142))	πD or $2\pi R$

Figure 2.32 Table of shapes and formulae

Volume

Sometimes it is necessary to work out the volume of an object, such as a cylinder or the amount of concrete needed. All that needs to be done is to work out the base area and then multiply that by the height.

For a concrete area, if a 1.2 m square needs 3 m of height then the calculation is:

$$1.2 \times 1.2 \times 3 = 4.32\,\text{m}^3$$

To work out the volume of a cylinder you need to know the base area × the height. The formula is:

$$\pi r^2 \times H$$

So if a cylinder has a radius (r) of 0.8 and a height of 3.5 m then the calculation is:

$$3.142 \times 0.8 \times 0.8 \times 3.5 = 7.038\,\text{m}^3$$

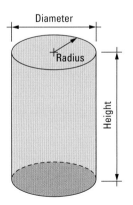

Figure 2.33 Cylinder

Pythagoras

Pythagoras' theorem is used to work out the length of the sides of right angled triangles. The theory states that:

In all right angled triangles the square of the longest side is equal to the sum of the squares of the other two sides (that is, the length of a side multiplied by itself).

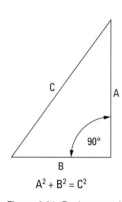

$A^2 + B^2 = C^2$

Figure 2.34 Pythagoras' theorem

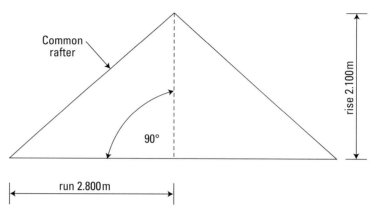

Figure 2.35 Section of a pitched roof

See Chapter 7 for more about geometry, trigonometry and Pythagoras.

Measuring materials

Using simple measurements and formulae can help you work out the amount of materials you will need. This is all summarised in the following table.

Material	Measurement
Timber	Can be sold by the cubic metre. To work out the length of material divide the cross section area of one section by the total cross section area of the material.
Flooring	To work out the amount of flooring for a particular area multiply the width of the floor by the length of the floor.
Stud walling	Measure the distance that the stud partition will cover then divide that distance by a specified spacing. This will give you the number of spaces between each stud.
Rafters and floor joists	Measure the distance between the adjacent walls then take into account that the first and last joist or rafter will be 50 mm away from the wall. Measure the total distance and then divide it by the specified spacing.
Fascias, barges and soffits	Measure the length and then add a little extra to take into account any necessary cutting and jointing.
Skirting	You need to work out the perimeter of the room and then subtract any doorways or other openings. This technique can be used to work out the necessary length of dado, picture rails and coving.
Bricks and mortar	Half-brick walls use 60 bricks per metre squared and one-brick walls use double that amount. You should add 5 per cent to take into account any cutting or damage. For mortar assume that you will need 1 kg for each brick.

Table 2.5

How to cost materials

Once you have found out the quantity of materials necessary you will need to find out the price of those materials. It is then simply a case of multiplying those prices by the amount of materials actually needed to find out approximately how much they will cost in total.

Materials and purchasing systems

Many builders and companies will have preferred suppliers of materials. Many of them will already have negotiated discounts based on their likely spending with that supplier over the course of a year. The supplier will be geared up to supply them at an agreed price.

In other cases builders may shop around to find the best price for the materials that match the specification. It is not always the case that the lowest price is necessarily the best. All materials need to be of a sufficient quality. The other key consideration is whether the materials are immediately available for delivery.

It is vital that suppliers are reliable and that they have sufficient materials in stock. Delays in deliveries can cause major setbacks on site. It is not always possible to warn suppliers that materials will be needed, but a well-run site should be able to anticipate the materials that are needed and put in the orders within good time.

Large quantities may be delivered direct from the manufacturer straight to site. This is preferable when dealing with items where consistency, for example of colour, is required.

Labour rates and costs

The cost of labour for particular jobs is based on the hourly charge-out rate for that individual or group of individuals multiplied by the time it would take to complete the job.

Labour rates can depend on the:

* expertise of the construction worker
* size of the business they work for
* part of the country in which the work is being carried out
* complexity of the work.

According to the International Construction Costs Survey 2012, the following were average costs per hour:

* Group 1 tradespeople – plumbers, electricians etc.: £30
* Group 2 tradespeople – carpenters, bricklayers etc.: £30
* Group 3 tradespeople – tillers, carpet layers and plasterers: £30
* general labourers: £18
* site supervisors: £46.

Quotes, estimated prices and tenders

As we have already seen, estimates, quotes and tenders are very different. We need to look at these in slightly more detail, as can be seen in the following table.

REED
TIP
...
A great career path can start with an apprenticeship. 80 per cent of the staff at South Tyneside Homes started off as apprentices. Some have worked their way up to job roles such as team leaders, managers and heads of departments.

Type of costing	Explanation
Estimate	This needs to be a realistic and accurate calculation based on all the information available as to how much a job will cost. An estimate is not binding and the client needs to understand that the final cost might be more.
Quote	This is a fixed price based on a fixed specification. The final price may be different if the fixed specification changes; for example if the customer asks for additional work then the price will be higher.
Tender	This is a competitive process. The customer advertises the fact that they want a job done and invites tenders. The customer will specify the specifications and schedules and may even provide the drawings. The companies tendering then prepare their own documents and submit their price based on the information the customer has given them. All tenders are submitted to the customer by a particular date and are sealed. The customer then opens all tenders on a given date and awards the contract to the company of their choice. This process is particularly common among public sector customers such as local authorities.

Table 2.6

Inaccurate estimates

Larger companies will have an estimating team. Smaller businesses will have someone who has the job of being an estimator. Whenever they are pricing a job, whether it is a quote, an estimate or a tender, they will have to work out the costs of all materials, labour and other costs. They will also have to include a **mark-up.**

It is vital that all estimating is accurate. Everything needs to be measured and checked. All calculations need to be double-checked.

It can be disastrous if these figures are wrong because:

● if the figure is too high then the client is likely to reject the estimate and look elsewhere as some competitors could be cheaper

● if the figure is too low then the job may not provide the business with sufficient profit and it will be a struggle to make any money out of the job.

KEY TERMS

Mark-up

– a builder or building business, just like any other business, needs to make a profit. Mark-up is the difference between the total cost of the job and the price that the customer is asked to pay for the work.

DID YOU KNOW?

Many businesses fail as a result of not working out their costs properly. They may have plenty of work but they are making very little money.

CASE STUDY

South Tyneside Homes

South Tyneside Council's Housing Company

Bringing all your skills together to do a good job

Marcus Chadwick, a bricklayer at Laing O'Rourke, talks about maths and English skills.

'Obviously you need your maths, especially being a bricklayer. From the dimensions on your drawings, you have to be able to work out how many materials you'll need – how many bricks, how many blocks, how much sand and cement you need to order. Eventually it ends up being rote learning, like the way you learn your times-tables. With a bit of practice, you'll be able to work out straight away, "Right, I need x number of blocks, I need 1000 bricks there, I need a ton of sand, therefore I need seven bags of cement." Though there are still times when I get the calculator out!

If you get it wrong and miscalculate it can delay the progression of the building, or your section. I'm the foreman and if I set out a wall in the wrong position then there's only one person to blame. So you check, then double-check – it's like the old saying, "Measure twice, cut once".

Number skills really are important; you can't just say "Well, I'm a bricklayer and I'm just going to work with my hands". But that all comes with time; I wasn't that good at maths when I left school, though it was probably a case of just being lazy. When you come into an environment where you need to start using it to earn the money, then you'll start to get it straight away.

Your communication skills are definitely important too. You need to know how to speak to people. I always find that your lads appreciate you more if you ask them to do something as opposed to tell them. That was your old 1970s mentality where you used to have your screaming foreman – "Get this done, get that done!" – it doesn't work like that anymore. You have to know how to speak to people, to communicate.

All the lads that work for me, they're my extended family. My boss knows that too and that's why I've been with him for seven years now – he takes me everywhere because he knows I've got a good relationship with all my workforce.

You'll also have the odd occasion where the client will come around to visit the site and you've got to be able to put yourself across, using good diction. That also goes for when you're ordering materials – because you deal with different regions across the country, you've got to be clearly spoken so they understand you, so things don't get messed up in translation.'

Purchasing or hiring plant and equipment

Normally, if a piece of plant or equipment is going to be used on a regular basis then it is purchased by the company. By maximising the use of any plant or equipment, the business will save on the costs of repeatedly hiring and the transport of the item to and from the site.

It also does not make sense to leave plant and equipment on a site if it is no longer being used. It needs to be moved to a new site where it can be used.

Many smaller construction companies have no alternative other than to hire. This is because they cannot afford to have an enormous amount of money tied up in the plant or equipment, whether this comes from earned profits or from a loan or finance agreement. Loans and finance agreements have to be paid back over a period of time.

The decision as to whether to purchase or to hire is influenced by a number of factors:

* The working lives of the plant or equipment – how long will it last? This will usually depend on how much it is used and how well maintained it is.

* The use of the plant or equipment – is the company going to get good use out of it if they buy it? If they are hiring it then it should only be hired for the time it is actually needed. There is no point in having the plant or equipment on site and paying for its hire if it is not being used.

* Loss of value – just like buying a brand new car, the value of new plant or equipment takes an enormous drop the moment you take delivery of it. Even if it is hardly used it is considered second-hand and is not worth anything like its price when it was new. The biggest falls in value are in the first few years that the company owns it. It then reaches a value that it will sit at for some years until it is considered junk or scrap.

* Obsolescence – what might seem today to be the most advanced and technologically superior piece of plant or equipment may not be so tomorrow. Newer versions will come onto the market and may be more efficient or cost-effective. It is probable that the plant or equipment will be obsolete, or outdated, before it ends its useful working life.

* Cost of replacement – investing in plant or equipment today means that at some point in the future they will have to be replaced. The business will have to take account of this and arrange to have the necessary funds available for replacement in the future.

* Maintenance costs – if the construction business owns the plant or equipment they will have to pay for any routine maintenance, repairs and of course operators. Hired equipment, such as diggers or cranes, is the responsibility of the hiring company. They pay for all the maintenance and although they charge for the operator, the operator is on their wage bill.

* Insurance and licences – owning plant or equipment often means additional insurance payments and the company may also have to obtain licences that allow them to use that type of equipment in a particular area. Hired plant and equipment is already insured and should have the relevant licences.

* Financial costs – if the decision is to buy rather than hire, the money that would have otherwise been sitting in a bank account, earning interest, has been spent. If the company had to borrow the money to buy the plant or equipment then interest charges are payable on loans and finance agreements.

PRACTICAL TIP

Many construction companies that know they are going to be working on a project for a long period of time will actually buy plant and equipment for that contract. Once the contract has been completed they will sell on the plant and equipment.

Planning the sequence of materials and labour requirements

One of the most important jobs when organising work that will need to be carried out on site is to calculate when, where and how much materials and labour will be needed at any one time. This is organised in a number of different ways. The following headings cover the main documents or processes that are involved.

Bill of quantities

This is used by building contractors when they quote for work on larger projects. It is usually prepared by a quantity surveyor. Fig 2.36 shows you what a bill of quantities looks like.

BILL OF QUANTITIES				
Contract		DWG No.		
DESCRIPTION	QUANTITY	UNIT	RATE	AMOUNT

Figure 2.36 Bill of quantities form

The form is completed using information from the working drawings, specification and schedule (this is called the take off). It describes each particular job and how many times that job needs to be carried out. It sets the number of units of material or labour, the rate at which they are charged and the total amount.

Programmes of work

A programme of work is also an important document, as it looks at the length of time and the sequence of jobs that will be needed to complete the construction. It has three main sections:

* A master programme that shows the start and finish dates. It shows the duration, sequence and any relationships between jobs across the whole contract.

* A stage programme – this is the next level down and it covers particular stages of the contract. A good example would be the foundation work or the process of making the building weather-tight. Alternatively it might look at a period of up to two months' worth of work in detail.

* A weekly programme – there will be several of these, which aim to predict where and when work will take place across the whole of the site. These are very important as they need to be compared against actual progress. The normal process is to review and update these weekly programmes and then update the stage and master programmes if delays have been encountered.

Stock systems and lead times

One of the greatest sources of delays in construction is not having the right materials and equipment available when it is needed. This means that someone has to work out not only what is needed and how many, but when. It is a balancing act because there are dangers in having the stock on site too early. If all the materials needed for a construction job arrived in the first week then this would cause problems and it is unlikely that there would be anywhere to store them. Materials need to be ordered to ensure that they are on site just before they are needed.

One of the problems is lead times. There is no guarantee that the supplier will have sufficient stock available when you need it. They need to be warned that you will need a certain amount of material at a certain time in advance. This will allow them to either manufacture the stock or get it from their supplier. Specialist materials have longer lead times. These may have to be specially manufactured, or perhaps imported from abroad. All of this takes time.

Once the quantities of materials have been calculated and the sequence of work decided, comparing that to the duration of the project and the schedule, it should be possible to predict when materials will be needed. You will need to liaise with your suppliers as soon as possible to find out the lead times they need to get the materials delivered to the site. This might mean that you will have to order materials out of sequence to the work schedule because some materials need longer lead times than others.

Planning and scheduling using charts

To plan the sequence of materials and labour requirements it is often a good idea to put the information in a format that can be easily read and understood. This is why many companies use charts, graphs and other types of illustrated diagram.

The most common is probably the Gantt chart. This is a series of horizontal bars. Each different task or operation involved in a project is shown on the left-hand side of the chart. Along the top are days, weeks or months. The planner marks the start day, week or month and the projected end day, week or month with a horizontal bar. It shows when tasks start and when they end. It will also be useful in showing when labour will be needed. It will also show which jobs have to be completed before another job can begin. An example of a Gantt chart can be seen in Fig 2.37.

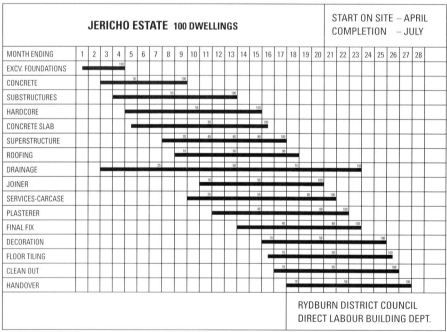

Typical programme for rate of completion on a housing development contract

Figure 2.37 Gantt chart

There are other types of bar chart that can be used to plan and monitor work on the construction site. It is important to remember that each chart relates to the plan of work:

● A single bar – this focuses in on a sequence of tasks and the bar is filled in to show progress.

PRACTICAL TIP

A Gantt chart shows that start dates for jobs are often dependent on the finish date of other jobs. If the early jobs fall behind then all of the other tasks that rely on that job will be delayed.

SINGLE BAR SYSTEM

	ACTIVITY	Week 1	Week 2	Week 3	Week 4
1	Excavate O/site				
2	Excavate Trenches				
3	Concrete Foundations				
4	Brickwork below DPC				

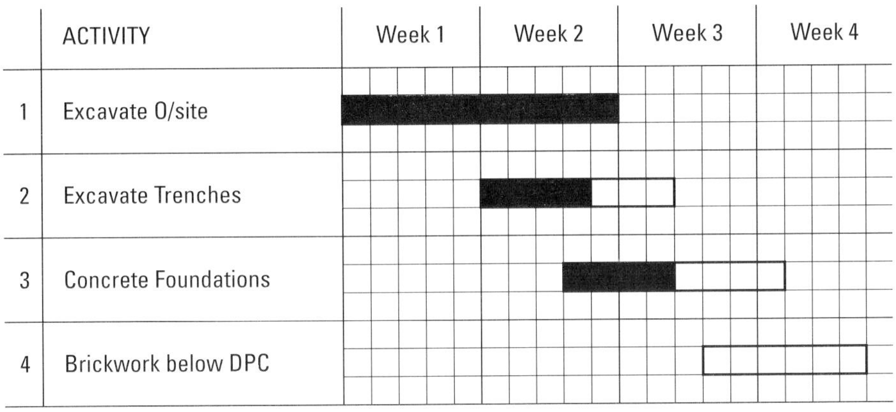

Figure 2.38 Single bar chart

● A two-bar – this tracks the amount of work that has been carried out against the planned amount of work that should have been carried out. In other words it shows the percentage of work that has been completed. It is there to alert the site manager that work may be falling behind and extra resources are needed.

TWO BAR SYSTEM

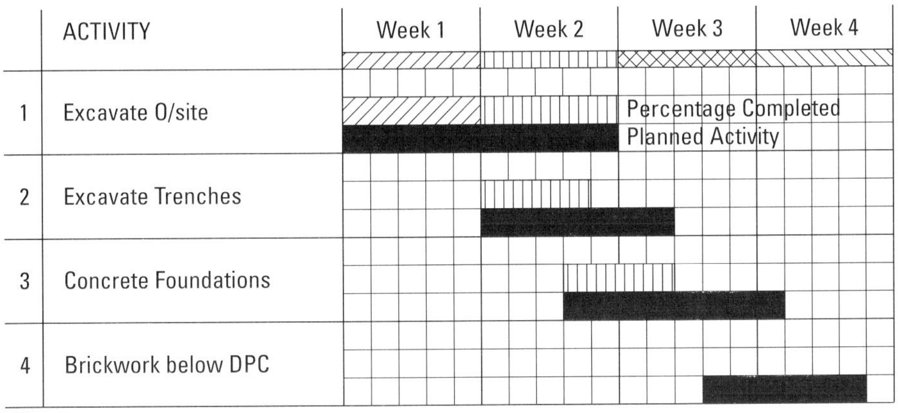

	ACTIVITY	Week 1	Week 2	Week 3	Week 4
1	Excavate O/site			Percentage Completed	
				Planned Activity	
2	Excavate Trenches				
3	Concrete Foundations				
4	Brickwork below DPC				

Figure 2.39 Two-bar system

* A three-bar – this shows the planned duration of the activity, the actual days that have seen work being done on that activity and the percentage of the activity that has been completed. This gives a snapshot view of how work has progressed over the course of a period of time.

THREE BAR SYSTEM

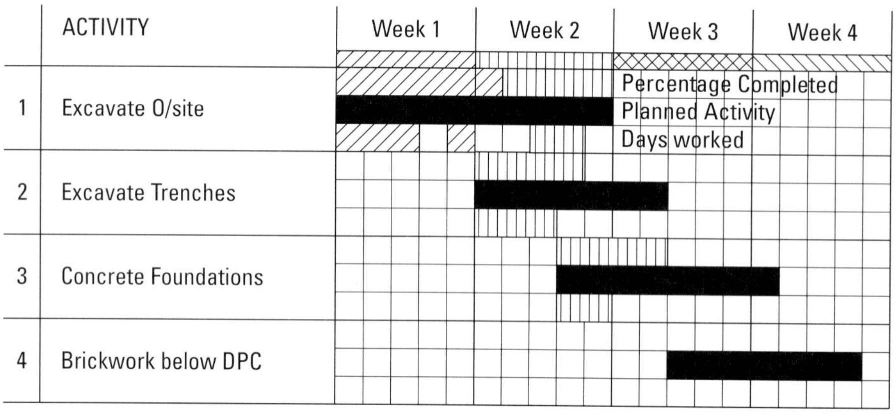

	ACTIVITY	Week 1	Week 2	Week 3	Week 4
1	Excavate O/site			Percentage Completed / Planned Activity / Days worked	
2	Excavate Trenches				
3	Concrete Foundations				
4	Brickwork below DPC				

Figure 2.40 Three-bar system

Calculating hours required

We have already seen that different types of construction workers attract different hourly rates of pay. The simple solution in order to work out the cost to complete particular work is to look at the programmes of work and the estimated time required to complete it. The next stage is to estimate how many workers will be needed to carry out that activity and then multiply that by the estimated labour cost per hour.

Added costs

When a construction company estimates the costs of work they have to incorporate a number of other different costs. A summary of these can be seen in the following table.

Added cost	Explanation
National Insurance contributions	Companies employing workers have to pay National Insurance contributions to the government for each employee.
Value Added Tax (VAT)	For businesses that are registered for VAT they have to charge a sales tax on any services that they provide. They collect this money on behalf of the government. The VAT is added to the final cost of the work.
Pay As You Earn (PAYE)	PAYE, or income tax, has to be paid on the income of all workers straight to the government.
Travel expenses	This is particularly relevant if workers on site have to travel a considerable distance in order to do their work. This can reasonably be passed on to the client.
Profit and loss	Many businesses will make the mistake of trying to estimate the costs of their work in the knowledge that they are competing with other businesses, trimming their estimates if they can. As we will see when we look at profitability, a business does need to make money from contracts otherwise it will not have sufficient funds to continue to operate.
Suppliers' terms and conditions	Suppliers will often have set payment terms, such as 30, 60 or 90 days. The construction company needs to have sufficient funds to pay their suppliers on the due date stated on the invoice. If they fail to do this then they could run into difficulties as the supplier may decide not to give them any more credit until outstanding invoices have been paid.
Wastage	It is rarely possible to buy materials and components that are an exact fit. Blocks and bricks, for example, will have a percentage damaged or unusable on each pallet. Materials can also be mis-cut, damaged or otherwise wasted on site. It is sensible for the construction company to factor in a wastage rate of at least 5 per cent. This is often required to allow for cutting and fitting when using stock lengths.
Penalty clauses	Many projects are time sensitive and need to be completed by specified dates. The contracts will state whether there are any penalties to be made if critical dates are missed. Penalty clauses are rather like fines that the construction company pays if they fail to meet deadlines.

Table 2.7

Total estimated prices

As we have seen, the total estimated price needs to incorporate all of the added costs. But there are two other issues that should not be forgotten:

* The cost of any plant and equipment hire – the length of time that these are necessary will have to be calculated together with an additional period in case of delays.

* Contingencies – it is not always possible to determine exact prices, especially in groundworks, and, in any case, not all construction jobs will run smoothly. It is therefore sensible to set aside funds should additional work be needed. This may be particularly true if additional work needs to be done to secure the foundations or if workers and equipment are on site yet it is not possible to work due to poor weather conditions.

Profitability

Setting an initial price for a business's services is, perhaps, one of the most difficult tasks. It needs to take account of the costs that are incurred by the business. It also needs to consider the prices charged by key competitors, as the optimum price that a business wishes to charge may not be possible if competitors are charging considerably lower prices.

It is difficult for a new business to set prices, because its services may not be well known. Its costs may be comparatively higher because it will not have the advantages of providing services on a large scale. Equally, it cannot charge high prices because neither the company nor its services are established in the market.

The process of calculating revenue is a relatively simple task. Sales revenue or income is equal to the services sold, multiplied by the average selling price. In other words, all a business needs to know is how many services have been sold, or might sell, and the price they will charge.

We have already seen that a business incurs costs that must be paid. Clearly these have a direct impact on the profitability of a business.

We can already see that there is a direct relationship between costs and profit. Costs cut into the revenue generated by the business and reduce its overall profitability.

Gross profit is the difference between a company's revenue and its costs. Businesses will also calculate their operating profit. The operating profit is the business's gross profit minus its **overheads.**

A business may also calculate its pre-tax profits, which are its profits before it pays its taxes. It may have one-off costs, such as the replacement of a piece of plant or equipment. These costs are deducted from the operating profit to give the pre-tax profit.

The business will then pay tax on the remainder of its profit. This will leave them their net profit. This is the amount of money that they have actually made over the course of a year or on a particular job.

Profits are an important measure of the success of a business. Like other businesses construction companies do borrow money, but profit is the source of around 60 per cent of the funds that businesses use to help them grow.

Businesses can look for ways to gradually increase their profit. They can look at each type of job they do and work out the most efficient way of doing it. This might mean looking for a particular mix of employees, or buying or hiring particular plant and equipment that will speed up the work.

KEY TERMS

Overheads

– these are expenses that need to be paid by the business regardless of how much work they have on at any one time, such as the rent of builders' yard.

GOOD WORKING PRACTICES

Like any business, construction relies on a number of factors to make sure that everything runs smoothly and that a company's reputation is maintained.

There needs to be a good working relationship between those who work for the construction company, and other individuals or companies that they regularly deal with, such as the local authority, and professionals such as architects and clients.

This is achieved by making sure that these individuals and organisations continue to have trust and confidence in the company. Any promises or guarantees that are made must be kept.

In the normal course of events communication needs to be clear and straightforward. When there are problems accurate and honest communication can often deal with many of them. It can set aside the possibility of misunderstandings.

Good working relationships

Each construction job will require the services of a team of professionals. They will need to be able to work and communicate effectively with one another. Each has different roles and responsibilities.

Although you probably won't be working with exactly the same people all the time you are on site, you will be working with on-site colleagues every day. These may be people doing the same job as you, as well as people with other roles and responsibilities who you need to work with to ensure that the project runs smoothly.

Working with unskilled operatives

It's important to remember that everyone will have different levels of skill and experience. You, as a skilled or trade operative, are qualified in your trade, or working towards your qualification. Some people will be less experienced than you; for example, unskilled operatives (manual workers) are entry level operatives without any formal training. They may, however, be experienced on sites and will take instructions from the supervisor or site manager. You should be patient with colleagues who are less experienced or skilled than you – after all, everyone has to learn. However, if you see them carrying out unsafe practices, you should tell your supervisor or charge-hand straight away.

Working with skilled employees

You'll also work with people who are more experienced than you. It's a good idea to watch how they work and learn from their example. Show them respect and don't expect to know as much as they do if they have been working for much longer. However, if you see them ignoring safety rules, don't copy them; speak to your supervisor.

Working with professional technicians

You might also work with professional technicians, such as civil engineers or architectural technicians. They will have extensive knowledge in their field but may not know as much as you about bricklaying or carpentry. For your relationship to run smoothly, you should respect each other's knowledge, share your thoughts on any issues, and listen to what each other has to say.

Working with supervisors

Supervisors organise the day-to-day running of the site or a team. Charge-hands supervise a specific trade, such as bricklayers or carpenters. They will be your immediate boss, and you must listen to their instructions and obey any rules they set out. These rules enable the site to be run smoothly and safely so it is in your interest to do what your supervisor says.

Working with the site manager

The site manager or site agent runs the construction site, makes plans to avoid problems and meet deadlines, and ensures all processes are carried out safely. They communicate directly with the client. They are ultimately responsible for everything that goes on at the construction site. Even if you don't communicate with them directly, you should follow the guidance and rules that they have put in place. It's in your interest to do your bit to keep the site safe and efficient.

Working with other professionals

You may also need to work with or communicate with other professionals. For example, a clerk of works is employed by the architect on behalf of a client. They oversee the construction work and ensure that it represents the interests of the client and follows agreed specifications and designs. A contracts manager agrees prices and delivery dates. These professionals will expect you to do the job that has been specified and to draw their attention to anything that will change the plans they are responsible for.

Hierarchical charts

A hierarchy describes the different levels of responsibility, authority and power in a business or organisation. The larger the business the more levels of management it will have. The higher up the management structure the more responsibility each person will have.

Decisions are made at the top and instructions are passed down the hierarchy. It is best to imagine most organisations as like a pyramid. The directors or owners of the business are at the top of that pyramid. A site manager may be part-way down and at the bottom of the pyramid are all the workers who are on site.

Trust and confidence

The trust of colleagues develops as a result of showing that the company is reliable, cooperative and committed to the success or goals of the colleague or client. Trust does not happen automatically but has to be earned through actions. An important part of this is building a positive relationship with colleagues. Over time, trust will develop into confidence. Colleagues will have confidence in the company being able to deliver their promises.

The reputation of a construction company is very important. Criticisms of the company will always do far more damage than the positive benefits of a successfully completed contract. This shows how important it is to get things right the first time and every time.

In an industry where there is so much competition, trust and confidence can mean the difference between getting the contract and being rejected before your bid has even been considered.

As the construction company becomes established they will build up a network of colleagues. If these colleagues have trust and confidence in the business they will recommend the business to others.

Ultimately, earning trust and confidence relies on the business being able to solve any problems with the minimum of fuss and delay. Fair solutions need to be identified. These solutions should not be against the interests of anyone involved.

Accurate communication

Effective communication in all types of work is essential. It needs to be clear and to the point, as well as accurate. Above all it needs to be a two-way process. This means that any communication that you have with anyone must be understood.

In construction work it is essential to keep to deadlines and follow strict instructions and specifications. Failing to communicate will always cause confusion, extra cost and delays. In an industry such as this it is unacceptable and very easy to avoid. Negative communication or poor communication can damage the confidence that others have in you to do your job.

It is important to have a good working relationship with colleagues at work. An important part of this is to communicate in a clear way with them. This helps everyone understand what is going on, what decisions have been made. It also means being clear. Most communication with colleagues will be verbal (spoken). Good communication results in:

* cutting out mistakes and stoppages (saving money)

* avoiding delays

* making sure that the job is done right the first time and every time.

The more complex a contract, the more likely it is that changes and alterations will be needed. The longer the contract runs for, the more likely it is that changes will happen. Examples are as follows:

* Alterations to drawings – this can happen as a result of several different factors. The architect or the client may decide at a fairly late stage that changes need to be made to the design of the project. This will require all documents that rely on information from the drawings to be amended. This could mean changes to the schedule, specification and work programmes and the need for materials and labour at particular times.

* Variation to contracts – although the construction company may have agreed with the client to carry out work based on particular drawings and specifications, changes to design and to the requirements may happen. It may be necessary to put in new estimates for additional work and to inform the client of any likely delays.

* Changes to risk assessments – it is not always possible to predict exactly what hazards will be encountered during a project. Neither is it possible to predict whether new legislation will come into force that requires extra risk assessments.

* Work restrictions – although the site will have been surveyed for access and cleared of obstacles such as low trees, problems may arise during the work. Local residents, for example, may complain about lighting and noise. This could reduce working hours on site. This could all have an impact on the schedule of work.

* Change in circumstances – this could cover a wide variety of different problems. Key suppliers may not be able to deliver materials or components on time. Tried and trusted sub-contractors may not be available. The client may run out of money or a problem may be unearthed during excavation and preparation of the site.

REED TIP

Open and frank communication means being able to say no if something is not possible. It's OK to say that something can't be done, rather than saying 'yes, yes, yes' and then being unable to complete a task.

DID YOU KNOW?

During the construction of the Olympic basketball site in London the whole site had to be evacuated when a Second World War bomb was found. It had to be removed by specialists before work could continue.

TEST YOURSELF

1. What is the system that is gradually taking over from CAD as the main way to produce construction drawings?

    a. CTIBM

    b. CIM

    c. BIM

    d. SIM

2. What does a block plan show?

    a. The construction site and its surrounding area

    b. Local boundaries and roads

    c. Elements and components

    d. Constructional details

3. Who might give you a delivery note?

    a. A postal worker

    b. An architect

    c. A contractor

    d. A supplier

4. Which type of construction drawings show the different faces or views of an object?

    a. Orthographic

    b. Section

    c. Elevation

    d. Plan

5. How many different types of pictorial projection are there?

    a. 3

    b. 4

    c. 5

    d. 6

6. Which of the following is usually a bid for a fixed amount of work in competition with other companies?

    a. Estimate

    b. Quotation

    c. Tender

    d. Invoice

7. To calculate the area of a room, which two measurements are needed?

    a. Length and height

    b. Height and width

    c. Length and width

    d. Length and circumference

8. If you had 5 workmen being paid £25 per hour and they were working for 4 hours, what would be the total labour cost?

    a. £125

    b. £250

    c. £500

    d. £600

9. Some contracts state that if a deadline is missed a fine has to be paid. What are these called?

    a. Terms and conditions

    b. Wastage

    c. Penalty clause

    d. Critical date payment

10. A business's operating profit is its gross profit minus which of the following?

    a. Tax

    b. Overheads

    c. Net profit

    d. Labour costs

Chapter 3

Unit CSA–L3Core08
ANALYSING THE CONSTRUCTION INDUSTRY AND BUILT ENVIRONMENT

LEARNING OUTCOMES

LO1: Understand the different activities undertaken within the construction industry and built environment

LO2: Understand the different roles and responsibilities undertaken within the construction industry and built environment

LO3: Understand the physical and environmental factors when undertaking a construction project

LO4: Understand how construction projects can benefit the built environment

LO5: Understand the principles of sustainability within the construction industry and built environment

INTRODUCTION

The aim of this unit is to:

● help you understand more about the construction industry and its place in society.

CONSTRUCTION INDUSTRY AND BUILT ENVIRONMENT ACTIVITIES

Half of all the non-renewable resources used across the globe are consumed by construction. Construction and the built environment are also linked with the pollution of drinking water, the production of waste and poor air quality.

Nevertherless, buildings create wealth. In the UK, buildings represent three-quarters of all wealth. Buildings are long-term assets. Today it is recognised that buildings should have the ability to satisfy user needs for extended periods of time. They must be able to cope with any changing environmental conditions. They also need to be capable of being adapted over time as designs and demands change.

There is an increasing move towards naturally lit and well-ventilated buildings. There is also a move towards buildings that use alternative energy sources.

The first part of this chapter looks at the type of work that has developed around the broader construction industry and built environment. It looks at the work that is undertaken and the different types of clients who use the construction industry.

Range of activities

The construction industry and the broader built environment is a highly complex network of different activities. While there are a great many small businesses that focus on one particular aspect of construction, they need to be seen as part of a far larger industry. Increasingly it is a global industry, with major business organisations operating not just in the UK but also in a wide variety of locations around the world. Their skills and expertise are in great demand wherever there is construction. The following table outlines some of the activities that are undertaken by the construction industry and the broader built environment.

Activity	Description
Building	This is the accepted and traditional activity of the construction industry. It involves building homes and other structures, from garden walls to entire housing estates or even Olympic villages.
Finishing	Finishing refers to a part of the industry that focuses on decorative work, such as painting and decorating. Once buildings are completed, in order to make them ready for habitation a broad range of professions are needed. Plumbers will install water and sanitation. Electricians will connect electrical services and equipment. Interior designers will create the desired look for the building.
Architecture	Architects and technicians design buildings for clients. The structures are designed to meet the needs of the client while ensuring that they conform to Building Regulations, local planning laws and decisions, as well as other legislation such as CDM Regulations and ensure they are sustainable.
Town planning	Town planning involves organising the broader built environment in a particular area. Town planners need to examine each planning application and see how it fits into the overall long-term future of the area. They need to ensure that the area meets the needs of future generations.
Surveying	This involves measuring and examining land on which building or other external work will take place. It can involve setting out the building. Surveyors use drawings by an architect to correctly position the building. They will be able to work out the area of the building and any volumes. Building surveyors check that the building is structurally sound, while quantity surveyors look after costs.
Civil engineering	Civil engineers are usually involved in major projects, such as road and railway building, the construction of dams, reservoirs and other projects that are not usually buildings. They are involved in what is known as infrastructure projects, such as transport links, networks and hubs.
Repair and maintenance	All buildings need professionals who are able to repair and maintain a broad range of features. From the foundations to the roof, carpenters, builders, electricians, plumbers and more specialist companies, such as pest control, can all be considered to be part of the repair and maintenance side of the industry. Pre-1919 buildings also have particular requirements and are maintained and repaired in a way that suits their construction and to avoid further damage or inappropriate work that will look out of place. This requires people who have specialist heritage skills.
Building engineering services	When buildings are occupied they need to be continually supported in terms of a rigorous checking and maintenance programme. This part of the industry can deal with lifts and escalators, lighting and heating, fire alarms and other inbuilt systems.
Facilities management	For larger commercial buildings or hospitals, schools, colleges and universities, systems need to be in place to replace parts of the building if they wear out or are damaged. This includes cleaning, air conditioning companies, painting and decorating, replacement of doors, windows and a host of other activities.
Construction site management	Construction sites can be complex and demanding places and someone needs to organise them and to monitor progress. Construction site management involves organising the delivery of materials, security, safety, the management of the workforce and contractors.
Plant maintenance and operation	Just as commercial buildings and dwellings need constant maintenance, so too do factories and other sites where products are made or processes are carried out. These individuals can be involved in the energy industry, at gas, oil and nuclear plants, or be responsible for maintaining factories that produce vehicles or food.
Demolition	Demolition experts are responsible for levelling sites in a safe and controlled way. They may have to demolish buildings that could contain asbestos or they may have to use controlled explosions.

Table 3.1

Types of work

There is also a wide range of work that is undertaken in different sectors. Some of this is very specialised work. Some companies will focus purely on that type of work, gaining a reputation and expertise in that area. The following table outlines the type of work that is undertaken within the construction industry.

Type of work	Description
Residential	This is any work connected with domestic housing or dwellings. It can include the building of new homes, extensions or renovations on existing homes and the construction of affordable accommodation for organisations such as housing associations.
Commercial	This is work related to any buildings used by businesses. It can include factories, office blocks, production units, industrial units or private hospitals.
Industrial	This is more specialist work, as it can involve construction, including civil engineering, of heavy industrial factories, such as oil refineries or plants for car manufacturing.
Retail	This can include building or refurbishing shops in high streets or the construction of out-of-town retail parks.
Recreational and leisure	Many of these projects are designed for use by communities, such as sports facilities, fitness clubs, leisure centres, swimming pools and other community sports projects. In the past decade the construction industry was involved in the various London 2012 Olympic facilities.
Health	This includes specialist building services to create hospitals and other health facilities, such as doctors' surgeries and care homes.
Transport infrastructure	This is another broad area of work that includes roads, motorways, bridges, railways, underground trains and tram systems, as well as airports, bus routes and cycle paths.
Public buildings	This is the building and maintenance of large buildings for local and central government. It can include offices, town halls, art galleries, museums and libraries.
Heritage	Heritage involves work on listed properties of historical importance. This is a specialist area, as Building Regulations, planning laws and Listed Status require any work to be carried out in sympathy with the original design of the building.
Conservation	This is an increasingly important area of work, as it involves the protection of natural habitats. It would involve work in National Parks, Areas of Outstanding Natural Beauty, animal sanctuaries and could also include construction work related to coastal erosion and flood defences.
Educational	This is the construction of schools, colleges, universities and other buildings used for educational purposes.
Utilities and services	This is work that is related to the installation, maintenance and repair of the key utilities, which include gas, electricity and water.

Table 3.2

Types of client

As we have seen, there is a huge range of activities and types of work in the broader construction industry and built environment areas. This means that there is a huge range of different potential types of client. Some are private individuals but at the other end of the scale they might be huge companies or government departments. The following table outlines the range of different types of client.

Type of client	Description
Private	These are usually individual owners of homes or buildings. They may be people who want work done on their own homes, or on their own business premises, such as a small shop. Many of the individual shop or business owners may be sole traders. These are individuals who run and own a small business.
Corporate	Corporate is a term that is used to describe larger companies or businesses. They can be individuals that run factories, larger shops, industrial units or some kind of service-based organisation, including banks, insurance companies and estate agents. Some construction companies have long-term contracts with corporate businesses, which have many branches around the country. There is a rolling programme of maintenance, upgrading and repair. The companies can be public limited companies (PLC), who are owned by shareholders with their shares traded on the Stock Exchange.
Government	The government can be a client on a local, regional or national basis. This part of the industry has become more complicated, as there are multiple levels of government across the UK. There is also a Scottish Parliament and a Welsh Assembly, in addition to the UK Parliament based in London. Local councils will be responsible for maintaining a wide range of services and they will also be involved in construction. This includes schools, roads, the maintenance of social housing and parks and leisure facilities. In addition to this there are government departments based in London with regional offices, such as the Ministry of Defence, which is responsible for facilities related to the armed forces, and the National Health Service, which is responsible for hospitals and other health provision. The government (including local authorities and non-departmental public bodies) must comply with strict procurement (buying) rules, which often involve tenders. They also have limited budgets, which could affect the building project's schedule.

Table 3.3

CONSTRUCTION INDUSTRY AND BUILT ENVIRONMENT ROLES AND RESPONSIBILITIES

As we have discussed, the construction industry and the built environment is a complex network of different activities. As the industry has developed over time it has become important for individuals to specialise and take on specific roles and responsibilities.

Roles and responsibilities of the construction workforce

The following tables show the broad range of different roles and briefly outline their responsibilities within the construction industry.

Engineering

Role	Responsibilities
Building services engineers	They are involved in the design, installation and maintenance of heating, water, electrics, lighting, gas and communications. They work either for the main contractor or the architect and give instruction to building services operatives.
Structural engineer	Structural engineers are involved in ensuring that construction work is strong enough to deal with its use and the external environment. So they will be involved in the shape, design and the materials used. They will not only deal with new construction work but also advise on older buildings or buildings that have been damaged.
Consulting/ building engineer	These individuals are involved in site investigation, building inspection and surveys. They get involved in a wide range of construction and maintenance projects.
Plant engineer	A plant engineer is responsible for maintaining and repairing a variety of machinery and equipment. They will also install and modify machinery and equipment in factories as part of an industrial or manufacturing process.
Site engineer	A site engineer is involved in setting out the plans for sewers, drains, roads and other services.
Specialist engineer	A good example of a specialist engineer is one that deals entirely with insulation. They will advise and install a range of energy conservation materials and equipment. A geotechnical engineer is another example. They carry out investigations into below foundation level and look at rock, soil and water.
Mechanical engineer	Mechanical engineers are primarily involved in installing and maintaining machinery and tools. It is a wide ranging profession but they will have overall responsibility for their particular area of work.
Demolition engineer	These engineers perform the task of tearing down old structures or levelling ground to make way for new buildings.
Infrastructure engineer	These engineers deal with the planning, construction and management of roads, bridges and similar structures.

Table 3.4

From the design and planning phase onwards

Role	Responsibilities
Client	The client, such as a local authority, commissions the job. They define the scope of the work and agree on the timescale and schedule of payments.
Customer	For domestic dwellings, the customer may be the same as the client, but for larger projects a customer may be the end user of the building, such as a tenant renting local authority housing or a business renting an office. These individuals are most affected by any work on site. They should be considered and informed with a view to them suffering as little disruption as possible.
Architect	They are involved in designing new buildings, extensions and alterations. They work closely with clients and customers to ensure the designs match their needs. They also work closely with other construction professionals, such as surveyors and engineers.
Estimator	Estimators calculate detailed cost breakdowns of work based on specifications provided by the architect and main contractor. They work out the quantity and costs of all building materials, plant required and labour costs.
Planner	Consultant planners such as civil engineers work with clients to plan, manage, design or supervise construction projects. There are many different types of consultant, all with particular specialisms.
Buyer	This individual works closely with the quantity surveyor. It is the buyer's job to source suitable materials as specified by the architect. They will negotiate prices and delivery dates with a range of suppliers.

Table 3.5

Surveying

Role	Responsibilities
Land agent	This is an individual who is authorised to act as an agent in the sale of land or buildings by the owner. Basically they are estate agents that sell plots of land.
Land surveyor	A land surveyor measures, records and then produces a drawing of the landscape. The data that they produce is used to plan out construction work.
Building surveyor	A building surveyor is responsible for making sure that both old and new buildings are structurally sound. They are involved in the design, maintenance, repair, alteration and refurbishment of buildings.
Quantity surveyor	Quantity surveyors are concerned with building costs. They balance maintaining standards and quality against minimising the costs of any project. They need to make choices in line with Building Regulations. They may work either for the client or for the contractor.

Table 3.6

REED TIP

Any work experience is relevant to your job applications. It doesn't have to be paid work – e.g. volunteering to help run Scout and Guide activities shows your sense of responsibility. Think of the times when others have had to rely on you.

CASE STUDY

South Tyneside Homes

South Tyneside Council's Housing Company

Your apprenticeship is just the start

Gary Kirsop, Head of Property Services, started at South Tyneside Homes as an apprentice 24 years ago.

'After becoming qualified, I had two options. I could have stayed working on the sites and become a site manager or technical assistant. I qualified as a building surveyor, doing my advanced craft at Sunderland College. After that, I went to Newcastle College to do my ONC and CHND, and eventually went on to finish a degree at Newcastle.

When I was a technical assistant I worked on education and public buildings, and spent a year in housing. As a technical assistant I was working on drawing (CAD), estimating small jobs to large jobs. Then an opportunity for Assistant Contracts Manager on capital works came up. Since then, I've also worked in disrepair and litigation, as well as two years with the empty homes department, and I've worked as a Construction Services Manager, responsible for the capital side, new homes, decent homes, and the gas team.

Four years ago, the Head of Property Services job came up and it's been a fantastic opportunity – my team has been one of the best in the country for performance. My department is responsible for repairs and maintenance, capital works, empty homes, and management of the operational side. We do responsive repairs for emergency situations, planned repairs, work for the "Decent Homes" programme where we bring properties up to standard, and we've recently built four new bungalows. Anything in construction, we have the skills and labour to do it in property services.

The full management team here in property services all started as apprentices, like me. It really helps that we understand the whole process from beginning to end.

So you can see that doing your apprenticeship is not only great in itself, but it also gives you skills for life and ongoing opportunities for education, training and your career.'

PHYSICAL AND ENVIRONMENTAL FACTORS AND CONSTRUCTION PROJECTS

Increasingly, people working in construction and the built environment are being asked to ensure that they minimise physical and environmental impacts when carrying out construction work. Construction has an enormous impact on the environment. Environmental measures will depend on the nature of the work and the site. For example, excavations that result in changes in the levels of land can cause problems with water quality and soil erosion. Many of these negative impacts can be reduced during the planning stage.

Physical and environmental factors

Physical factors relate to the impact that any new construction project will have on any existing structures and their occupants. Any new construction project is going to have a negative impact on home owners and businesses. There will be increased traffic on roads and a host of other considerations.

Once the construction has been completed there may be longer term impacts. A prime example would be building a new housing development in an area that lacks good roads, sufficient schools or access to health facilities. During the planning and development stage these factors will be looked at to see what the knock-on effects might be in the short and long term.

Environmental factors concern the impact that a construction project has on the natural environment. This would include any possible impacts on trees and vegetation, wildlife and habitats. It can also have an impact on the air quality or noise levels in the area.

Physical factors and the planning process

There is a wide variety of different physical factors that have to be taken into consideration during the planning process. These are outlined in the following table.

Physical factor	Explanation
Planning requirements	The majority of new developments or changes to existing buildings do require consent or planning permission. The local planning authority will make a decision whether any such construction will go ahead. Each authority has a development framework that outlines how planning is managed. This includes the change of use of a building or a piece of land.
Building Regulations	There are 14 technical parts of the Building Regulations covering everything from structural safety to electrical safety. They also outline standards of quality of work and materials used. All new developments and major changes to existing buildings must comply with Building Regulations.
Development or land restrictions	This is a complicated area, as there are often many restrictions on building and the use of land. One of the most complex is restrictive covenants, which are created in order to protect the interests of neighbours. They might restrict the use of the land and the amount of building work that can take place.
Building design and footprint	The footprint is the physical amount of space or area that the proposed development takes up on a given plot of land. There may be limits as to the size of this footprint. In terms of building design, certain areas may have restrictions as the local authority may not approve the construction of a building that is out of character, or that would adversely affect the overall look of the area.
Use of building or structure	Each building or structure will have a Use Class, such as 'residential', 'shops' or 'businesses'. Redeveloping an existing building and not changing the use to which it is put, for example renovating a building from a butcher to a chemist, does not usually require planning permission. However, changing from a bank to a bar would require planning permission. Certain uses, due to their unique nature, do not fall into any particular Use Class and planning permission is always required. A good example would be a nightclub or a casino.
Impact on local amenities	During the construction phase it is likely that roads or access may have to be blocked, which could impact on local businesses. In the longer term additional traffic and the need for parking may have an impact on local amenities, as will the demand for their use.
Impact on existing services and utilities	Any new development or major change in use of an existing structure may put extra strain on services and utilities in the area. A new housing development, for example, would require power cables to be run to the site. It would also need excavation work to connect it to the sewers and underground pipes run onto the site for potable water. All of this is potentially disruptive and may require considerable investment by the utility or service provider.
Impact on transportation infrastructure	Major new developments will have a huge impact on the roads and public transport in an area. Permission for major developments often comes with the requirement to improve access routes, build new roads and the requirement to make a contribution to improvements in the infrastructure. New developments can radically change the flow of traffic in an area and may have a knock-on effect in terms of maintenance and repair in the longer term.
Topography of the proposed development site	The term topography refers to the location of the site and how dominant it will be in the local landscape. Obviously a development that is situated on a hill or ridge is far more obvious and will have a longer lasting impact on the local area. If the development is considered to be too obtrusive or visible then it may be deemed as inappropriate to situate the development on that site.
Greenfield or brownfield site	A greenfield site is an area of land that has never been used for non-agricultural purposes. A brownfield site is usually former industrial land, or land that has been used for some other purpose and is no longer in use. There is more information on greenfield and brownfield sites in the next section of this chapter.

Table 3.7

Environmental factors and the planning process

Just as there are physical factors, there are also different environmental factors that need to be considered. Some of the major ones are detailed in the following table.

Environmental factor	Explanation
Topography of the development site	As mentioned in the previous table, the topography of the development site can have a marked impact on the local environment. It may dominate what is otherwise a predominantly natural environment, perhaps with woodland or rolling hills.
Existing trees and vegetation	Sites may have to be cleared in order to provide the necessary space for the footprint of the structure. It may be prohibited to remove or otherwise interfere with certain trees and vegetation, as they may be protected. The normal course of events is to minimise the impact on existing plant life and to have a replanting phase after the site has been developed.
Impact on existing wildlife and habitats	Any potential impact on wildlife and plants that are under threat could mean that the site would not receive the go ahead. An environmental impact study will identify whether there are any specific dangers that will affect the natural habitat of the area, or endanger any local species of wildlife.
Size of land and building footprint	There is a formula that determines the usually permitted footprint of a piece of land compared to the actual size of the plot of land. For example, a 4-bedroom house on an average housing estate would take up approximately 1/12th of an acre (11.5m × 29m).
Access to the building or structure	It is not only the building plot that needs to be considered in terms of its environmental impact. Access to the site is another concern. Existing roads may have to be widened, perhaps a roundabout installed. Alternatively new roads may have to be built across other plots of land. For pedestrian traffic footpaths may also be necessary. These can either be alongside existing roads or built alongside new roads, requiring even more space. There may be existing footpaths and this could mean that access needs to be provided through the site or the footpaths diverted.
Supply of services to the building or structure	Running above ground services and utilities to the site may also present a problem as far as its impact on the environment is concerned. It may not be possible to allow features such as pylons or street lights to dominate the landscape.
Natural water resources	New developments can affect the biodiversity of an area by impacting on natural waterways. Local wildlife and plants rely on this resource. In addition to this, construction could either pollute or affect the quality of the local water.
Land restrictions	There may be land restrictions that limit either the use or the size of any development. Developments will not be allowed to adversely affect surrounding properties and owners. There are conservation areas, scheduled monuments, archaeological sites and scheduled or listed buildings. These are all protected and construction on or near them is either prohibited or severely limited.
Future development and expansion	Although the intention may be to restrict the environmental impact of the site in the first phase of development, in the future this might not be possible. Major housing development is often carried out in phases and the size of the development will gradually increase as demand increases. It is therefore important when permission is initially given that the likelihood of future development and expansion is taken into account.

Table 3.8

Figure 3.1 Trees on a proposed site may need to be protected during construction work

HOW CONSTRUCTION PROJECTS BENEFIT THE BUILT ENVIRONMENT

The construction industry is one of the UK's largest employers. It is a hugely diverse industry. Construction projects can have a massive impact on the built environment. They can rejuvenate whole areas; improve the housing stock, amenities and the general life and well-being of the local population. The built environment describes the overall look and layout of a specific area. Each new construction project and its architectural design will have an impact on that built environment and the broader, natural environment. If it is carefully and sympathetically planned and organised it can have a positive impact on the way people live, work and interact with one another.

Each new development has enormous environmental, social and economic consequences. Increasingly it has a role to play in ensuring that our built environment has a strong and sustainable future.

Land types available for development and their advantages and disadvantages

In March 2012 the National Planning Policy Framework was published, which aims to review planning guidance across the UK. The idea was to encourage the building of domestic dwellings. It stated that there would be a policy to try to use as many brownfield sites as possible, but that greenfield sites in rural areas would no longer be protected at any cost. Where development was necessary it would take place, as there was a huge demand for homes, shops and workplaces.

The first targets for development would be sites that had been used in the past for other purposes.

Greenfield land or sites
Greenfield sites are usually either agricultural or amenity land. Given the fact that there is a housing crisis in the UK and that land needs to be allocated to build millions of new homes, greenfield sites are very much under consideration.

The problem in doing this is that there is huge resistance, particularly in rural areas, to losing greenfield sites for the following reasons:

* Once a greenfield site has been developed it is extremely unlikely that it will ever return to agricultural use. Any loss of agricultural land means a reduction in the amount of food that can be produced in the UK. There might also be a drop in employment in the local area as fewer farm workers are needed.

* Natural habitats of wildlife and plants are destroyed forever.

* Greenfield or amenity land, if lost, means that the land can no longer be used for leisure and recreation.

* Developments on greenfield sites can have a negative impact on the local transport infrastructure and will increase the amount of energy used because things are further away from town centres.

* The loss of green belts of agricultural land around cities, towns and villages means that each separate area loses its identity and in effect becomes a suburb of a larger town or city.

Figure 3.2 Building on greenfield and greenbelt land is a controversial issue

Brownfield land or sites

Brownfield sites are pieces of land that have been previously developed. They were probably used for either industrial or commercial purposes, but are now derelict and abandoned.

Figure 3.3 Brownfield sites have previously been built on

Brownfield sites can be found in areas where there is a high demand for new homes. It has been estimated that there are more than 66,000 hectares of brownfield sites in England alone. At least a third of this land can be found in the southeast of England, where there is the highest demand for housing. Around 60 per cent of new housing is being built on brownfield sites. This is a trend that is likely to accelerate over the next 10 years.

Brownfield sites are not just used for housing projects but are also sites for commercial buildings, as well as recreational sites and newly planted woodland.

Reclaimed land

There are areas, particularly around the coast and in estuaries, which for many years have been bogs or salt marshes. These damp grasslands can be gradually drained of water and eventually provide agricultural land or, in some cases, land suitable for housing developments. With global warming and climate change threatening to permanently flood huge areas of the UK, it may seem strange to consider humans reversing the process.

The area is converted by digging flood relief channels and drainage ditches to encourage the water to flow out and away from the land. To protect the land during this process banks are built to keep out river and seawater. It is a long and involved process but can provide possible land for redevelopment. This process has been successful in many different parts of the world, notably in the Fens in East Anglia, on the Netherlands coast, where pumping stations reclaim land from the sea, and in the Middle and Far East where huge projects have reclaimed vast areas of land.

Figure 3.4 Reclaiming land enables it to be put other uses

Contaminated land

Many brownfield sites, particularly those once used for industrial purposes, are contaminated with varying levels of hazardous waste and pollutants. Before any development can take place an environmental consultant will organise the analysis of soil, ground water and surface water to identify any risks.

Special licences are required to reclaim brownfield sites and this can be a very expensive process for developers. The main way of dealing with brownfield sites is a process known as remediation. This involves the removal of any known contaminants to a level that will not affect the health of anyone living or working on the site both during construction and after building is complete.

Not all brownfield sites are, therefore, suitable or cost-effective. In some cases the cost of removing the contaminants exceeds the value of the land after it has been developed. There are new ways of dealing with contaminants:

Figure 3.5 Contaminated land must be cleaned before use

* Bioremediation – this uses bacteria, plants, fungi and micro-organisms to destroy or neutralise contaminants.

* Phytoremediation – plants are encouraged to grow on the site and the contaminants are taken up into the plant and stored in their leaves and stems.

* Chemical oxidation – this involves injecting oxygen or oxidants into contaminated soil and water to destroy contaminants.

Social benefits of construction development

The construction industry and the built environment do provide a range of potential benefits, particularly to local areas. These are examined in the following table.

Social benefit	Explanation
Regeneration of brownfield sites	Disused land, usually former industrial sites, and have been developed for new housing and commercial sites. In London, virtually the whole of the 2012 Olympic village was built on brownfield sites.
Local employment	Construction sites need the skills of local construction workers and offer opportunities for small businesses. Long-term projects offer long-term employment for local people.
Improved housing	New developments and refurbishment of older properties provide greener and more energy efficient dwellings. This has a long-term positive impact for the environment and the reduction in the use of non-renewable resources.
Improvements to local infrastructure	A new development of any size often comes with the requirement for the developers to contribute towards the building of new roads and other infrastructure projects for the area. New developments, in order to work, need access roads, transport and other facilities.
Improvements to local amenities	Modern housing developments and commercial properties need to have amenities near them in order to make them viable in the longer term. This means the building of schools, hospitals, health centres and shops.

Table 3.9

Figure 3.6 Sustainable developments aim to be pleasant places to live

KEY TERMS

Landfill

– 170 million tonnes of waste from homes and businesses are generated in England and Wales each year. Much of this has to be taken to a site to be buried.

Biodiversity

– wherever there is construction there is a danger that the wildlife and plants could be disturbed or destroyed. Protecting biodiversity ensures that at risk species are conserved.

SUSTAINABILITY

Carbon is present in all fossil fuels, such as coal or natural gas. Burning fossil fuels releases carbon dioxide, which is a greenhouse gas linked to climate change.

Energy conservation aims to reduce the amount of carbon dioxide in the atmosphere. The idea is to do this by making buildings better insulated and, at the same time, making heating appliances more efficient. It also means attempting to generate energy using renewable and/or low or zero carbon methods.

According to the government's Environment Agency, sustainable construction is all about using resources in the most efficient way. It also means cutting down on waste on site and reducing the amount of materials that have to be disposed of and put into **landfill.**

In order to achieve sustainable construction the Environment Agency recommends:

* reducing construction, demolition and excavation waste that needs to go to landfill

* cutting back on carbon emissions from construction transport and machinery

* responsibly sourcing materials

* cutting back on the amount of water that is wasted

* making sure construction does not have an impact on **biodiversity.**

What is meant by sustainability?

In the past buildings have been constructed as quickly as possible and at the lowest cost. More recently the idea of sustainable construction has focused on ensuring that the building is not only of good quality and that it is affordable, but that it is also energy efficient.

Sustainable construction also means having the least negative environmental impact. So this means minimising the use of raw materials, energy, land and water. This is not only during the build period but also for the lifetime of the building.

Figure 3.7 Eco houses are becoming more common

Construction and the environment

In 2010, construction, demolition and excavation produced 20 million tonnes of waste that had to go into landfill. The construction industry is also responsible for most illegal fly tipping (illegally dumping waste). In any year there are at least 350 serious pollution incidents caused as a result of construction.

Figure 3.8 Always dispose of waste responsibly

Regardless of the size of the construction job, everyone in construction is responsible for the impact they have on the environment. Good site layout, planning and management can help reduce this impact.

Sustainable construction helps to encourage this because it means managing resources in a more efficient way, reducing waste and reducing your **carbon footprint.**

Finite and renewable resources

We all know that resources such as coal and oil will eventually run out. These are examples of finite resources.

Oil is not just used as fuel – it is used in plastic, dyes, lubricants and textiles. All of these are used in the construction process.

Renewable resources are those that can be produced by moving water, the sun or the wind. Materials that come from plants, such as biodiesel, or the oils used to make some pressure-sensitive adhesives, are examples of renewable resources.

The construction process itself is only part of the problem. It is also the longer term impact and demands that the building will have on the environment. This is why there has been a drive towards sustainable homes and there is a Code for Sustainable Homes, which is a certification of sustainability for new build housing.

The future

Sustainability also means ensuring that future generations do not suffer from the ill-considered activities of today's generation. The following table outlines some of the present dangers and concerns.

KEY TERMS

Carbon footprint

– this is the amount of carbon dioxide produced by a project. This not only includes burning carbon-based fuels such as petrol, gas, oil or coal, but includes the carbon that is generated in the production of materials and equipment.

DID YOU KNOW?

Search on the internet for 'sustainable building' and 'improving energy efficiency' to find out more about the latest technologies and products.

Global warming

– a rise in temperature of the earth's atmosphere. The planet is naturally warmed by rays, some being reflected back out into space. The atmosphere is made up of gases (some are called greenhouse gases) which are mainly natural and form a kind of thermal blanket. The human-made gases are believed to make this blanket thicker, so less of the heat escapes back into space. Over the past 100 years, our climate has seen some rapid changes. This is believed to be linked to changes in the makeup of the atmosphere and land use.

Climate change

– the burning of fossil fuels (coal, gas, wood, oil) has resulted in an increase in the amount of greenhouse gases. This has pushed up global temperatures. Across the world, millions do not have enough water, species are dying out and sea levels are rising. In the UK we see extreme events such as flooding, storms, sea level rise and droughts. We have wetter warmer winters and hotter drier summers.

Present or future concern	Explanation
Global shortages	Many naturally found resources will eventually run out and they will have to be replaced with alternatives. Acting now to discover, develop or use alternatives will delay this. Construction is at the forefront of finding alternatives and looking at different construction materials and methods.
Needs of future generations	Buildings constructed today must to be useful and affordable for future generations. At the same time, materials and construction methods should not leave a bad legacy that future generations have to deal with.
Global warming	The construction industry has been criticised over its contribution to global warming. A lack of co-ordination between different parts of the industry has produced poor quality, energy-inefficient buildings. The government is keen to ensure that the industry trains people about the principles of sustainable design and efficient technologies. These steps need to be put in place to inform decisions at the design stage of a building.
Climate change	Construction projects need to take into account the effects of climate change and consider ways to reduce the project's impact on the environment. This means minimising carbon emissions, using sustainable (or renewable) energy and reducing water consumption.
Extinction of species and vegetation	Global warming and climate change has an impact on animals and plants. On a local level, this is also a problem as construction can destroy natural habitats. Increasingly, this is closely monitored and environmental impact studies are used to prevent this from happening.
Destruction of natural resources	There are strict planning laws that aim to prevent the industry from destroying or harming natural habitats. Ancient woodland, sites of scientific importance and other sites of interest are all protected. It is also the case that development in areas that are likely to flood or cause flooding are prohibited or controlled.

Table 3.10

Figure 3.9 Climate change may be a serious problem over the next decades

Social regeneration

Construction projects are often used to regenerate areas of the UK that have lacked investment in the past. As industry develops and changes over time, whole areas that would once have been extremely busy in the past now have empty industrial units and high unemployment levels. As the area loses jobs housing deteriorates, as does the local infrastructure, as there is no money in the local economy.

Redeveloping these waste sites is seen as a way in which a whole area can be regenerated or reborn. Construction projects bring jobs relating to the project but they also bring the promise of longer term jobs. These areas have relatively cheap land and lower rents. Also the workforce expects lower rates of pay. This attracts businesses to relocate to the new buildings created by construction developers. This brings work, improved housing, and improvements to the local infrastructure and amenities.

Sustainability and its benefits

Energy efficiency is all about using less energy to provide the same result. The plan is to try to cut the world's energy needs by 30 per cent before 2050. This means producing more energy efficient buildings. It also means using energy efficient methods to produce the materials and resources needed to construct buildings.

Alternative methods of building

The most common type of construction in the UK is brick and blockwork. However there are plenty of other options:

* timber frame – using pre-fabricated timber frames which are then clad

* insulated concrete formwork – where a polystyrene mould is filled with reinforced concrete

* structural insulated panels – where buildings are made up of rigid building boards rather like huge sandwiches

* modular construction – this uses similar materials and techniques to standard construction, but the units are built off site and transported ready-constructed to their location.

Figure 3.10 Insulated concrete formwork   Figure 3.11 Modular construction

There are alternatives to traditional flooring and roofing, all of which are greener and more sustainable. Green roofing (both living roofs and roofs made from recycled materials) has become an increasing trend in recent years. Metal roofs made of steel, aluminium and copper often use a high percentage of recycled material. They are also lightweight. Solar roof shingles, or solar roof laminates, while expensive, decrease the cost of electricity and heating for the dwelling. Some buildings even have a living roof which consists of a waterproof membrane, a drainage layer, a growing material and plants such as sedum. This provides additional insulation, absorbs air pollution, helps to collect and process rainwater and keeps the roof surface temperature down.

Just as roofs are becoming greener, so too are the options for flooring. The use of renewable resources such as bamboo, eucalyptus and cork is becoming more common. A new version of linoleum has been developed with **biodegradable**, **organic** ingredients. Some buildings are also using floorboards and joists made from non-timber materials that can be coloured, stained or patterned.

Figure 3.12 Solar roof tiles provide their own solar power

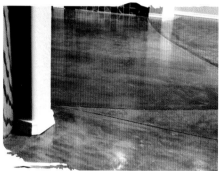

Figure 3.13 A stained concrete floor can be a striking feature

An increasing trend has been for what is known as off-site manufacture (OSM). European businesses, particularly those in Germany, have built over 100,000 houses. The entire house is manufactured in a factory and then assembled on site. Walls, floors, roofs, windows and doors with built-in electrics and plumbing all arrive on a lorry. Some manufacturers even offer completely finished dwellings, including carpets and curtains. Many of these modular buildings are designed to be far more energy efficient than traditional brick and block constructions. Many come ready fitted with heat pumps, solar panels and triple-glazed windows.

Architecture and design

The Code for Sustainable Homes Rating Scheme was introduced in 2007. Many local authorities have instructed their planning departments to encourage sustainable development. This begins with the work of the architect who designs the building.

Figure 3.14 A timber-framed HUF haus is assembled off site

Local authorities ask that architects and building designers:

* ensure the land is safe for development – that if it is contaminated this is dealt with first

* ensure access to and protection for the natural environment – this supports biodiversity and tries to create open spaces for local people

* reduce the negative impact on the local environment – buildings should keep noise, air, light and water pollution down to a minimum

* conserve natural resources and cut back carbon emissions – this covers energy, materials and water

* ensure comfort and security – good access, close to public transport, safe parking and protection against flooding.

Figure 3.15 Eco developments, like this one in London, are becoming more common

DID YOU KNOW?

Local planning authorities now require that all new developments generate at least 10 per cent of their energy from renewable sources.

Using locally managed resources

The construction industry imports nearly 6 million cubic metres of sawn wood each year. Around 80 per cent of all the softwood used in construction comes from Scandinavia or Russia. Another 15 per cent comes from the rest of Europe, or even North America. The remaining 5 per cent comes from tropical countries, and is usually sourced from sustainable forests. However there is plenty of scope to use the many millions of cubic metres of timber produced in managed forests, particularly in Scotland.

Local timber can be used for a wide variety of different construction projects:

* Softwood – including pines, firs, larch and spruce – for panels, decking, fencing and internal flooring.

* Hardwood – including oak, chestnut, ash, beech and sycamore – for a wide variety of internal joinery.

Eco-friendly, sustainable manufactured products and environmentally resourced timber

There are now many suppliers that offer sustainable building materials as a green alternative. Tiles, for example, can be made from recycled plastic bottles and stone particles.

There is a National Green Specification database of all environmentally friendly building materials. This provides a checklist where it is possible to compare specifications of environmentally friendly materials to those of traditionally manufactured products, such as bricks.

Simple changes to construction, such as using timber or ethylene-based plastics instead of PVCU window frames is a good example.

Finding locally managed resources such as timber makes sense in terms of cost and in terms of protecting the environment.

The Timber Trade Federation produces a Timber Certification System. This ensures that wood products are labelled to show that they are produced in sustainable forests.

PRACTICAL TIP

www.recycledproducts.org,uk has a long list of recycled surfacing products, such as tiles, recycled wood and paving and detials of local suppliers

Building Regulations

In terms of energy conservation, the most important UK law is the Building Regulations 2010, particularly Part L. The Building Regulations:

* list the minimum efficiency requirements

* provide guidance on compliance, the main testing methods, installation and control

* cover both new dwellings and existing dwellings.

DID YOU KNOW?

The Forest Stewardship Council has a system that verifies that wood comes from well-managed forests. The Programme for the Endorsement of Forest Certification Schemes promotes sustainable managed forests throughout the world. (www.fsc.org and www.pefc.co.uk)

A key part of the regulations is the Standard Assessment Procedure (SAP), which measures or estimates the energy efficiency performance of buildings.

Local planning authorities also now require that all new developments generate at least 10 per cent of their energy from renewable sources. This means that each new project has to be assessed one at a time.

Energy conservation

By law, each local authority is required to reduce carbon dioxide emissions and to encourage the conservation of energy. This means that everyone has a responsibility in some way to conserve energy:

* Clients, along with building designers, are required to include energy efficient technology in the build.

* Contractors and sub-contractors have to follow these design guidelines. They also need to play a role in conserving energy and resources when actually working on site.

* Suppliers of products are required by law to provide information on energy consumption.

In addition, new energy efficiency schemes and building regulations cover the energy performance of buildings. Each new build is required to have an Energy Performance Certificate. This rates a building's energy efficiency from A (which is very efficient) to G (which is least efficient).

Some building designers have also begun to adopt other voluntary ways of attempting to protect the environment. These include: BREEAM (Building Research Establishment Environmental Assessment Method, a voluntary measurement rating for green buildings) and the Code for Sustainable Homes (a certification of sustainability for new builds).

High, low and zero carbon

When we look at energy sources, we consider their environmental impact in terms of how much carbon dioxide they release. Accordingly, energy sources can be split into three different groups:

* high carbon – those that release a lot of carbon dioxide

* low carbon – those that release some carbon dioxide

* zero carbon – those that do not release any carbon dioxide.

Some examples of high carbon, low carbon and zero carbon energy sources are given in the tables below.

**energy®
saving
trust**

Figure 3.16 The Energy Saving Trust encourages builders to use less wasteful building techniques and more energy efficient construction

High carbon energy source	Description
Natural gas or LPG	Piped natural gas or liquid petroleum gas stored in bottles
Fuel oils	Domestic fuel oil, such as diesel
Solid fuels	Coal, coke and peat
Electricity	Generated from non-renewable sources, such as coal-fired power stations

Table 3.11

Low carbon energy source	Description
Solar thermal	Panels used to capture energy from the sun to heat water
Solid fuel	Biomass such as logs, wood chips and pellets
Hydrogen fuel cells	Converts chemical energy into electrical energy
Heat pumps	Devices that convert low temperature heat into higher temperature heat
Combined heat and power (CHP)	Generates electricity as well as heat for water and space heating
Combined cooling, heat and power (CCHP)	A variation on CHP that also provides a basic air conditioning system

Table 3.12

Zero carbon energy	Description
Electricity/wind	Uses natural wind resources to generate electrical energy
Electricity/tidal	Uses wave power to generate electrical energy
Hydroelectric	Uses the natural flow of rivers and streams to generate electrical energy
Solar photovoltaic	Uses solar cells to convert light energy from the sun into electricity

Table 3.13

It is important to try to conserve non-renewable energy so that there will be sufficient fuel for the future. The idea is that the fuel should last as long as is necessary to completely replace it with renewable sources, such as wind or solar energy.

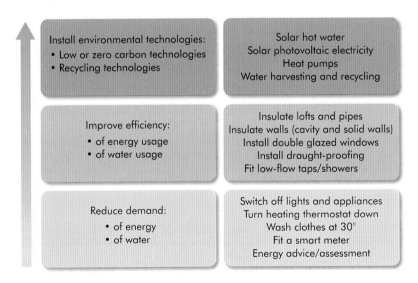

Figure 3.17 Working towards reducing carbon emissions

Alternative energy sources

There are several new ways in which we can harness the power of water, the sun and the wind to provide us with new heating sources. All of these systems are considered to be far more energy efficient than traditional heating systems, which rely on gas, oil, electricity or other fossil fuels.

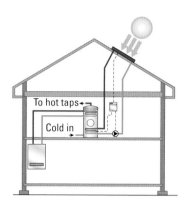

Figure 3.18 Solar thermal hot water system

Solar thermal

At the heart of this system is the solar collector, which is often referred to as a solar panel. The idea is that the collector absorbs the sun's energy, which is then converted into heat. This heat is then applied to the system's heat transfer fluid.

The system uses a differential temperature controller (DTC) that controls the system's circulating pump when solar energy is available and there is a demand for water to be heated.

In the UK, due to the lack of guaranteed solar energy, solar thermal hot water systems often have an auxiliary heat source, such as an immersion heater.

Biomass (solid fuel)

Biomass stoves burn either pellets or logs. Some have integrated hoppers that transfer pellets to the burner. Biomass boilers are available for pellets, woodchips or logs. Most of them have automated systems to clean the heat exchanger surfaces. They can provide heat for domestic hot water and space heating.

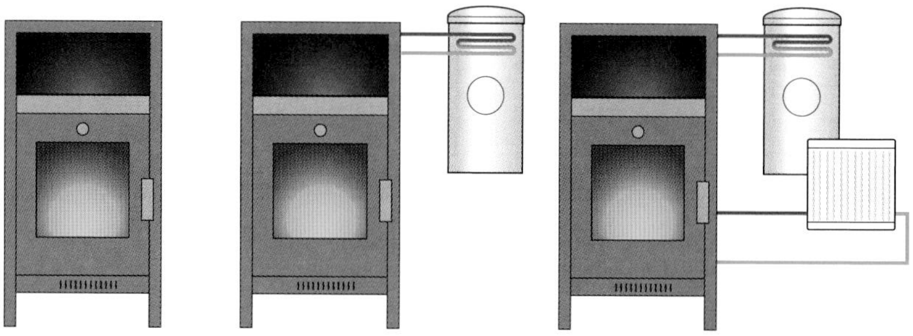

Stove providing room heat only

Stove providing room heat and domestic hot water

Stove providing room heat, domestic hot water and heating

Figure 3.19 Biomass stoves output options

KEY TERMS

Heat sink

– this is a heat exchanger that transfers heat from one source into a fluid, such as in refrigeration, air conditioning or the radiator in a car.

Geothermal

– relating to the internal heat energy of the earth.

Heat pumps

Heat pumps convert low temperature heat from air, ground or water sources to higher temperature heat. They can be used in ducted air or piped water **heat sink** systems.

There are different arrangements for each of the three main systems:

* Air source pumps operate at temperatures down to minus 20°C. They have units that receive incoming air through an inlet duct.

* Ground source pumps operate on **geothermal** ground heat. They use a sealed circuit collector loop, which is buried either vertically or horizontally underground.

* Water source pump systems can be used where there is a suitable water source, such as a pond or lake. Energy extracted from the water is used as heat.

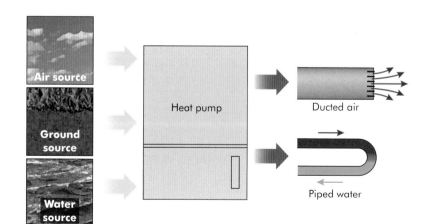

Figure 3.20 Heat pump input and output options

The heat pump system's efficiency relies on the temperature difference between the heat source and the heat sink. Special tank hot water cylinders are part of the system, giving a large surface-to-surface contact between the heating circuit water and the stored domestic hot water.

Combined heat and power (CHP) and combined cooling heat and power (CCHP) units

These are similar to heating system boilers, but they generate electricity as well as heat for hot water or space heating (or cooling). The heart of the system is an engine or gas turbine. The gas burner provides heat to the engine when there is a demand for heat. Electricity is generated along with sufficient energy to heat water and to provide space heating.

CCHP systems also incorporate the facility to cool spaces when necessary.

Wind turbines

Freestanding or building-mounted wind turbines capture the energy from wind to generate electrical energy. The wind passes across rotor blades of a turbine, which causes the hub to turn. The hub is connected by a shaft to a gearbox. This increases the speed of rotation. A high speed shaft is then connected to a generator that produces the electricity.

Solar photovoltaic systems

A solar photovoltaic system uses solar cells to convert light energy from the sun into electricity.

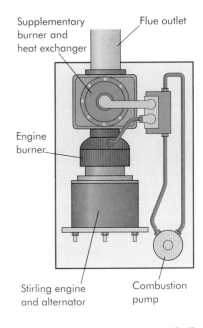

Figure 3.21 Example of a MCHP (micro combined heat and power) unit

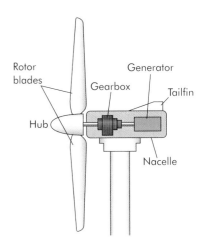

Figure 3.22 A basic horizontal axis wind turbine

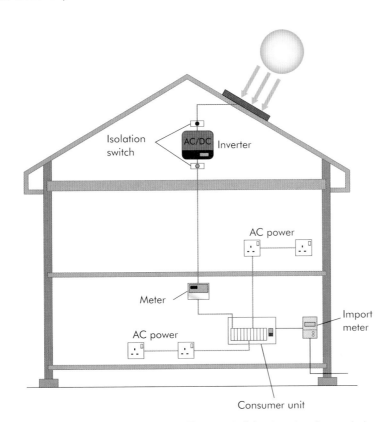

Figure 3.23 A basic solar photovoltaic system

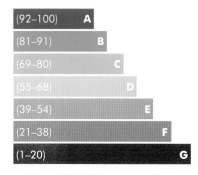

Figure 3.24 SAP energy efficiency rating table. The ranges in brackets show the percentage energy efficiency for each banding

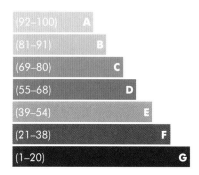

Figure 3.25 SAP environmental impact rating table

Energy ratings

Energy rating tables are used to measure the overall efficiency of a dwelling, with rating A being the most energy efficient and rating G the least energy efficient.

Alongside this, an environmental impact rating measures the dwelling's impact in terms of how much carbon dioxide it produces. Again, rating A is the highest, showing it has the least impact on the environment, and rating G is the lowest.

A Standard Assessment Procedure (SAP) is used to place the dwelling on the energy rating table. This will take into account:

* the date of construction, the type of construction and the location
* the heating system
* insulation (including cavity wall)
* double glazing.

The ratings are used by local authorities and other groups to assess the energy efficiency of new and old housing and must be provided when houses are sold.

Preventing heat loss

Most old buildings are under-insulated and will benefit from additional insulation, which can be for ceilings, walls or floors.

The measurement of heat loss in a building is known as the U Value. It measures how well parts of the building transfer heat. Low U Values represent high levels of insulation. U Values are becoming more important as they form the basis of energy and carbon reduction standards.

By 2016 all new housing is expected to be Net Zero Carbon. This means that the building should not be contributing to climate change.

Many of the guidelines are now part of Building Regulations (Part L). They cover:

* insulation requirements
* openings, such as doors and windows
* solar heating and other heating
* ventilation and air conditioning
* space heating controls
* lighting efficiency
* air tightness.

Building design

UK homes spend £2.4bn every year just on lighting. One of the ways of tackling this cost is to use energy saving lights, but also to maximise natural lighting. For the construction industry this means:

* increased window size
* orientating window angles to make the most of sunlight – south facing windows maximise sunlight in winter and limit overheating in the summer
* window design – with a variety of different types of opening to allow ventilation.

Solar tubes are another way of increasing light. These are small domes on the roof, which collect sunlight and direct it through a tube (which is reflective). It is then directed through a diffuser in the ceiling to spread light into the room.

Waste water recycling

Water is a precious resource, so it is vital not to waste it. To meet the current demand for water in the UK, it is essential to reduce the amount of water used and to recycle water where possible.

The construction industry can contribute to water conservation by effective plumbing design and through the installation of water efficient appliances and fittings. These include low or dual flush WCs, and taps and fittings with flow regulators and restrictors. In addition, rainwater harvesting and waste water recycling should be incorporated into design and construction wherever possible.

Statutory legislation for water wastage and misuse

Water efficiency and conservation laws aim to help deal with the increasing demand for water. Just how this is approached will depend on the type of property:

* For new builds, the Code for Sustainable Homes and Part G of the Building Regulations set new water efficiency targets.

* For existing buildings, Part G of the Building Regulations applies to all refurbishment projects where there is a major change of use.

* For owners of non-domestic buildings, tax reduction schemes and grants are available for water efficiency projects.

In addition, the Water Supply (Water Fittings) Regulations 1999 set a series of efficiency improvements for fittings used in toilets, showers and washing machines, etc.

Reducing water wastage

There are many different ways in which water wastage can be reduced, as shown in the table below.

Method	Explanation
Flow reducing valves	Water pressure is often higher than necessary. By reducing the pressure, less water is wasted when taps are left running.
Spray taps	Fixing one of these inserts can reduce water consumption by as much as 70 per cent.
Low volume flush WC	These reduce water use from 13 litres per flush to 6 litres for a full flush and 4 litres for a reduced flush.
Maintenance of terminal fittings and float valves	Dripping taps or badly adjusted float valves can cause enormous water wastage. A dripping tap can waste 5,000 litres a year.
Promoting user awareness	Users who are on a meter will certainly see a difference if water efficiency is improved, and their energy bills will be reduced if they use less hot water.

Table 3.14

Captured and recycled water systems

There are two variations of captured and recycled water systems:

* Rainwater harvesting captures and stores rainwater for non-potable use (not for drinking).

* Greywater reuse systems capture and store waste water from baths, washbasins, showers, sinks and washing machines.

Rainwater harvesting

In this system, water is harvested usually from the roof and then distributed to a tank. Here it is filtered and then pumped into the dwelling for reuse. The recycled water is usually stored in a cistern at the top of the building.

Greywater reuse

The idea of this system is to reduce mains water consumption. The greywater is piped from points of use, such as sinks and showers, through a filter and into a storage tank. The greywater is then pumped into a cistern where it can be used for toilet flushing or for watering the garden.

Waste management

The expectation within the construction industry is increasingly that working practices conserve energy and protect the environment. Everyone can play a part in this. For example, you can contribute at home by turning off hose pipes when you have finished using water.

Simple things, such as keeping construction sites neat and orderly, can go a long way to conserving energy and protecting the environment. A good way to remember this is Sort, Set, Shine, Standardise:

Sort – sort and store items in your work area, eliminate clutter and manage deliveries.

Set – everything should have its own place and be clearly marked and easy to access. In other words, be neat!

Shine – clean your work area and you will be able to see potential problems far more easily.

Standardise – using standardised working practices you can keep organised, clean and safe.

Reducing material wastage

Reducing waste is all about good working practice. By reducing wastage disposal and recycling materials on site, you will benefit from savings on raw materials and lower transportation costs.

Let's start by looking at ways to reduce waste when buying and storing materials:

* Only order the amount of materials you actually need.

* Arrange regular deliveries so you can reduce storage and material losses.

* Think about using recycled materials, as they may be cheaper.

* Is all the packaging absolutely necessary? Can you reduce the amount of packaging?

* Reject damaged or incomplete deliveries.

* Make sure that storage areas are safe, secure and weatherproof.

* Store liquids away from drains to prevent pollution.

By planning ahead and accurately measuring and cutting materials, you will be able to reduce wastage.

Statutory legislation for waste management

By law, all construction sites should be kept in good order and clean. A vital part of this is the proper disposal of waste, which can range from low risk waste, such as metals, plastics, wood and cardboard, to hazardous waste, for example asbestos, electrical and electronic equipment and refrigerants.

DID YOU KNOW?

If waste is not properly managed and the duty of care is broken, then a fine of up to £5,000 may be issued.

Waste is anything that is thrown away because it is no longer useful or needed. However, you cannot simply discard it, as some waste can be recycled or reused while other waste will affect health or the quality of the environment.

Legislation aims not only to prevent waste from going into landfill but also to encourage people to recycle. For example, under the Environmental Protection Act (1990), the building services industry has the following duty of care with regard to waste disposal:

* All waste for disposal can only be passed over to a licensed operator.

* Waste must be stored safely and securely.

* Waste should not cause environmental pollution.

The main legislation covering the disposal of waste is outlined in the table below.

Legislation	Brief explanation
Environmental Protection Act (1990)	Defines waste and waste offences
Environmental Protection (Duty of Care) Regulations (1991)	Places the responsibility for disposal on the producer of the waste
Hazardous Waste Regulations (2005)	Defines hazardous waste and regulates the safe management of hazardous waste
Waste Electrical and Electronic Equipment (WEEE) Regulations (2006)	Requires those who produce electrical and electronic waste to pay for its collection, treatment and recovery
Waste Regulations (2011)	Introduces a system for waste carrier registration

Table 3.15

Safe methods of waste disposal

In order to dispose of waste materials legally, you must use the right method.

* **Waste transfer notes** are required for every load of waste that is passed on or accepted.

* **Licensed waste disposal** is carried out by operators of landfill sites or those that store other people's waste, treat it, carry out recycling or are involved in the final disposal of waste.

* **Waste carriers' licences** are required by any company that transports waste, not just waste contractors or skip operators. For example, electricians or plumbers that carry construction and demolition waste would need to have this licence, as would anyone involved in construction or demolition.

* **Recycling** of materials such as wood, glass, soil, paper, board or scrap metal is dealt with at materials reclamation facilities. They sort the material, which is then sent to reprocessing plants so it can be reused.

* **Specialist disposal** is used for waste such as asbestos. There are authorised asbestos disposal sites that specialise in dealing with this kind of waste.

Recycling metals

Scrap metal is divided into two different types:

- **Ferrous** scrap includes iron and steel, mainly from beams, cars and household appliances.

- **Non-ferrous** scrap is all other types of metals, including aluminium, lead, copper, zinc and nickel.

Recycling businesses will collect and store metals and then transport them to **foundries**. The operators will have a licence, permit or consent to store, handle, transport and treat the metal.

Recycling plastics

Different types of plastic are used for different things, so they will need to be recycled separately. Licensed collectors will pass on the plastics to recycling businesses that will then remould the plastics.

Recycling wood and cardboard

Building sites will often generate a wide variety of different wood waste, such as off-cuts, shavings, chippings and sawdust.

Paper and cardboard waste can be passed on to an authorised waste carrier.

Disposing of asbestos

Asbestos should only be disposed of by specialist contractors. It needs to be double wrapped in approved packaging, with a hazard sign and asbestos code information visible. You should also dispose of any contaminated PPE in this way. The standard practice is to use a red inner bag with the asbestos warning and a clear outer bag with a carriage of dangerous goods (CDG) sign.

Asbestos waste should be carried in a sealed skip or in a separate compartment to other waste. It should be transported by a registered waste carrier and disposed of at a licensed site. Documentation relating to the disposal of asbestos must be kept for three years.

Disposing of electrical and electronic equipment

The Waste Electrical and Electronic Equipment (WEEE) Regulations were first introduced in the UK in 2006. They were based on EU law – the WEEE Directive of 2003.

Normally, the costs of electrical and electronic waste collection and disposal fall on either the contractor or the client. Disposal of items such as this are part of Site Waste Management Plans, which apply to all construction projects in England worth more than £300,000.

- For equipment purchased after August 2005, it is the responsibility of the producer to collect and treat the waste.

- For equipment purchased before August 2005 that is being replaced, it is the responsibility of the supplier of the equipment to collect and dispose of the waste.

- For equipment purchased before August 2005 that is not being replaced, it is the responsibility of either the contractor or client to dispose of the waste.

Disposing of refrigerants

Refrigerators, freezer cabinets, dehumidifiers and air conditioners contain **fluorinated gases**, known as chloro-fluoro-carbons (CFCs). CFCs have been linked with damage to the Earth's **ozone layer**, so production of most CFCs ceased in 1995.

Refrigerants such as these have to be collected by a registered waste company, which will de-gas the equipment. During the de-gassing process, the coolant is removed so that it does not leak into the atmosphere.

Key benefits of using sustainable materials

In summary:

- Using locally sourced materials not only cuts down on the transportation costs but also the pollution and energy used in transporting that material. At the same time their use provides employment for local suppliers.

- In choosing sustainable materials rather than materials that have to go through complex production processes or be shipped in from other parts of the world, construction should be more efficient and have a lower general impact on the environment.

- The use of energy saving materials will have a long-term and lasting impact on the use of energy for the duration of the property's life.

- Not only will the construction industry have a lower carbon footprint, but also everything they build will have been constructed using lower carbon technologies and materials.

- Protecting the local natural environment from damage by construction work or surrounding infrastructure is only part of the environmental consideration. In choosing sustainable materials to use in construction projects the natural environment is protected elsewhere, by reducing quarrying, tree-felling and the use of scarce resources

- Recycling as much construction waste as possible, particularly from demolition, means that the industry will make less contribution to landfill. Most materials except those that are toxic or hazardous can be repurposed.

TEST YOURSELF

1. What area of the construction and built environment industry would be involved in examining planning applications regarding the long-term future of an area?

   a. Town planners

   b. Surveyors

   c. Civil engineers

   d. Construction site managers

2. What is the term used to describe transport routes such as roads, motorways, bridges and railways?

   a. Services

   b. Infrastructure

   c. Commercial

   d. Utilities

3. Which of the following is an example of a corporate client?

   a. Small business owner

   b. Local authority

   c. Government department

   d. Insurance company

4. What is another term that can be used to describe a land agent?

   a. Land surveyor

   b. Quantity surveyor

   c. Estate agent

   d. Building inspector

5. Which job role involves overseeing construction work on behalf of an architect or client to represent their interests on site?

   a. Clerk of works

   b. Main contractor

   c. Sub-contractor

   d. Building control inspector

6. Which of the following is an example of a renewable energy resource?

   a. Plants

   b. Sun

   c. Wind

   d. All of these

7. What does the National Green Specification Database provide?

   a. Methods on how to recycle

   b. A list of all recycling sites

   c. A list of environmentally friendly building materials

   d. A list of components required for building jobs

8. Which part of the Building Regulations focuses on energy conservation?

   a. Part B

   b. Part G

   c. Part H

   d. Part L

9. Which of the following is an example of biomass?

   a. Coal

   b. Peat

   c. Coke

   d. Logs

10. In addition to providing heating, which of the following also provides cooling?

    a. CCHP

    b. CHP

    c. MCHP

    d. HPCP

Chapter 4

Unit CSA L1Occ12

ERECT AND DISMANTLE ACCESS EQUIPMENT AND WORKING PLATFORMS

LEARNING OUTCOMES

LO1/2: Know how to and be able to prepare for erecting access equipment and working platforms

LO3/4: Know how to and be able to inspect access equipment and working platform components and identify defects

LO5/6: Know how to and be able to erect and work from access equipment and working platforms

LO7/8: Know how to and be able to dismantle and store access equipment and working platform components

INTRODUCTION

The aims of this chapter are to:

* help you select appropriate access equipment and working platforms for the work to be carried out

* help you use access equipment and working platforms following health and safety guidelines

* show you how to dismantle and store access equipment and working platforms appropriately.

PREPARING FOR ERECTING ACCESS EQUIPMENT AND WORKING PLATFORMS

Painters and decorators often have to work at height to reach the tops of walls, ceilings and other awkward spaces. Access equipment and working platforms are very helpful for these tasks. However, they must be chosen, used, dismantled and stored correctly – or else accidents can occur.

Before you use access equipment and working platforms, you need to be aware of:

* the potential hazards

* relevant legislation and regulations

* instructions or guidance from the manufacturer

* any PPE you may need

* which equipment is best to use for the situation

* how to protect your working area from damage.

DID YOU KNOW?

A third of all falls from height involve either ladders or step ladders.

Hazards, health and safety and risk assessment

The main hazards of erecting and dismantling access equipment and working platforms are falls from height. Falls can be the result of:

* equipment that is faulty, damaged or broken, e.g. bent locking bars on a step ladder

* equipment that hasn't been inspected regularly

* equipment that has not been checked before use, e.g. to check that the last use did not leave any damage

* equipment that has not been set up correctly, e.g. placing a ladder on an uneven surface

* equipment that is being used for the wrong sort of job, e.g. using a ladder for heavy or lengthy work.

Other possible hazards include:

* slips and trips, e.g. from wearing inappropriate footwear, or grease on ladder rungs

- cuts and abrasions, e.g. getting your hands or fingers trapped when erecting or using equipment

- using equipment incorrectly, e.g. overreaching, standing on too high a rung.

The risks of erecting/dismantling and using access equipment are very similar, as you can see from Table 4.1 below.

Risks and hazards of erecting, using and dismantling access equipment
• Injury caused by incorrect manual handling of heavy or awkward equipment or components
• Slips and trips over components or tools
• Cuts and abrasions when operating moving parts
• Injury caused by equipment not erected correctly, e.g. collapsing
• Falls from height
• Objects falling from height, causing injury to others
• Slips when climbing
• Trips over tools on working platforms
• Cuts and abrasions when operating moving parts
• Faulty equipment causing falls or cuts and abrasions
• Incorrect use of access equipment resulting in injury

Table 4.1 Risks of erecting, using and dismantling access equipment

Hazard identification records

Hazard identification records, such as a health and safety file or hazard book, are a tool to quickly highlight common hazards on a work site or for a particular task. These not only help to reduce accidents but are also a legal requirement. In the event of an accident, the record will show which hazards on site had been identified and any steps taken to prevent the accident from happening.

Refresh your memory about hazards, health and safety and risk assessments in Chapter 1.

PPE

If you don't wear the correct PPE when erecting or dismantling access equipment and working platforms, then the risk of injuring yourself is higher. You may need:

- gloves to protect your hands from cuts and abrasions

- a hard hat for whenever there is a risk of head injury

- a high vis jacket so you are visible to your colleagues and members of the public

- steel cap boots to protect your feet from any falling debris or parts of scaffolding.

The PPE you choose will also depend on the policy of your workplace or college.

More information on PPE and the laws concerning it can be found in Chapter 1.

REED TIP

Every workplace, college or training centre is a different working environment. They may each have different outcomes from their risk assessments, and so each will have its own policies about health and safety and PPE that you should follow.

DID YOU KNOW?

Fifty per cent of accidental deaths in construction are caused by falls from height.

PRACTICAL TIP

It is the responsibility of everyone on site to ensure that appropriate PPE is worn.

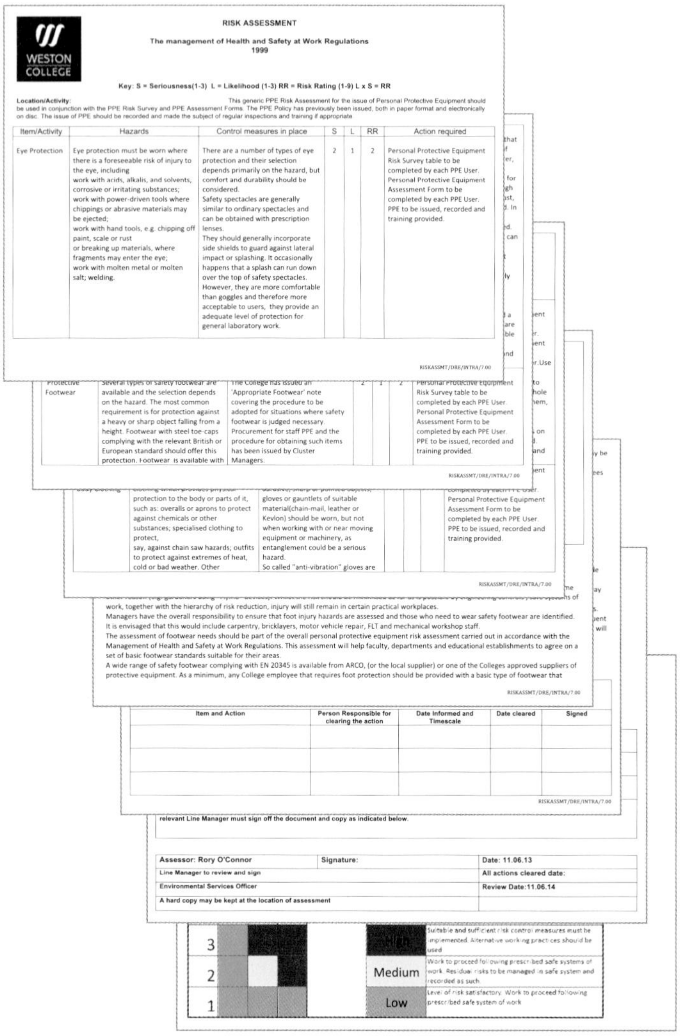

RISK ASSESSMENT

The management of Health and Safety at Work Regulations 1999

Key: S = Seriousness(1-3) L = Likelihood (1-3) RR = Risk Rating (1-9) L x S = RR

Location/Activity: This generic PPE Risk Assessment for the issue of Personal Protective Equipment should be used in conjunction with the PPE Risk Survey and PPE Assessment Forms. The PPE Policy has previously been issued, both in paper format and electronically on disc. The issue of PPE should be recorded and made the subject of regular inspections and training if appropriate

Item/Activity	Hazards	Control measures in place	S	L	RR	Action required
Eye Protection	Eye protection must be worn where there is a foreseeable risk of injury to the eye, including work with acids, alkalis, and solvents, corrosive or irritating substances; work with power-driven tools where chippings or abrasive materials may be ejected; work with hand tools, e.g. chipping off paint, scale or rust or breaking up materials, where fragments may enter the eye; work with molten metal or molten salt; welding.	There are a number of types of eye protection and their selection depends primarily on the hazard, but comfort and durability should be considered. Safety spectacles are generally similar to ordinary spectacles and can be obtained with prescription lenses. They should generally incorporate side shields to guard against lateral impact or splashing. It occasionally happens that a splash can run down over the top of safety spectacles. However, they are more comfortable than goggles and therefore more acceptable to users, they provide an adequate level of protection for general laboratory work.	2	1	2	Personal Protective Equipment Risk Survey table to be completed by each PPE User. Personal Protective Equipment Assessment Form to be completed by each PPE User. PPE to be issued, recorded and training provided.

RISKASSMT/DRE/INTRA/7.00

Figure 4.1 A sample risk assessment

Accident Record
Report No:...........................

1: About the person who had the accident

Name: ....

Address: ....

....

....

Postcode: ....

Occupation: ....

2: About the person filling in this record

▼If you did not have the accident write your name, address and occupation below

Name: ....

Address: ....

....

....

Postcode: ....

Occupation: ....

3: About the accident (continue on the back of form if required)

▼When did it happen?

Date: ....

Time: ....

▼Write how the accident happened. Give cause if known.

....

....

▼If the person in the accident was injured give details.

....

....

Signature:

....

Date: ....

For employer only

4: Complete this box if accident is recordable under Reporting of Injuries, Diseases and Dangerous Occurences Regulations 1995 (RIDDOR).

▼How was is it reported: ....

Signature:

....

Date reported: ....

Figure 4.2 A sample accident report form

Sources of health and safety guidance

The Work at Height Regulations 2005

The Work at Height Regulations make it compulsory for an employer to carry out a risk assessment before any work at height is started. The starting point of the regulations is that work at height should be avoided wherever possible. Access equipment should be used if there is no other option but to work at height.

The regulations also have specific requirements for access equipment and working platforms. These can be found in the schedules at the end of the regulations, for example, *Schedule 6 Requirements for ladders* and *Schedule 3 Requirements for working platforms*. Key points from the regulations have been included throughout the rest of this chapter.

If you would like to see what the regulations look like, you can find them here: *www.legislation.gov.uk/uksi/2005/735/contents/made*. For some more information on the Work at Height Regulations, see page 4.

DID YOU KNOW?

A free pdf of the CDM 2007 ACoP is available here: *http://www.hse.gov.uk/pubns/priced/l144.pdf*

Approved Codes of Practice (ACoPs)

The ACoP for the Construction (Design and Management) Regulations 2007 (CDM Regulations) provides useful advice and guidance for people working in the construction industry. It aims to improve health and safety when managing of construction projects. If you follow the CDM ACoP then you should be following the law that it relates to. It covers the practical requirements and general duties that must be followed on all construction sites. There are also ACoPs for other regulations that may be relevant to your work.

Manufacturers' instructions

In the workplace, always listen to and follow directions from your employer. These will include using any safety equipment or training you are given. In addition, when working with access equipment, you must always follow the instructions that are provided by the manufacturer. These may appear on the equipment itself or may come as a separate guide on how to put together, use, dismantle and store access equipment or working platforms.

Choosing suitable access equipment and working platforms

There are a number of different types of access equipment and working platforms for various situations and uses. This may depend on the location you're working in, the height at which you're working and the length of time you'll need to use the equipment.

For example, ladders are best used for short-term work (less than 20 minutes) and only when doing light work (not carrying anything heavier than 10 kg), such as applying paint or preparing surfaces. Some equipment is better suited to use indoors, such as a step ladder. Other equipment is designed mostly to be used outside, such as scaffolding.

Access equipment and working platforms	Internal use	External use
Ladders	✓	✓
Step ladders	✓	
Mobile towers		✓
Trestle platforms	✓	
Tubular scaffolding		✓
Hop-up	✓	
Proprietary staging	✓	
Podiums	✓	

Table 4.2 Internal and external access equipment and working platforms

In general, however, almost any access equipment can be used either inside or outside, but it depends on the size and needs of the project. For instance, if working inside a large building, you may find yourself using full tubular scaffolding, and if you were doing a small job outside, such as painting a door frame, you might use a hop-up or step ladder.

Types of access equipment and working platforms

Ladders

Ladders are one of the most common pieces of access equipment. Ladders must be leant up against something because they do not have their own **prop**, like a step ladder does. The steps of the ladder are called the **rungs**, and the long, upright parts that hold the rungs together are called the **stiles**. Ladders can be made of wood, aluminium or fibreglass. Wooden ladders are becoming less common as they are heavier and more expensive to buy.

There are different types of ladders that you may come across:

* Single ladders – these have only one section, i.e. they cannot be reduced or extended in size

* Double ladders – a type of extension ladder; they are made up of two sections that slide along each other and lock into place

* Extension ladders – an adjustable ladder with two or three sections that can be released to make the ladder longer; they can also be used as a single ladder and, when made of aluminium, they are light and portable

* Pole ladders – these are single ladders often used to access scaffolding platforms; they are made of timber and have **tie rods** under every rung

* Roof ladders – specifically designed for working on roofs, they have a hook that sits over the top of the roof to keep it stable.

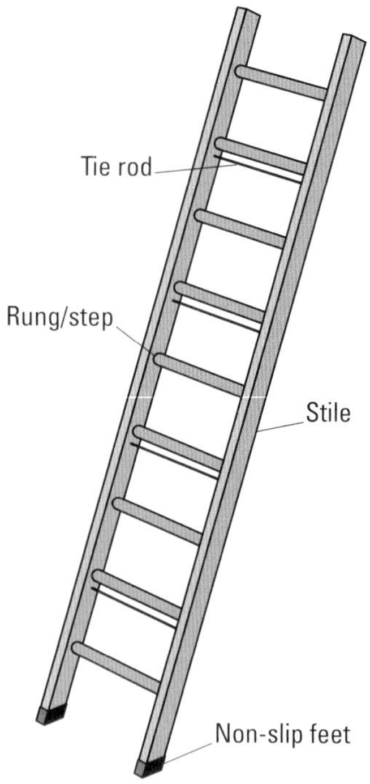

Tie rod

Rung/step

Stile

Non-slip feet

Figure 4.3 A ladder

Figure 4.4 An extension ladder

Figure 4.5 A pole ladder

Figure 4.6 A roof ladder

Step ladders and platform steps

Unlike a standard ladder, step ladders have a prop, so they can stand on their own without being rested against another surface. Step ladders are very regularly used for various construction tasks. They should only be used for quick, lightweight tasks that will be finished within about 30 minutes. A longer job will require more sturdy equipment. Like ladders, step ladders are made of wood, aluminium or fibreglass.

Platform steps are a shorter type of step ladder that has a small platform at the top and a bar or rail to hold onto. Sometimes this rail is used for hanging paint kettles.

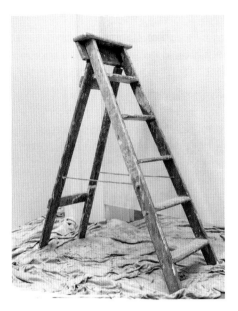

Figure 4.7 A step ladder

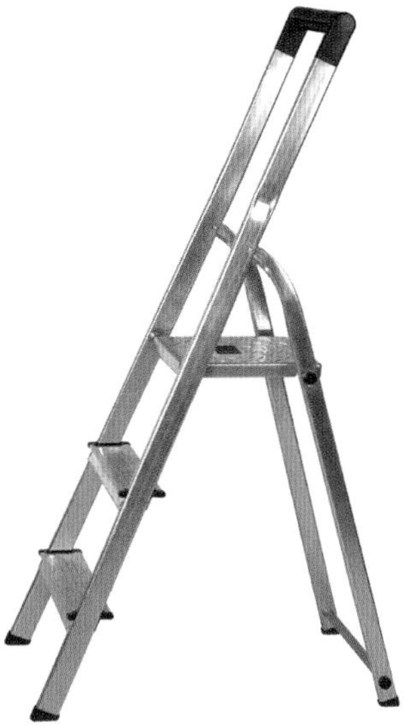

Figure 4.8 A platform step ladder

PRACTICAL TIP

Go to the HSE website and print off its quick guide pocket card, *Top tips for ladder and step ladder safety: www.hse.gov.uk/pubns/indg405.pdf*

DID YOU KNOW?

Platform steps are also known as swingback steps.

Mobile towers

A mobile access tower can be a safe and useful way to access work at height. They are widely used, but must be put together and used correctly, otherwise accidents can occur. Mobile towers are always constructed on lockable wheels so that they can be moved around the site without being taken apart. They are also free-standing, meaning that the scaffolding is not attached to or dependent on the building for support. Their parts are constructed of aluminium or fibreglass.

For painting and decorating, mobile towers tend to be used when there are two people working on a task, such as repairing fascia board or guttering, or applying a textured finish to a ceiling.

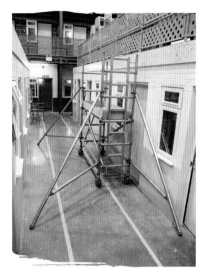

Figure 4.9 A mobile tower

Trestle platforms

A trestle platform has two parts: the frame (or **trestle**) that supports the stage (or platform), and the platform itself. They are used for jobs that will likely take more than a few minutes as they are more secure than a ladder or step ladder.

A-frame trestles are shaped like the letter A and are used in pairs to support a scaffold board. Their height cannot be adjusted. Steel trestles on the other hand can be raised and lowered, and are generally considered more stable.

KEY TERMS

Trestle

– often found in pairs, these hold up a flat board or platform, acting like the four legs of a table. They include A-frames and steel trestles.

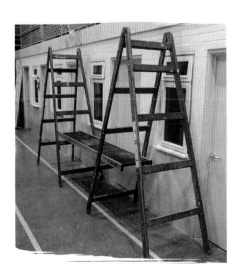

Figure 4.10 An A-frame trestle

Figure 4.11 A steel trestle

DID YOU KNOW?

Mobile access towers are also referred to as mobile scaffold towers, tower scaffolds, proprietary towers or just towers.

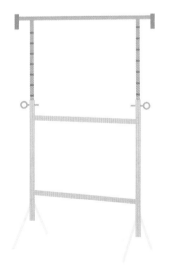

Figure 4.13 A podium

Proprietary staging

Proprietary staging is an attached working platform of staging boards specifically designed to help safe movement at heights. They are also known as crawling boards, lightweight stagings, and sometimes even by the brand-name 'Youngman boards'.

Proprietary staging is attached to trestles and tied in if working above 2 m. The painter and decorator would use them for spanning doorways and gaining access near the roof, e.g. when painting guttering. When working above 2 m, handrails and kickboards must also be used.

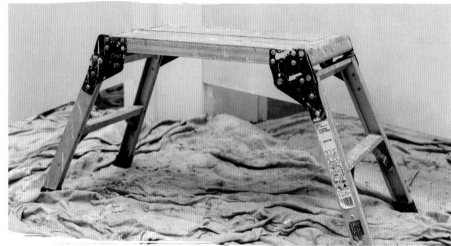

Figure 4.12 A hop-up

Podiums and hop-ups

Hop-ups and podiums are commonly used for painting and decorating. They are designed to be used for low-level access work. Hop-ups tend to be about 40–50 cm in height, whereas podiums can have a platform height of up to around 1–1.5 m. Podiums have 1 m-high guard rails to protect the user from falling, and are similar to mobile towers in that they can be moved. Some podiums are specifically designed for working on staircases. They are a safe alternative to step ladders.

Mobile Elevated Working Platforms (MEWPs)

Even though you may not have to erect or dismantle all types of access equipment, it's possible that an employer will still expect you to use them. Some of the following MEWPs require training from a hire company or from your employer before you can use them, and some kinds will require you to hold a licence or card.

A MEWP is a driveable, mechanical piece of equipment that is powered by diesel or electricity. It is raised up and down by the person working at height in the basket. MEWPs come with additional risks:

* Overturning – because they are tall and narrow structures, if moved onto an uneven surface, or perhaps in extreme weather, the machine could fall over and throw the operator from the basket.

* Entrapment – if the basket were raised too high or the platform moved without checking, the operator could become stuck between the basket and a fixed structure.

* Collision – because they are a mechanical, moving object, they could accidentally hit a pedestrian, another vehicle, or overhead cables.

MEWPs should never be moved at the base while there are people on the raised platform.

Cherry picker

These are also known as boom lifts. A cherry picker is a platform that is raised vertically from the floor, but can also move forwards and backwards.

Scissor lift

A scissor lift moves only in a vertical direction, i.e. up and down, not forward and backward in the air. The platform sits on top of an accordion-like mechanism (see Fig 4.15).

See Table 1.10 on page 24 for a summary of equipment, features and safety checks.

See Table 1.10 on page 24 for a summary of equipment, features and safety checks.

Figure 4.14 A cherry picker

Figure 4.15 A scissor lift

DID YOU KNOW?

Some of these pieces of access equipment are known by different names in different parts of the country. For example, platform steps may be known as hop-ups.

Protecting the work and its surrounding area

Tools and equipment used for erecting and dismantling access equipment need to be looked after. Any damage to components could mean that the equipment cannot be put together properly or may no longer be safe when in use.

It's not just your own work that you are protecting the area from – other site activities going on around you could also cause damage, such as other people using your equipment, site traffic, and mobile access equipment. Tools and equipment should be properly stored so that they do not create a slip or trip hazard. Leaving access equipment outside and unprotected could expose them to bad weather and damage. By keeping your tools and equipment clean and dry, you will make them perform better and last longer.

Keeping your work area tidy is always good practice. It ensures that you do not create hazards for yourself or other people on site. Any waste should be correctly disposed of.

When erecting, using or dismantling access equipment, it is a good idea to use screens, barriers or timber hoardings so that people walking by or underneath are not at risk of bumping the equipment or from being hit by falling items. Putting up notices will also help to remind others on site that you are working at height.

PRACTICAL TIP

If several people are working at height fall prevention means guard rails and solid working platforms. To arrest a fall, nets and airbags can be used. For individuals, restraints and ropes can also be used if correct instruction or training is received.

INSPECTING ACCESS EQUIPMENT AND WORKING PLATFORM COMPONENTS AND IDENTIFYING DEFECTS

Components of access equipment and working platforms

Each of the types of access equipment listed above has specific parts and features that should be checked before using the equipment. See the table below for an explanation of these.

Component	Use
Stiles	The vertical parts of a ladder or step ladder which hold the rungs together.
Rungs	The horizontal parts of a ladder. They can be round or rectangular in shape.
Tie rods	Steel rods that sit beneath the rungs and stop the rungs on timber ladders from breaking and the stiles from separating. Tie rods should be at least under the second rung and then spaced regularly up the ladder. They should also appear under the first and fourth tread of steps.
Ropes	Two lengths of rope or cord is attached to each side of a step ladder to stop it from opening too far. However, the HSE has provided recent guidance that fixed metal stays are now preferred to ropes as they are more secure.
Pulleys	A wheel that controls the movement of a rope or cord. They are found on some types of scaffold and on rope-operated ladders.
Treads	The parts of a step ladder that you step on. Treads should be a minimum of 90 mm deep. Remember: never stand on the top step.
Hinges	Fixed to both sides of a step ladder, hinges stop the ladder from giving way.
Swingbacks	A type of step ladder that doesn't have a platform and the steps go right to the top of the stiles. It has two parts: the back frame and the front stepping part.
Locking bars	Fitted to aluminium step ladders, they hold the back frame and front steps together to prevent the ladder from collapsing. They must be fully extended and locked into place.
Non-slip inserts	Attached to the ends of the stiles. They reduce movement and slipping and can also protect the floor surface.
Scaffold boards	Made of timber and aluminium, they are used on fixed scaffolding and are the part that is walked on.
Platform staging	Made of aluminium, platforms are usually used on mobile access towers, providing a tough and non-slip surface.

Table 4.3 Components of access equipment and their use

REED TIP

Good communication is about listening – not just talking – and understanding the different ways people prefer to communicate.

Scaffolding

While you won't be erecting or dismantling scaffolding, you may well be using it. You will need to know how to use it safely and also how to visually inspect it and look for any faults that may need to be reported.

Tubular scaffolding

Tubular scaffolding is a structure of upright tubes (standards) and horizontal tubes (ledgers) made from either steel or aluminium. The scaffolding boards which form the working platform are supported by transoms which extend between the ledgers.

Scaffolding can be either independent (free-standing, but tied to a building) or dependent (attached to the brickwork of a building). See Fig 4.16 to help you identify the different parts of a scaffold.

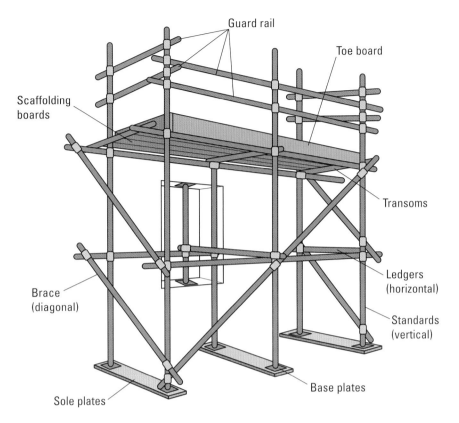

Figure 4.16 Parts of a scaffold

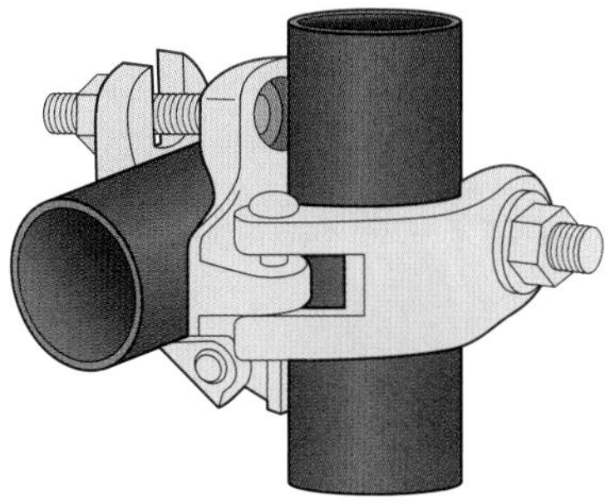

Figure 4.17 A universal coupler

The scaffold is held together by various fittings including:

* universal couplers (Fig 4.17) which connect two tubes at right angles (one vertical, one horizontal)
* right angle couplers (Fig 4.18) which connect two tubes at right angles (both horizontal)
* swivel couplers (Fig 4.19) which connect two tubes at any angle
* sleeve couplers (Fig 4.20) which connect two tubes end to end
* base plates (Fig 4.21) which spread the load under the standards.

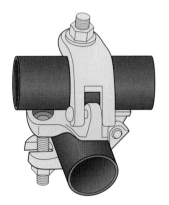

Figure 4.18 A right angle coupler

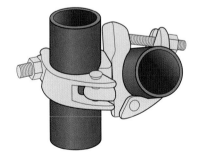

Figure 4.19 A swivel coupler

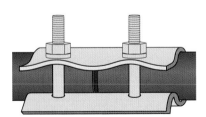

Figure 4.20 A sleeve coupler

Figure 4.21 A base plate

Slung scaffold

This is a type of dependent scaffold that is suspended from the ceiling of a building, usually in large spaces such as a theatre, factory or train station. The platforms have toe boards and guard rails all the way around so that it is safe to work above an area that is in constant use. Compare these with birdcage scaffolds in the case study on page 118.

CASE STUDY

South
Tyneside

South Tyneside Council's
Housing Company

Further training enhances your job prospects

Glen Richardson is a final year apprentice at South Tyneside Homes.

'You learn new skills on the job every day because every job is different. Obviously you're learning new skills in college too, but if you're given the chance to do any further training with your employer, then that's something extra you can put on your CV. It will be transferrable to other jobs too.

I've done a manual handling course on how to pick up heavy and awkward objects correctly and how to store them without injuring myself. It's definitely improved the way I work.

Even more useful was the scaffolding qualification I got. I can now erect and dismantle scaffolding, which is something you're not automatically allowed to do without the proper training. This means that I get to work on bigger jobs and buildings, and it will definitely help if I ever need to look for another job.'

Figure 4.22 A scaffolding tag

Carrying out inspections

Inspections of your access equipment should be carried out both before it is put together and once it is in use. It is essential to spot any problems or faults in the components. Inspecting equipment is an important part of ensuring accidents don't occur.

It is important not only to make sure the equipment is safe before it is used, but also while it is in use. This is because components can move around and loosen because of the weight on them, be exposed to weather when working outdoors, or get damaged while the equipment is being worked on. Even if equipment is safe for use, it must still be used correctly to avoid accidents.

Before using scaffolding, consider the following:

* Does it look safe?
* Are there any signs or tags attached to tell you whether it is unsafe or unfinished?
* Are the boards clear of debris and bulky work materials?
* Are the guardrails, scaffold boards and toe boards all in place?
* Is there a safe and suitable way of accessing the scaffold?
* Does it look complete, i.e. can you see all the parts identified in Fig 4.16?
* Are the right fittings being used?

If any faults to scaffolding have been discovered, then a tagging system is used to alert others to the problem so that the component or structure is not used by someone else. **Note:** you should *never* adjust fixed scaffolding unless you have been properly trained and are a competent, **carded scaffolder**.

When carrying out an inspection on scaffolding or working platforms, you should fill in an inspection report. An example of this from the HSE is shown below.

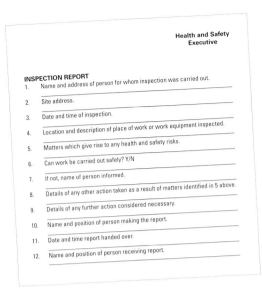

Health and Safety Executive

INSPECTION REPORT
1. Name and address of person for whom inspection was carried out.
2. Site address.
3. Date and time of inspection.
4. Location and description of place of work or work equipment inspected.
5. Matters which give rise to any health and safety risks.
6. Can work be carried out safely? Y/N
7. If not, name of person informed.
8. Details of any other action taken as a result of matters identified in 5 above.
9. Details of any further action considered necessary.
10. Name and position of person making the report.
11. Date and time report handed over.
12. Name and position of person receiving report.

Figure 4.23 A sample inspection report

> **PRACTICAL TIP**
>
> Note that only a nominated and trained person (a certified scaffolder) can do a full scaffold inspection.

Pre-erection inspections

Mobile towers and podiums

* Check castors and wheels are swivelling freely.

* Check that there is a brake and that it is working.

Step ladders

* Check that treads, bolts, screws, and hinges are in good condition and not loose.

* Check the retaining cords and hinges are the same length and are not damaged or frayed.

* Check there are no splits or cracks in the stiles.

* Check that treads are secure and do not have any splits.

* If using an aluminium step ladder, check that stiles and treads are not warped, twisted or badly dented.

> **DID YOU KNOW?**
>
> You should never paint a wooden ladder or step ladder as it could hide any damage.

Platforms

* Check that any platforms or scaffold boards are not split.

* Check they are not warped or twisted.

* Check that boards do not have significant weaknesses, such as large knots in the wood.

> **DID YOU KNOW?**
>
> If a ladder is damaged, it must be destroyed and not repaired.

Ladders

See practical task 2, *Pre-use inspection of ladders* for the correct pre-use checks.

Inspection intervals

Some access equipment and components need to be checked more often than others. The HSE provides guidance on these inspection intervals as follows.

Scaffolds and working platforms above 2 m high:

* should first be inspected after installation (being put into position) or assembly (being put together); note that mobile towers or platforms can still be moved around after they've been assembled without further inspection

* should be inspected after any exceptional circumstances which are likely to jeopardise the safety of the equipment, for example, after a major weather event such as a storm, after vandalism, or after an incident

* should be inspected at least every seven days and there should be a record of this inspection.

All other working platforms:

* should be inspected in the same way as scaffolding, except that regular inspections need only take place at suitable intervals rather than every seven days.

Guard rails, toe boards, barriers and any other fall protection:

* should be inspected in the same way as all other working platforms.

Ladders and step ladders:

* should be inspected at suitable intervals

* should be inspected after exceptional circumstances that might affect the equipment's safety

* should be checked on each occasion before use, though a report on this inspection is not needed.

ERECTING AND WORKING FROM ACCESS EQUIPMENT AND WORKING PLATFORMS

To set up and work safely from access equipment and working platforms, you must be aware of how to:

* identify a secure base

* correctly load a working platform

* use safe manual handling techniques.

Refer back to Chapter 1, pages 20–1 where techniques for safe lifting are covered in detail. You will need to bear these in mind when erecting and dismantling access equipment because it can at times be heavy and/or awkward to move.

The table below shows how each type of equipment should be secured, loaded and handled.

Access equipment	Securing the base	Correct loading and use	Manual handling and erecting
Ladders	• Place on firm ground, or use a board. • Place on level, clean and solid ground with good grip for ladder feet. • Tie both stiles of ladder at suitable point, preferably near the top (see Fig 4.24). • Or, if not possible, wedge against something solid, e.g. another wall. • Set up at angle of 75° (or 1 m for every 4 m up). • Don't rest against weak or flexible surfaces, e.g. plastic guttering.	• Keep three points of contact with ladder (two feet, one hand). • Don't do work involving both hands. • Both hands are holding on when climbing. • The ladder is close enough to the work, i.e. no overreaching. • Take one rung at a time when going up or down. • No standing on top third of the ladder, or top two or three rungs. • Avoid carrying loads over 10 kg. • Never move the ladder while in use.	• Short ladders can be carried alone, holding it vertically against the shoulder, grasping a lower rung, and using other hand to hold stile avoid going near power lines or take extra care if necessary to work near them. • Long ladders should be carried by two people, one at each end, held on the shoulder. • To erect a long ladder, one person stands on bottom rung while second person takes the top rung and 'walks' down the ladder rung by rung with their hands.

Access equipment	Securing the base	Correct loading and use	Manual handling and erecting
Step ladders	• Make sure the legs are fully open. • All four legs must be in contact with ground. • Placed on a level and stable base. • Tie the step ladder if high risk work.	• Do not work side-on. • Make sure you have a handhold. • No standing on top two or three rungs (knees should be below top step). • The ladder is close enough to the work, i.e. no overreaching. • Take one rung at a time when going up or down. • Don't use to access higher levels, e.g. a roof. • Avoid carrying loads over 10 kg. • Person climbing should be facing the work. • Never move the ladder while in use.	• For set up, lean it forward a little, holding onto the stiles, then pull the back support away from the front. Make sure any locks are engaged.
Mobile towers	• Place on firm and level ground. • Brakes must be on while in use. • Surface must be strong enough to take the load at each of the four points. • Use boards underneath if needed to avoid sinking or tilting. • If using indoors, height of tower should be a base width to height ratio of 1:3.5 (but check manufacturer's guidelines). • If using outdoors, height of tower should be a base width to height ratio of 1:3 (but check manufacture guidelines).	• Do not move while people are working on the tower. • Do not move while tools, equipment and materials are on the tower. • Do not use outdoors in high winds. • For use outdoors, tie in against the structure if possible. • Be aware of overhead power lines – do not use if lines are near the tower. • Do not overreach – if you can't reach, then move the tower correctly. • To access the working platform, climb only on the ladder *inside* the tower. • Never add a ladder/step ladder to the platform for extra height.	• When moving, always push from the bottom of the tower to avoid toppling. • If you need to increase the height, the width of the base can be increased by adding outriggers to stabilise the tower (see Fig 4.26). • Anyone erecting a mobile tower must be fully trained and competent or under supervision of an experienced person.
Trestle platforms	• Set each trestle on a firm, level base. • Ensure A-frames are opened fully. • Steel trestles should be no further than 1.2 m apart. • Rest steel trestles on top of a flat scaffold board.	• Use only one working platform. • If there is a risk of falling, guard-rails, barriers and toe boards should be used. • Scaffold boards used as a platform must be the same length and thickness, and boards on an A-frame trestle should be no less than 450 mm wide. • Trestle should be stable when in use. • Access to the platform should also be safe, e.g. a securely tied ladder. • The working platform boards should be checked for splitting, twisting, warping and knots that may weaken them.	• Lift the trestle into the correct place, hold it evenly and pull both sides away from each other. Make sure all parts are opened as far as they can go.

Access equipment	Securing the base	Correct loading and use	Manual handling and erecting
Proprietary staging (crawling boards)	• Secure or tie in your platform boards with ropes into the trestles, if working at heights over 2 m.	• Do not overload. • If working over 2 m heights, there must be a handrail and kickboard. • Use the correct access equipment to reach your boards, e.g. a pair of steps to erect it and reach it when in use. • When using with trestles, the overhang should be no more than four times the thickness of the board.	• Proprietary staging boards are heavy and awkward to lift. You will need two people to erect or dismantle.
Podiums/ hop-ups	• If on wheels, use brakes to lock them in. Place on even ground.	• If on wheels, do not pull yourself and the podium along the working area. • Select the right height equipment so that work is conducted on the platform itself. • Do not move while people are working. • Do not move while tools, equipment and materials are on the platform. • Only one person should use at a time. • Close and lock the gate before starting work.	• Platform should be locked into place. • Handrails should be secured.

Table 4.4 Securing, loading and handling access equipment and working platforms

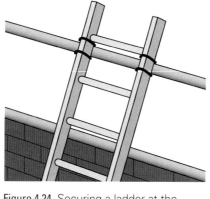

Figure 4.24 Securing a ladder at the top – for working on only. For access, a ladder must be tied 1 m or 3 rungs above the working platform.

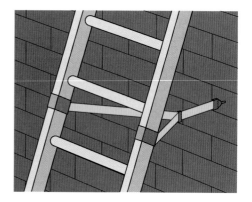

Figure 4.25 Securing a ladder at the base

Figure 4.26 An outrigger for mobile tower base

PRACTICAL TIP

'Footing' the ladder is the last alternative to securing the base, if it is not possible to tie the stiles or wedge it. Footing is where a second person stands on the bottom rung with both feet. It is not ideal because it does not stop the ladder from slipping sideways. When tying the ladder off initially, you will need someone to foot the ladder.

Always follow the manufacturer's instructions when erecting access equipment and working platforms. Each make of equipment could be slightly different from others you have worked with before.

DISMANTLING AND STORING ACCESS EQUIPMENT AND WORKING PLATFORMS

Sequence for dismantling

Ladders

Dismantling a ladder is basically the same as erecting it, but in reverse.

1. The bottom rung should be footed.

2. Lift the ladder carefully away from the surface it is leaning on.

3. Put both hands on the stiles.

4. Walk slowly backwards while moving your hands from rung to rung.

5. Lay the ladder on the ground.

Platforms, platform steps and trestles

You can dismantle platform steps as follows:

1. Lean the steps forwards.

2. Undo the lock (if there is one).

3. Ensure the rope can move freely.

4. Move the back frame towards the front frame.

5. Lift the steps into a secure position.

Storage requirements

Ladders

Timber ladders will decay over time if stored outside. If they can't be stored under cover, then they should be covered or placed in a position that is protected from the wind and rain. Keep them away from heat sources such as boilers.

Do not hang ladders from their rungs or stiles. Instead, store horizontally on a rack, with the weight on the stiles only. They should be supported along their length so that they don't sag.

Ladders made from aluminium will corrode if store near set lime or cement.

Step ladders, trestles and platform steps

These should always be stored in an upright position in a covered area to avoid weathering. They should preferably be kept off the ground so they do not get exposed to damp.

REED TIP

Helping out on maintenance jobs around the home with your family at the weekend still counts as experience you can put on your CV or application form. Remember, however, that any work-based evidence as part of your qualification has to be countersigned by an experienced qualified operative.

CASE STUDY

South
Tyneside *Homes*

South Tyneside Council's
Housing Company

Working with a birdcage scaffold

Birdcage scaffolding is often used in buildings such as churches or theatres.

The birdcage scaffold is a type of independent scaffold with an enclosed working area so that there are no gaps for a worker to fall through. It is raised to just below the ceiling and forms a complete platform, from corner to corner, like an extra floor. It is quite expensive to erect and so tends to be used for jobs that will take a long time.

When the Hull New Theatre was being refurbished, there was a lot of painting detail on the ceiling to be restored. More than 2,600 square metres of ceiling, wall panels, balconies and box seating in the auditorium were redecorated with the use of a birdcage scaffold and a large team of painters.

As a similar alternative to birdcage scaffolding, it is possible to use a slung scaffold. This is so that it is still possible to have a congregation or audience coming into the space below during long-term inner roof and ceiling work. It can also mean faster construction times, e.g. so that the seating can be added at the same time as the ceiling work.

Figure 4.27 A birdcage scaffold

PRACTICAL TASK

1. PREPARE TO ERECT ACCESS EQUIPMENT AND WORKING PLATFORMS

OBJECTIVE

To be able to interpret guidance information, produce a hazard identification record, select and use the right PPE, and protect the work and its surrounding area from damage.

SCENARIO

You and one other person are preparing a new timber fascia for the frontage of a shop inside a shopping centre. The work can only be completed outside normal shop opening hours, therefore you will not have access to the shop itself. Due to the lack of access to a power supply, you will only be able to use hand tools.

The timber fascia is 4.5 m long and 2 m high from ground level. You will be preparing surfaces for painting. The maximum amount of time you have to complete the surface preparation is 2 hours.

STEP 1 Decide what sort of access equipment you will need to work at the highest point.

- Think about the type of work you will be doing and how long you will need to be working at height. Think about how high you will need to reach and whether you will need to move and reach your access equipment. What sorts of access equipment might be suitable?

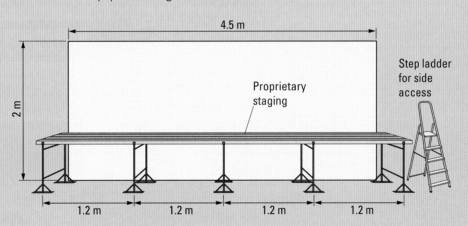

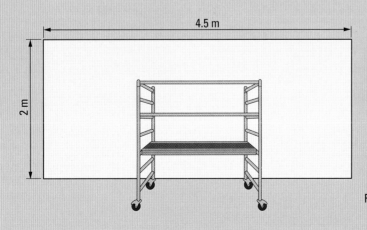

Figure 4.28 Diagram of possible working solutions for access equipment

PRACTICAL TIP

As you are working in a limited space/width, a set of steel trestles with lightweight staging platforms would be wide enough to span the whole area.

Remember that working platforms must be supported every 1.2 m. You would also require a form of access to the platform itself, e.g. a step ladder.

Because of the limited time available to complete the work, you might consider using an aluminium mobile tower which would limit the time needed to erect, dismantle and re-erect a steel trestle type of working platform.

Remember that even though you need to reach up to 2 metres, you do not need to stand above 2 metres to do this. If you would be working for more than 10–15 minutes at a time in one spot, then a step ladder would not be the best option for the entire job.

You will need something higher than a hop-up to reach the 2 m point comfortably.

PRACTICAL TIP

Your lecturer may give you a print out of a **hazard recording sheet**, or you can download a template at: *www.planetvocational.com/subjects/build*

STEP 2 Fill out a hazard identification record.

- Think about the environment: are there likely to be other workers on site? Are there likely to be members of the public in the area?
- Think about the hazards of erecting and working with the type of access equipment you have chosen, e.g. competence of persons using the access equipment, uneven ground condition, falls from height, slips and trips, cuts and abrasions.
- Think of specific problems for the environment you're in and the equipment you've chosen, e.g. collection and storage of the equipment, security of the equipment for each separate hazard.
- For each separate hazard, what can you do to avoid or control the risk?

STEP 3 Choose your PPE.

- Using the same hazard recording sheet, consider which PPE you will need to both erect and use the access equipment you've chosen.

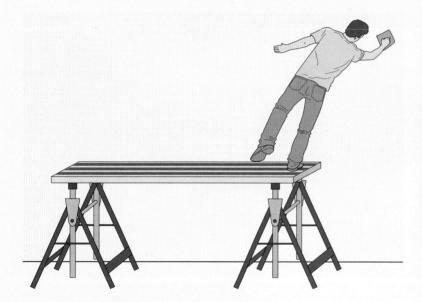

Figure 4.29 Incorrect use of access equipment

STEP 4 Look at the manufacturer's instructions for the access equipment you've chosen.

- Manufacturer's instructions can usually be found in the form of a label stuck to the equipment itself. From a hire company, the company must provide an instruction leaflet, which is in addition to the manufacturer's instruction label found on the equipment itself.

- Do you understand them? Do you need any extra pieces of PPE you haven't thought about? Is the equipment suitable for the job, e.g. does it give a maximum span and weight?

- Add any extra hazard and PPE information you've found to your hazard record.

Figure 4.30 Manufacturer's instructions

STEP 5 Protect the work and surrounding area from damage.

- Produce a sketch of the intended work area showing how you protect the work and the work area.

- Think about the working environment. Are there other operatives working on site? Are there vehicles in the area that could come into contact with your access equipment? Are there members of the public who might walk into or underneath your equipment? Is anyone at risk of falling objects from your access equipment? Will you need to put up any signs, barriers or hoardings to protect the area?

- Also consider what PPE you might need when protecting the area. You can add this to your hazard recording sheet.

PRACTICAL TIP

Your lecturer may give you a print out of a **sketch record sheet**, or you can download a template at: *www.planetvocational.com/subjects/build*

DID YOU KNOW?

To remind yourself of the safety signs commonly used on construction sites and what they mean, go to page 33.

2. PRE-USE INSPECTION OF LADDERS

OBJECTIVE

To be able to inspect and record findings about the condition of two types of access ladders prior to erection and use, and select relevant PPE to be used when checking access ladders in the workplace.

TOOLS AND EQUIPMENT

A step ladder

An extension ladder

Socket set

Spanners

Screwdrivers

PPE

Ensure you select PPE appropriate to the job and site where you are working. Refer to the PPE section of chapter 1.

STEP 1 Lay your extension ladder flat on the floor, without extending it. Open up your step ladder to its standing position.

PRACTICAL TIP

When inspecting access equipment, you will still be on site. Even though you may not be using the equipment or carrying out work tasks, you must still wear the PPE required by your college or employer, e.g. hi-vis jackets and hard hats.

STEP 2 Check for the following on your ladder:

❑ Are the stiles straight and undamaged?

❑ Are all of the feet present?

❑ Are the feet worn, damaged or caked in dirt?

❑ Are all the rungs there?

❑ Are any of the rungs bent, worn, loose or damaged?

❑ Is the ladder clean of mud, grease etc.?

❑ If you're using a purely wooden ladder, are there enough tie rods? (Note: there should be a rod for every other rung on a ladder that has wooden rungs and stiles.)

On your step ladder, check the following:

❑ Are all of the feet present?

❑ Are the feet worn, damaged or caked in dirt?

❑ Is the platform split, bent or buckled?

❑ Are the steps or treads clean of mud or grease?

STEP 4 Check for the following on both your ladder and step ladder:

❑ Do the moving parts move easily?

❑ Are there any loose bolts or screws?

❑ Does the locking mechanism on the extension fully engage?

❑ Are the locking mechanisms worn or damaged?

STEP 5 Remove any dirt or grease from the ladder and tighten any loose parts.

Note that very minor adjustments are fine, but anything more serious than this, consult your supervisor. The ladder may need replacing.

STEP 6 If there are any defects that you think could make the ladders unusable, you must report your concerns to the charge hand, foreman, supervisor or employer, and do not use until rectified.

Record all comments and findings using the **ladder condition checklist record**.

STEP 3 Extend the ladder to its full length and lean it against a wall.

PRACTICAL TIP

Remember: you should never make any alterations, remove or add anything to a ladder or other access equipment.

PRACTICAL TIP

Your lecturer may give you a print out of a **ladder condition checklist record**, or you can download a template at: *www.planetvocational.com/subjects/build*

PRACTICAL TASK

3. ERECT, USE, DISMANTLE AND STORE AN EXTENSION LADDER

Remember to always ask yourself the following questions before using the access equipment:

- Is the access equipment of the correct *type* for the intended work?
- Have you read the manufacturer's instructions?
- Is it *fit* for purpose?
- Is it the correct *size*?
- Can you use it by yourself or do you need *help* to erect, use and dismantle the access equipment?
- Is the *grounding* level and firm?

OBJECTIVE

As part of a team of two people, erect and use an extension ladder in the recognised and safe manner, dismantle, handle and store the ladder without causing any minor or long-term damage.

TOOLS AND EQUIPMENT

A class 1 extension ladder

Ladder ties

A ladder stabilising device

PPE

Ensure you select PPE appropriate to the job and site where you are working. Refer to the PPE section of chapter 1.

STEP 1 Place the ladder on the ground facing the intended resting position (in a safe position, away from doors etc.).

Figure 4.31 Preparing to raise a ladder

STEP 2 While another person is using their feet to stop the ladder moving, or by placing the ladder against the bottom of the wall, walk the ladder to an upright position.

Figure 4.32 Footing the ladder

STEP 3 Pull the bottom of the ladder out from the wall until you reach an angle of 75° (or a ratio of 4:1).

STEP 4 Secure (tie) both stiles of the ladder near the base, or at the top, or half way down if possible. If tying the ladder is not possible, you should use a ladder stability device, or wedge it against a wall. As a last resort, you can use another person to 'foot' the ladder.

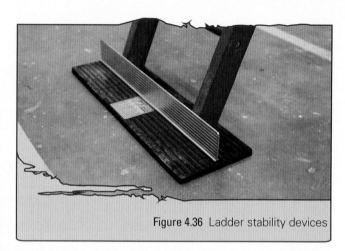

Figure 4.36 Ladder stability devices

Figure 4.33 Tying the ladder near the base

PRACTICAL TIP

Remember that if you're tying a ladder at the top as pictured, that it is safe for working on, but shouldn't be used for gaining access to a roof or working platform.

PRACTICAL TIP

There are two positions for your hands when ascending and descending a ladder. You can steady yourself by moving one hand to the stile of the ladder (side) whilst the other hand holds on to the rungs as you ascend or descend. This method should be used when you are carrying small types of resources such as paint tins.

Figure 4.34 Tying the ladder at the top

Figure 4.37 Ascending a ladder while carrying an object

Figure 4.35 Tying the ladder partway down

STEP 5 Climb the ladder safely, always facing the rungs and maintaining three points of contact.

If you are using the ladder for access only, you should use both hands on the rungs.

STEP 6 Carry out the work you need to do on the ladder. When working on the ladder, take care not to overreach. A good guideline is to keep your belt buckle between the two stiles.

PRACTICAL TIP

Note: it is always best practice to use two people to carry long access equipment such as an extension ladder. You may not be able to see what is coming around the corner or any obstacles on site, such as doorways.

STEP 7 Carry the ladder safely.

To move the ladder to the storage area you should get help from another person to lift and carry it. If you have to move the ladder by yourself, place the ladder at approximately 75° over your shoulder, and hold the ladder with both hands. This should only be done for short distances.

Figure 4.38 Correct manual handling of a timber ladder by two people

STEP 8 Store the ladder off the ground, if possible, horizontally on at least two ladder hooks, in a dry and secure environment.

TEST YOURSELF

1. What is the biggest risk when using access equipment and working platforms?

   a. Slips

   b. Falls from height

   c. Electrocution

   d. Cuts and abrasions

2. Why might you erect a screen around your access equipment?

   a. To stop passers-by damaging your work

   b. To protect others from any falling debris

   c. To stop vehicles and passers-by from running into access equipment

   d. All of the above

3. A tie rod is:

   a. a steel bar underneath rungs on a ladder

   b. a step on a ladder

   c. a piece of cord to tie a ladder to a structure

   d. the sides of a ladder that hold the rungs

4. Which of the following is NOT part of a ladder?

   a. Rung

   b. Stiles

   c. Toe board

   d. Tread

5. What should you check mobile scaffold towers for?

   a. Wheels that spin freely

   b. Twists

   c. Knots

   d. Splits

6. What does the 1 in 4 rule mean?

   a. You should check a ladder every fourth time you use it

   b. That the ladder is at a 75° angle to the surface it leans against

   c. One in four decorators will fall off a ladder

   d. You should work on a ladder for one quarter of an hour

7. When working on a step ladder you should NOT:

   a. work side-on to the surface

   b. carry loads of more than 10 kg

   c. use it to access higher levels

   d. all of the above

8. How should ladders be stored?

   a. Upright

   b. Hanging from the rungs

   c. Laid flat on a rack

   d. Outside

9. When moving a mobile scaffold you should always:

   a. push or pull from the bottom

   b. have the wheels locked

   c. leave your tools on the platform

   d. make sure someone is still standing on the platform

10. If you spot a problem with a scaffold you should:

   a. try to fix it

   b. tell your supervisor

   c. do nothing

   d. climb up to have a closer look

Chapter 5

Unit CSA–L3Occ141

PREPARE SURFACES FOR PROTECTION AND DECORATION

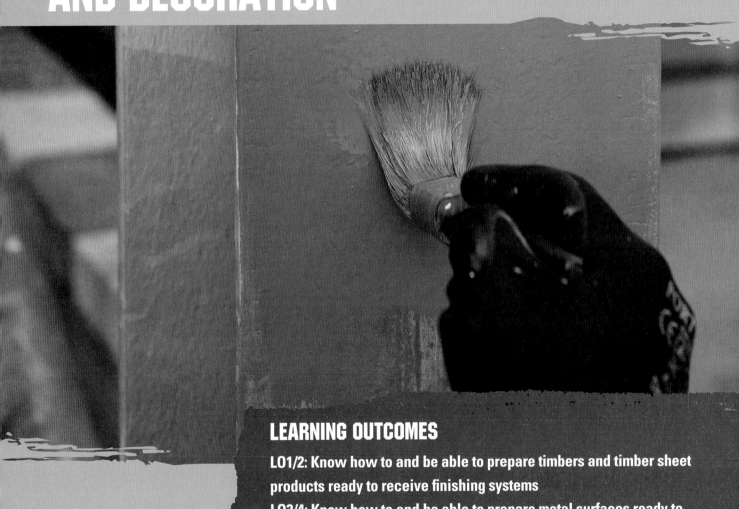

LEARNING OUTCOMES

LO1/2: Know how to and be able to prepare timbers and timber sheet products ready to receive finishing systems

LO3/4: Know how to and be able to prepare metal surfaces ready to receive finishing systems

LO5/6: Know how to and be able to prepare trowelled finishes and plasterboard ready to receive finishing systems

LO7/8: Know how to and be able to remove previously applied paint and paper ready to receive finishing systems

LO9/10: Know how to and be able to rectify surface conditions

LO11/12: Know how to and be able to repair and make good surfaces

INTRODUCTION

The aim of this chapter is to:

* help you learn how to prepare a wide range of surfaces for protection and decoration.

Hazards, health and safety and risk assessment

When preparing surfaces, the main hazards include:

* asthma or other respiratory complaints from breathing in contaminants, e.g. mould, dust particles, or substances such as wood preservatives or other chemicals

* getting dust or other irritants in your eyes when sanding/rubbing down wood, flaking paint and metal

* chemical burns, dermatitis, and other skin problems from working with materials such as paint strippers and solvents

* inhalation of toxic fumes or dust from old lead paint that has broken down, resulting in respiratory complaints, headaches or dizziness

* fire and risk of burns from working with flammable materials such as solvents or LPG equipment

* the risk of falling from height or the risk from falling objects (see Chapter 4 for more detailed information on working at height)

* electric shock or fire when working with electricity, e.g. preparing surfaces around power sockets, removing light fittings and working with electric tools

* trip hazards, e.g. materials and equipment lying around, such as dust sheets

* cuts and abrasions from sharp equipment, such as scrapers and knives

* asbestos from old insulation and coatings.

There are several precautions you should take to avoid these risks and hazards:

* Wear appropriate PPE.

* Do not smoke near flammable materials.

* Make sure there is good ventilation, i.e. keep doors and windows open.

* Carry out safety checks on access equipment.

* Keep a clean and tidy work area.

* Use washing facilities provided.

* Wash hands regularly (but NOT with white spirit or other solvents), especially before eating.

* Switch off electricity before removing light fittings and ensure electrical equipment is kept in good working order.

* Tape down dust sheets so you don't trip on them.

* Wet abrade instead of dry abrade to reduce dust.

* Identify any risk of asbestos and have it removed by a licensed contractor only.

PRACTICAL TIP

The Control of Substances Hazardous to Health (COSHH) Regulations also apply to the use of wood preservatives.

Chapter 1 has more details on general health and safety issues in a construction environment.

PPE

It is very important to wear the appropriate PPE when preparing surfaces. In particular, you should wear: gloves to protect your hands from chemicals; goggles or safety glasses to protect your eyes from dust particles or flaking materials; a dust mask to avoid breathing in particles and fumes, and overalls to protect your skin from chemicals and solvents.

Protecting the work and its surrounding area

Before surface preparation work starts, move any furniture and belongings away from the surface, e.g. into the centre of a room, and cover with dust sheets. Fixtures, fittings and soft furnishings, such as curtains, should also be removed. The floor should be covered from wall to wall to ensure that no water, solvents, primers or paints can leak down into the floor covering or floorboards. When working with water, particularly during surface preparation, use plastic dust sheets. Even if the carpets have been taken up, you should use dust sheets to stop old wallpaper sticking to the floor.

Keep your work area clean and tidy, as this will reduce any damage to your surfaces, tools and materials, either from your activities or from other people living or working on site. A tidy work area is a safe work area.

If you are working outdoors, it is even more important to protect your work as bad weather may damage surfaces and the materials you are using. Waste should be disposed of appropriately, especially toxic substances such as solvents. Leaving excessive waste, such as wet wallpaper that has been scraped off, to pile up can result in damage to floor surfaces, even if dust sheets are down.

> **PRACTICAL TIP**
>
> If you're going to be working with paint, including primers, then use a 'Wet paint' sign so that others don't touch drying paint. Otherwise you might have to prepare and paint the same area again, and the person who touched it won't thank you either!

Tools and equipment for preparing surfaces

Table 5.1 lists the different tools and equipment you will need for preparing surfaces, rectifying surface conditions and repairing defects.

> **REED TIP**
> •••
>
> Take care of your tools and they will last you a long time. On many sites you will be judged by the quality of your tools and their appearance.

Figure 5.1 A scraper

Figure 5.2 A putty knife

Figure 5.3 A chisel knife

Figure 5.4 A nail punch

Figure 5.5 A hot air gun

Figure 5.6 A dusting brush

Figure 5.7 A roller tray

Figure 5.8 A paint kettle

Figure 5.9 A wire brush

Figure 5.10 A filling knife

Figure 5.11 A filling board

Figure 5.12 Roller sleeves

Figure 5.13 A single arm roller frame

Figure 5.14 Rubbing blocks

Figure 5.15 A moisture meter

Figure 5.16 An orbital sander

Figure 5.17 A palm sander

Figure 5.18 A needle gun

Figure 5.19 A craft knife

Figure 5.20 A pointing trowel

Figure 5.21 A hawk

Figure 5.22 A pole sander

Figure 5.23 A caulking blade

Figure 5.24 A chisel

Figure 5.25 Shave hooks

Figure 5.26 A fibre brush

Figure 5.27 A bottle-type gas torch

Figure 5.28 A steam stripper

Tools, equipment and materials	Description use
Scraper (Fig 5.1)	Also known as a stripping knife, the scraper is used to remove old wallpaper and flaking paint. Some scrapers have changeable blades and long handles which are good for difficult jobs, e.g. many layers of old wallpaper. With the razor edges, it is important to apply pressure evenly, otherwise the plaster underneath can be damaged.
Putty knife (Fig 5.2)	Also known as a stopping knife or glazing knife, one side of the blade is straight and the other side is curved. It is used to push putty or stopper into small holes or cracks, as well as for scraping excess putty from windows.
Chisel knife (Fig 5.3)	Similar to a scraper, but with a narrower blade, a chisel knife can be used for various surface preparation tasks, such as scraping hard-to-reach areas, and small filling jobs.
Knotting brush	A knotting brush would usually be included on the lid of the knotting solution you buy. It is a short, round brush about one inch wide and is used to apply the shellac or patent knotting solution.
Nail punch (Fig 5.4)	A nail punch is a metal rod used for pushing nails into a timber surface prior to stopping and painting so they do not stick out. The punch is usually hit with a hammer to drive the nails in.
Hot air gun (Fig 5.5)	Also known as hot-air strippers, these are used for stripping off paint. They are safer to use than traditional blowtorches because they use hot air instead of a naked flame, and they are less likely to burn woodwork or crack glass. There are models that have temperature settings for extra control.
Hammer	Hammers are used for a variety of preparation tasks, e.g. removing wire clips, picture hooks and nails sticking out of woodwork. Hammers are also used for driving in nails, either on their own or with the help of a nail punch. A scaling hammer is used for removing rust and millscale (see page 147) when preparing a metal surface for painting. These are mostly used on industrial sites. Chipping hammers are used to rake out mortar, hack off plaster or chip away concrete. There are manual versions and hand-held power tools.
Dusting brush (Fig 5.6)	A dusting brush is used for removing dust, debris, grit or any loose and flaky material before any paint is applied. This will make sure that your finish is smooth and that no bits of debris find their way back into your paint kettle. A vacuum cleaner with a duster brush attachment can be used instead.
Roller tray (Fig 5.7)	Roller trays are used for holding paint and other coatings for use with paint rollers. There is a well that holds most of the paint at the bottom and a raised, textured tray for loading the roller. There are different sizes to suit different types of roller, and they are usually made of plastic, but sometimes metal.
Brushes	There are many different brush sizes and types. The filling, or bristle part, can be made of pure bristle (from the hair of wild pigs), man-made fibres (e.g. nylon), natural fibres (e.g. dried grass or plants) or mixtures of these. Synthetic bristles are best used with water-based paints and natural hair bristles are better for oil-based paints. There are some brush types that are multi-purpose. The width of brush you choose will depend on what you're painting. Wider brushes will allow you to apply paint more quickly, but you should choose a brush that is slightly narrower than the surface you're painting, e.g. a 3.5-inch door frame would be painted with a 3-inch brush. Brushes can also come in different shapes. Most are square-cut and are very versatile, but there are also brushes cut with a slight angle to the bristles which makes it easier to reach into corners. There are even special brushes for reaching into very awkward spaces, such as behind radiators.
Paint pots/kettles (Fig 5.8)	Also known as paint cans, a kettle is made of either metal or plastic and is used for holding the right amount of paint in for a job. Paint is poured into the kettle from the bigger tin of paint. The kettle has a handle for holding onto or the handle can be used with a kettle hook so that it can be attached to ladders. They should always be cleaned thoroughly after use so that fresh paint is not contaminated with flakes or different colours.

Tools, equipment and materials	Description use
Wire brush (Fig 5.9)	When preparing a metallic surface, you will need a wire brush to remove old flaking paint, loose rust and corrosion. They can also be used to clear loose bits from brickwork. They are made with bristles of steel or bronze. Bronze bristles will not cause any sparks, so they are the better choice for a high fire risk area. A powered rotary wire brush is a more powerful option for the same job.
Filling knife (Fig 5.10)	Similar in shape to a scraper, but made of thinner and more flexible metal, a filling knife is used to apply filler to cracks and holes. Its flexibility allows for more control when working with the filler.
Filling board (Fig 5.11)	A filling board, similar to an artist's paint palette, is used to hold and mix large amounts of fillers or stoppers. They are usually made of timber board or plastic and can have a pole attached to the base or a thumb hole to hold onto the board.
Buckets	Buckets are useful for holding water and other mixtures when stripping off wallpaper and cleaning surfaces before painting.
Sponges	A large sponge is useful for applying water and other mixtures to walls when removing wallpaper.
Rollers (Figs 5.12 and 5.13)	Rollers are used to apply large areas of paint quickly on flat surfaces. A roller comes in two parts: the frame, and the sleeve, which is detachable. There are several types of roller widths and thicknesses, each designed for a different purpose, such as for painting pipes and radiators. Roller sleeves can be made of foam or sponge, lambswool, synthetic fibres, and mohair. They can be short, medium or long piles. Short pile rollers are best for flat surfaces such as plastered walls because they give a smooth and even finish. Medium and long pile rollers are used on more uneven, rough or exterior surfaces. Synthetic sleeves are cheaper than wool and last longer, but should not be used for solvent-based paints. Lambswool sleeves are good for solvent-based paints and various textures. Mohair sleeves are used for gloss or eggshell paints, and foam or sponge can be used for gloss. Woven fabric rollers are resistant to shedding, which is important on smooth surfaces. They can be used with all types of paint.
Rubbing blocks (Fig 5.14)	Rubbing blocks are made from rubber, cork or wood. They are used to hold sandpaper, making it easier to use. The paper can be simply wrapped around the block and held in place, or the block may have clips or teeth to hold onto the sandpaper.
Sterilising fluid, fungicidal wash	Used for getting rid of mould and other growths from a surface before repainting. Often used together with a fungicidal paint to avoid mould growing back. Fungicide is poisonous – don't forget your PPE.
Stain block (proprietary and non-proprietary)	Used to both fix and avoid surface stains such as grease marks, dyes, nicotine and damp.
Barrier cream	Barrier creams can be applied to protect the skin, before using solvents, such as methylated spirits. This can be useful when using solvents for washing down, and can prevent drying out of skin by avoiding contact with the solvent.
Stiff/scrubbing brush	A stiff-bristled nylon brush used for removing rust and paint, or to remove moss or mould from rendered exterior walls.
Moisture meter (Fig 5.15)	Used to find out how much water is in the surface you are preparing, e.g. in wood or plaster, before it can be painted. The meter will give a reading either as a percentage of moisture content or on a scale of 0 to 100, with zero being dry and 100 being saturated.
Orbital sander (Fig 5.16)	An electrical tool for sanding which has abrasive paper attached to a pad. The pad moves in a circular (orbital) motion. Used for preparing and smoothing timber, metal and surfaces that have been painted before. Slower but easier to use than a belt sander (used for large areas such as floorboards). Best used on small surfaces. They produce a very fine finish.

Tools, equipment and materials	Description use
Palm sander (Fig 5.17)	A light, handheld sander, powered by electricity. Used for dry abrading before or between coatings. Particularly helpful for fiddly surfaces such as skirting boards.
Rotary sander	A mechanical tool powered by electricity or compressed air, it has a round sanding head. It is used for sanding paint, varnish, wood and removing rust. It can be quite a powerful tool and hard to control, which can cause an uneven surface if you sand in one place for too long. A lambswool attachment can be added for polishing.
Needle gun (Fig 5.18)	Used for removing rust and descaling, it has a number of steel needles which strike and retract from the surface continuously. The needles will adjust automatically to the surface, even if it is uneven. It is particularly useful for working around small, awkward areas, such as ornamental ironwork. Powered by compressed air.
Chisel gun	Used for descaling heavy rust and millscale, a chisel is set inside an electric tool (often this is a different head for a chipping hammer or needle gun) and thrust forwards and backwards. The surface may be damaged by the impact of the chisel, so it should only be used for small areas.
Lint-free cloths	Used for removing traces of dust before painting, and for polishing or waxing wood. They are lint-free so as not to leave any fibres behind which may show up in the finish.
Wall brush	A wide brush used for painting emulsion onto large areas and for applying water to a wall surface when removing wallpaper.
Craft knife (Fig 5.19)	Has a razor sharp, retractable, foldable or fixed blade. Useful for scoring and cutting, and scraping small bits of old paint. Often a scraper will be sufficient.
Pointing trowel (Fig 5.20)	A trowel is a bricklayer's tool used for repairing large cracks and holes in exterior walls. It spreads, levels and shapes the stopper, e.g. plaster, mortar or cement.
Wetting in brush	Used before applying the filler when making good render or plaster to make sure the crack doesn't dry out too quickly. A standard paintbrush is fine for this purpose.
Hawk (Fig 5.21)	A hawk is used with a trowel to hold the filling material when repairing larger cracks in walls. It is held in the left hand (if you're right-handed), while the other hand holds the trowel.
Pole sander (Fig 5.22)	An abrasive pad is attached to the end of a pole to help with sanding in hard to reach, high places. Also known as a sanding pole.
Caulking blades (Fig 5.23)	A stiff, plastic blade with a handle made of wood or plastic, used for filling in between plasterboard joints.
Chisel (Fig 5.24)	A hand tool with a flat cutting edge, it is used for chipping away hard materials, such as wood, brick or metal, as well as removing hard putty around wooden casement windows.
Shave hooks (Fig 5.25)	A sharp hand tool with a bevelled edge. It comes in three different shapes and is used for scraping paint from mouldings, alongside other paint removers.
Metal containers	These can be used for the storage of oily rags or dirty/used white spirit – this will reduce the amount of fresh solvent needed.
Fibre brush (Fig 5.26)	Used for cleaning and removing marks from metal or wood, or for brushing down exterior render before painting to remove dirt and loose render.
Transformer	Used to convert electrical current from 230V to 110V. A 110V transformer should be used on construction sites.
Extension cable	When operating power tools, you may need to move a slight distance away from the power source. A long extension cable will allow you to do this, but avoid trailing leads around the site as they can be a trip hazard.

Tools, equipment and materials	Description use
LPG burning off equipment (Fig 5.27)	Heat can be used to soften up old paint which can then be scraped off. An LPG torch is a portable and lightweight tool powered by butane or propane. It has different size nozzles to produce different jets of flame. The bottle is refillable and will last for a few hours, depending on how much flame is produced. The torch itself can either be attached to the bottle or as a detachable hose attached to a larger bottle. There are also disposable bottle types, which are even lighter to use, but cost more than a refillable bottle and are less powerful. There are various safety precautions that you must take when using an LPG torch as there is obviously a high fire risk. Note: they should *never* be used to remove lead paints.
Non-combustible panel	A fire-resistant fibre board, they are especially useful in buildings with a high fire risk. The panel or board should be used under a door or window for burning off paint with a hot air gun.
Steam stripper (Fig 5.28)	Usually powered by electricity, a steam stripper has a small water tank that heats the water like a kettle and sends steam through a hose to a plate with holes in it. The plate is then laid flat on the wallpaper and the steam seeps into the covering, which softens the paper and the glue, making it easier to scrape off.
Cartridge gun/cage	A cartridge gun or cage is used for applying decorator's caulk or silicone sealer.

Table 5.1 Tools and equipment for preparing surfaces and correcting defects

Materials for preparing surfaces

Materials	Description/use
Solvents	Solvents are used for cleaning and preparing metal surfaces. They remove grease from surfaces prior to painting. White spirit is an example of a solvent. Solvents are toxic so care should be taken to wear correct PPE and follow correct health and safety procedures (i.e. good ventilation) when using them.
Shellac/knotting solution	Knotting solution is used to treat knots or stains on timber to prevent them from **bleeding** through paint.
Stoppers	Stoppers are similar to fillers but are made from a stiffer material such as plaster or cement. They are best used for filling deep holes and gaps.
Single-pack fillers	Single-pack fillers are fast-drying fillers used for repairing and filling small holes on timber.
Two-pack fillers	Two-pack fillers come in 'two packs' – one containing a filler and the other a hardener. They are used on wood and give a hard-wearing finish.

Table 5.2 Materials for preparing surfaces and correcting defects

KEY TERMS

Bleeding

– when a stain (e.g. nicotine stains) or a knot comes through and stains the surface.

Solvent-based cleaners

Methylated spirit

Methylated spirit is used for general cleaning of dirt, particularly for removing grease and adhesives. Highly flammable, methylated spirit should not be used around naked flames or sparks and there should be no smoking nearby. It is also harmful if inhaled or if it comes into contact with the skin.

White spirit

White spirit is also used for cleaning and thinning. It can be helpful in cleaning up silicone, degreasing, and cleaning paintbrushes after use with solvent-based paints. It cannot be used to clean knotting solution from brushes. Both white spirit and methylated spirit are useful in preventing rust and corrosion. The advantage of using them is that they dry very quickly so that a surface can be painted straight after use.

Acetone

Acetone is a solvent and cleaner that can be used for removing marks from felt-tip pens, crayon and permanent markers. It can be used on metal and glass and is very good at removing grease. Note that it should not be used on plastic. Acetone is highly flammable, can cause eye irritation, skin irritation or dermatitis, and breathing it in can cause headaches, dizziness, and nausea, so it should be used in a well-ventilated area.

Detergents

Instead of solvent-based cleaners, detergents can be used along with warm water to remove dirt. They must be rinsed off so there is no residue, then left to dry properly before paint can be applied.

Sugar soap

Sugar soap is particularly helpful in washing down paintwork before repainting, and is very good at removing stains, such as from nicotine. It can also be used to help with stripping wallpaper or removing grease. Sugar soap comes in liquid or in powder form. It can cause some skin irritation, so it's best to wear gloves when using it. In powder form, be careful not to inhale it.

PREPARING TIMBERS AND TIMBER SHEET PRODUCTS

Uses and properties of timbers and timber sheeting

Timbers can include softwoods, hardwoods and sheet materials. They provide an absorbent surface and must be well prepared so that they do not take on moisture, which would cause them to rot or expand. Timbers have various uses including:

- structural, e.g. roofing, floor joists, fencing
- first fix, e.g. door and window frames, flooring, stud partitions, staircases
- second fix, e.g. kitchen units, skirting boards, architraves and other timber mouldings
- decorative, e.g. furniture, fireplace surrounds, banisters and high-quality joinery.

Softwood

Softwood timbers come from fast-growing, **coniferous** trees and are mainly used for indoor construction. They must be protected by a surface coating, and need to be properly prepared before paint is applied. Softwoods are used on components such as skirting boards, door and window frames, dado rails, picture rails and architraves.

Hardwood

Hardwood timbers come from **deciduous** trees, which have broad, flat leaves. Hardwoods have a more complex and dense cell structure than softwoods and are generally stronger. You can often see growth rings in the grain of hardwoods so they are usually varnished rather than painted and are used for more decorative purposes.

Timber sheet products

Sheet materials, such as plywood, chipboard, medium-density fibreboard (MDF), hardboard and blockboard, are wood products that are made from wood layers or fibres that are stuck together into sheets or boards. MDF is often used for skirting boards, architraves, door frames, picture and dado rails because it is a cheaper material.

Properties and characteristics of timber

Each type of timber has its own properties and characteristics:

* tactility – how easy it is to work with the timber to make different things

* porosity – how much air and water can pass through the timber

* aesthetics – how attractive the timber looks

* insulation – how much heat the timber retains

* hardness – how resistant to wear and tear the timber is

* strength – how well the timber holds together under pressure and use

* flexibility – how easily the timber bends or expands without breaking

* absorption – how much liquid a timber can take in (the more **porous** a timber, the more water it will absorb)

* adhesion – how easily a timber will stick to another surface

* capillarity (or capillary action) – how much water is absorbed or drawn up, which can rot and weaken the surface

* cohesion – how strong timber is in terms of resisting fracture

* surface tension – the property of the fluid on a surface to resist an external force. This property is caused by cohesion.

KEY TERMS

Porous

– a material that has tiny holes (pores) for air or liquid to pass through. The bigger the holes, the higher the porosity and the absorbency. Porous surfaces may need extra protection from water damage.

Type of timber	Applications	Surface properties	Physical properties
Softwoods			
Pine (also known as Redwood) (Fig 5.29)	Used for a variety of internal and external work including kitchen carcasses, mouldings, frames, doors, flooring and furniture	• Pale and yellowish in colour • Good for painting and staining • Can be glued and nailed • Minimal shrinkage • Resinous, can have a lot of knots • Can achieve a smooth finish	• Moderate strength • Flexible • Moderate durability • Not very resistant to insects or decay • Limited life outdoors
Cedar (Fig 5.30)	Often used for external features such as decking, fencing, garage doors and outdoor furniture	• Varies from straw colour to dark brown or red • Silvery in colour when exposed to weather • Very good for gluing, nailing and screwing • Receives paint, stain and polish well	• Durable and long-lasting, especially when treated • Tends not to warp or twist
Spruce (also known as Whitewood) (Fig 5.31)	Like pine, it is used similarly for internal carpentry and joinery; can be used outside only if treated	• Yellow–white to red–white colouring • Will darken with age • Finishes well with stain, varnish or paint • Good for gluing, screwing and nailing	• Strong and hard • Slightly durable • Poor resistance to insects and decay • Can suffer some shrinkage • Lightweight

Type of timber	Applications	Surface properties	Physical properties
Hardwoods			
Oak (Fig 5.32)	Used for high-quality joinery such as doors and wood panelling, heavy construction, flooring, cabinets, furniture; can be used for exterior joinery such as gates and fencing.	• Golden brown in colour • **Porous**, coarse texture • Takes wax, stain and polish well • Good for gluing, nailing and screwing	• Very strong wood, particularly English oak • Durable • Can suffer from shrinkage and splitting • Resists bending • Sap wood is susceptible to fungal attack, insect attack and decay • Will corrode steel fittings and stain from iron
Beech (Fig 5.33)	Used for internal work such as doors, flooring, tool handles and furniture.	• Pale in colour, sometimes pinkish • Holds stain and polish well • Good for nailing and screwing • Can be glued • Can achieve smooth finish	• Hardwearing • Close grained • Susceptible to attack from furniture beetles • Warps easily • Will decay if exposed to water
Mahogany (Fig 5.34)	Used for high-quality joinery, furniture and veneers	• Dark reddish-brown in colour • Rough grain with interesting patterns • Good for nailing, screwing and gluing • Stains well	• Resistant to decay • Strong • Uniform pore structure
Timber sheeting			
Medium density fibreboard (MDF) (Fig 5.35)	Used for ready-moulded skirting boards and other mouldings, furniture and panelling; for internal use only	• Smooth, even surface • Good for painting • Cannot use nails, only screws	• Made of sawdust • Made using dry process • Moisture and fire resistant versions available
Plywood (Fig 5.36)	Used for building construction, panelling and furniture making; can be used for exterior work	• Marked clearly if only to be used for interior work • Can delaminate in long-term wet weather • Suitable for painting • Smooth surface	• Very strong • Made of layers (veneers) glued with alternating grain • Available in different grades • Some grades have medium resistance to moisture • Should not warp or bend
Hardboard (or high-density fibreboard, HDF) (Fig 5.37)	Used for furniture backing, packaging, floor covering, door panels, curved surfaces	• Laminated plastic surface • Can be painted and varnished • Can be smooth on one side, textured on the other or both sides smooth	• Made of wet, compacted wood fibres • Light • Low cost • Flexible • High density • Some resistance to moisture
Blockboard (Fig 5.38)	Used for interior worktops, tables and shelves	• Similar to plywood, but inside layer formed from strips of timber • Can take paint, varnish or laminates • Can be nailed or screwed	• Strong enough to bear weight • High resistance to twisting and warping

Table 5.3 Timber types, their uses and properties

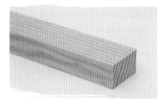

Figure 5.29 Pine

Figure 5.30 Cedar

Figure 5.31 Spruce

Figure 5.32 Oak

Figure 5.33 Beech

Figure 5.34 Mahogany

Figure 5.35 Medium density fibreboard (MDF)

Figure 5.36 Plywood

Figure 5.37 Hardboard

Figure 5.38 Blockboard

Implications of poor joinery design on integrity and durability of coating systems

The result of poor joinery is that the surface will break down when weathered. For example, if a window has sharp edges on the sill or window rails, paint will not adhere properly and it will leave the window exposed to the elements (e.g. wind and rain). This usually results in the window rotting and degrading. It is therefore important that timber products are prepared properly.

Common defects found in timber and timber sheet products

Table 5.4 lists the common defects found in timber when preparing architraves, skirting boards and window and door frames.

Defect	Description	How to make good
Knots	Knots can be both a defect and a feature of wood. They occur where the branches were joined to the trunk. Knots can mean that the timber is either weakened or more difficult to work with. Knots give out resin/sap that will seep through and stain the finish.	Coat knots with shellac knotting solution Prime
Resin exudation	Where sap has seeped out of a knot.	Abrade Apply knotting solution Prime
Glue residue	Bits of excess glue left on the surface after using wood adhesive.	Dry abrade any adhesive Spot prime with a water-based or oil-based adhesive

Defect	Description	How to make good
End grain	Where timber has been cut at a 90° angle to the grain – you will be able to see the rings. End grain absorbs moisture more easily and so it should be sealed with a primer.	Apply stopper Abrade Prime if bare timber is exposed
Cracks	Splits in the timber.	
Open joints	Gaps that have appeared in timber that has been previously joined together.	
Nail holes	Nail holes are commonly seen in plaster where people have hung pictures. Because they are small, they are easily repaired.	Apply all-purpose filler Dry abrade Spot prime
Projecting nail heads	These can be seen sticking out of plasterboard and timber where the nail is attaching the substrate to a surface. They should be punched in and filled so the surface is flat.	Punch in nail Apply all-purpose filler Abrade Spot prime
Infestation	An infestation occurs when insects, that cause damage, are present in large numbers. There are several types of insects that can cause infestations in wood, such as termites and wood-boring beetles, which are most common. Damage caused by wood-boring beetles can be seen as small holes in the timber that are formed when the adult beetle leaves the timber to mate and as the bore dust that it leaves behind containing its larvae. The larvae can live for three years if the moisture content of the timber is greater than 12 per cent. The infestation must be treated prior to painting.	Cut back timber 300 mm from the affected area before treating Remove any weakened/crumbling timber Apply treatment solution to replacement timber and replace Apply treatment solution to all surfaces of affected timber.
Wet rot	Wet rot grows in wood that has been exposed to a moisture content of more than 24 per cent. It is often caused by leaks from gutters and pipes. It is a brown or white fungus that eats away at the timber and is usually restricted to the wet area of the timber. If the source of moisture is removed the wet rot will go away.	Remove the source of moisture Rake out the rotten timber and allow wood to dry. Apply a coat of wood preservative and spot prime with wood primer Insert wood screws into the timber Fill with a two-pack filler (more than once if necessary) Abrade the surface
Dry rot	Dry rot is a serious form of timber decay. It is a fungus that eats away at the timber for food and causes the wood to lose its strength. It has the appearance of cotton wool. It is also caused by exposure to a high moisture content (of between 18 and 22 per cent) and poor ventilation. These conditions create high humidity which can speed up the growth of the fungus. It is more damaging than wet rot because it spreads further to other timbers.	Remove the source of moisture Cut out and replace the rotten timber (and remove rotten timber to be taken away and burnt) Apply a coat of wood preservative to surrounding timber and to the replacement timber Strip back and treat the adjacent plastered areas with steriliser and **biocide** before replastering
Moulds and fungi	Mould is a type of fungus caused by spores in the atmosphere which grow and feed on organic matter found in various surface finishes. It often grows in damp, poorly ventilated areas, especially in old buildings, in corners, behind furniture or curtains where air flow is poor.	Mould must be removed properly before paint can be applied. Sterilise and wash down the area with a fungicidal solution. Scrape mould off Wait a week and retreat if necessary Coat with paints that contain a fungicide.

Table 5.4 Common defects found in timber

Moisture content

Natural timber contains about 80 per cent moisture and is air-dried or **kiln-dried** to reduce its moisture content to around 15 to 20 per cent. If it isn't dried it will absorb or get rid of moisture causing it to expand and contract. This can cause damage to the wood (e.g. warping and wet rot/dry rot) and to paint coverings (e.g. delamination).

Before painting timber you need to check the moisture content using a moisture meter to ensure it is less than 20 per cent. The meter will give a reading as a percentage of moisture content.

Rectifying defects in timber and timber sheet products

Solvent wiping (or degreasing)

Timber surfaces need to be free from grease and oil before they can be painted. Any grease left on the surface will prevent the first coat of paint from adhering. For the removal of slight grease, use sugar soap or detergent and warm water to wash down the surface. For more stubborn grease, you may need to use a solvent (see Cleaning agents on pages 167–8), such as white spirit or turpentine (turps).

Dry abrading

Abrading is the smoothing or rubbing down of a surface to get rid of any flaws. It also provides a key for the new coating to stick to, giving a smoother, better finish. Wood is dry abraded (i.e. without the use of water or solvents).

Knotting

Bare timber often has **knots**, which occur where the branches were joined to the tree. They appear as darker areas, circular in the middle, and contain **resin** that can bleed from the knots over time. To stop the resin from coming out of the knots and staining the finished surface, a thin coat of knotting solution should be applied. If a knot is large and too resinous, it can be drilled out and plugged with another piece of wood or wood filler. Remember that knotting solution is highly flammable.

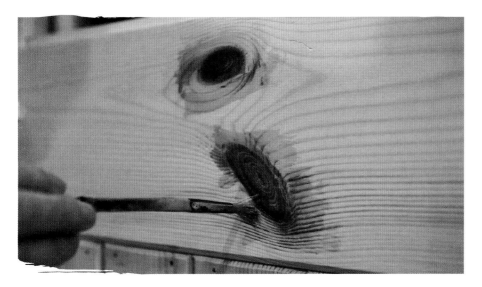

Figure 5.39 Applying knotting solution

Priming

Priming is the very first coat applied to a bare surface. The surface must be properly prepared before the primer is applied, otherwise it will not stick properly. Primers are designed for different types of surface, such as for wood, to give the best result. However, a universal primer is suitable for most jobs. Some primers can also double as the undercoat, i.e. you can apply it once as the primer, then apply it again as the undercoat. Primers should be applied by brush, or with a roller for large flat surfaces such as fresh plaster. For more detail on types of primers, see pages 145–5, 151 and 159 below.

PRACTICAL TIP

When abrading you will create a great deal of sawdust so you must always wear safety glasses and keep your nose and mouth covered.

KEY TERMS

Abrading

– the scraping away or wearing down of a surface using friction. Also known as 'sanding' or 'rubbing down'.

Knots

– a dark, resinous circle in timber where branches were joined to the tree.

Resin

– a thick, sticky substance that comes from trees and timber.

Flush

– when something is completely level or even with another surface, i.e. not sticking out.

DID YOU KNOW?

Primer is sometimes referred to as the 'base coat'.

PRACTICAL TIP

If you're not sure which primer to use on a particular surface, check on the tin. Paint manufacturers will always list the type of use, e.g. indoor/outdoor, plaster/wood/metal, and will also have instructions for application, drying time and cleaning.

Filling

When a surface is not in good condition, e.g. if it has patches, holes or cracks, then a filler may need to be used before the surface is abraded and primed. There are different ways of filling a gap, depending on its nature and size.

Flush filling

Flush filling is used on small cracks or dents in the surface. The filler is scraped back level with the surface using a filling knife or scraper.

Back filling

Back filling is used on deep holes or gaps. The filler is pushed into the back of the hole and allowed to dry, then more filler is applied in layers until the gap is flush or proud.

Proud filling

Proud filling is where the filler is pushed into the gap but is left to stand out from the main surface. This is because fillers can shrink as they dry, and you would need to fill the gap again. If, once it is dry, it is still standing proud, it will need to be sanded back so that it is **flush** with the surface.

Knife filling

Knife filling is the application of filler or stopper to a small area using a filling knife.

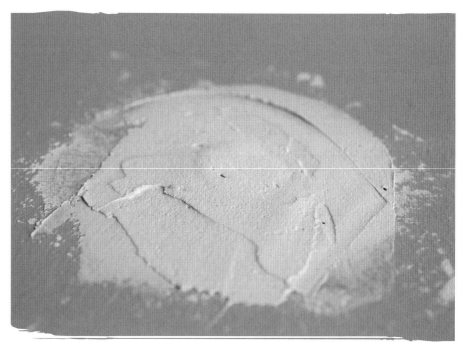

Figure 5.40 Proud filling

Table 5.17 on page 168 is a quick guide to the types of fillers and stoppers that can be used on various substrates.

Stopping

Stoppers are similar to fillers, but are made from a stiffer material such as plaster or cement. They are best used for filling deep holes and gaps using a trowel. Fillers are made of a smoother paste and are better for smaller jobs.

PRACTICAL TIP

When using a stopper for timber, you may need to use a tinted product to match the colour of the wood as closely as possible if they are to be varnished.

When working with open-grained timber, you are likely to use plastic woods, two-pack stoppers or putty.

- Putty is made from linseed oil and is used for filling holes in timber and setting glass into windows. It can become defective and start to crumble with age.
- Plastic woods are a mixture of resin and wood flour and tend to be used when your final coating will be a varnish rather than paint. These also come in a two-pack.
- Two pack stoppers are used on bare surfaces as they may affect coatings. They are particularly strong, meaning you can screw into them, and they set quickly without shrinking.

When stopping and making good surfaces, bear in mind that you may be working with solvent-based products, so your skin should be protected and you should work in a well-ventilated area. When abrading stoppers, ensure you are wearing safety goggles/glasses and a dust mask.

Sinking nail heads

Old surfaces will often have nails and hooks attached to them. If these are sticking out, they should either be removed (if they are not serving a purpose) or sunk in. To sink a nail head, use a nail punch and a hammer. Make sure the nail punch is covering the whole nail head, then hammer the nail in until it is flush with the surface. You have to fill the hole, sand it, prime, and finally paint. The filler and primer help to seal the surface and stop any staining or corrosion from the nail. If the surface is a new or uncoated interior timber, then it is best to use a water-based primer that is quick-drying.

Preparation processes for timber and timber sheet products

The table below shows the different processes needed for the preparation of new and previously painted timber.

Surface type	Preparation processes used	
New soft or hardwood and sheet materials	<ul><li>Dust off surface (abrading may scratch the surface)</li><li>Punch in nails</li><li>Fill gaps, end grain and sunken nail heads (use coloured stopper if varnishing or staining)</li><li>Apply knotting solution</li><li>Prime external timber with solvent-based wood primer or preservative</li><li>Prime internal timber with acrylic primer or solvent-based wood primer</li></ul>	
Previously painted timber	<ul><li>Degrease and rinse</li><li>Fill cracks and holes</li></ul>	<ul><li>Dry abrade</li><li>Dust down</li></ul>

Table 5.5 Preparation processes for timber

Abrasives for the preparation of timber

Abrasive	Material	Properties	Uses
Glass paper	Glass particles on a paper or cloth backing	Comes in Strong, Coarse, Medium and Fine grades. Tends to clog up easily, so has a short life. Can also scratch the surface too much.	Dry, hand or mechanical abrading of plaster or wood for a rough finish
Garnet paper	Natural garnet attached to a paper backing	Comes in grades of 40 to 240. Garnet has a tendency to fracture when used which creates new cutting edges.	Dry, hand or mechanical abrading of wood Fine finishing of wood
Aluminium oxide	Bauxite mineral stuck to paper backing	Comes in discs, belts or sheets in grades of 40 to 240. Long lasting as it doesn't clog up or wear down quickly.	Dry, hand or mechanical abrading of wood

Table 5.6 Types of abrasives for the preparation of timber

Figure 5.41 Sanding with the grain

Abrasive papers come in sheets, belts, discs or rolls, depending on whether it will be used by hand or on a power tool. When choosing the right grading – how fine or coarse the paper is – the lower the number, the more coarse the paper; the higher the number, the finer the paper.

● When sanding a rough surface where the flaws or debris are sticking out, use a coarse paper.

● When sanding down between coatings, use a medium abrasive paper.

● When finishing work, use a fine paper to get a smooth finish.

It is important to choose the correct abrasive. One that is too fine will take more time, use more paper, and may not successfully remove all of the flaws. Alternatively, an abrasive that is too rough will not leave a smooth enough surface and scratches may be noticed in the finish.

Primers for use on timbers

Primers are applied as a 'base coat'. They are needed to ensure that the surface coating adheres properly. They can be water-borne or solvent-borne.

● Solvent-borne primers give a glossy finish. They contain higher levels of VOCs and take longer to dry.

● Water-borne primers contain lower levels of VOCs and dry more quickly. They are generally not as durable but are continuously improving.

KEY TERMS

VOCs (volatile organic compounds)

– a material found in many paints and coatings that helps them to dry more quickly. They evaporate into the atmosphere and are bad for the environment.

DID YOU KNOW?

There are many new coating technologies being developed by paint manufacturers. This is in response to tighter environmental regulations and a need for products that give a better finish. For example, there is a lot of research and development in the area of low **VOC** paints, particularly for gloss finishes. Another new product you may come across is 'new work undercoat' and 'new work gloss' which are highly pigmented for better coverage on rough, textured or porous surfaces.

Primer	Description and use
Solvent-borne	
Aluminium	Used especially for resinous timbers due to its good **opacity** and self-knotting. It is particularly good for surfaces that are likely to bleed, such as coal-tar and old bitumen-coated surfaces. Cleaned and thinned with white spirit. Due to its dark colour an extra undercoat may be needed for light coloured finishes. Drying time 4–6 hours; overcoat in 24 hours.
White	A wood primer for soft and hardwoods. Can be used on interior or exterior wood.
Pink	Pink primer used to be used more commonly as a timber primer because of the red lead content in old paints. Pink primer is traditionally used on softwoods (aluminium leaf primer is used for hardwoods). It doesn't affect the finished coat.
Water-borne	
Acrylic	Can be used as primer and undercoat. Usually in white. Quick-drying, can overcoat in two hours, easy and cheap to thin and clean, non toxic. Used on timber, board, paper and dry plaster. Not used on metal.

Table 5.7 Primers for use on wood

PREPARING METAL SURFACES

Metal surfaces

Ferrous metals
Ferrous metals have iron content, such as cast iron, wrought iron and stainless steel. Iron and steel can rust when they come into contact with water and oxygen, e.g. when outdoors in the wind and rain. This rust, or **corrosion**, is formed by the iron content, so the more iron there is in a metal, the more rust will occur and the weaker the metal will become. Ferrous metal needs to be well prepared and protected to avoid rust and flaking.

Non-ferrous metals
Non-ferrous metals do not contain any iron, such as aluminium, copper, lead, brass, galvanised steel, bronze and zinc. They are not as strong as ferrous metals and are usually used only for decorative purposes. When non-ferrous metals corrode, the corrosion provides a layer of protection to the metal underneath. Although non-ferrous metals do not suffer damage (i.e. rust) from corrosion as ferrous metals do, they still need to be prepared and protected from the weather.

Properties and characteristics of metal
Each type of metal will have its own properties and characteristics, which can be described as follows.

Surface properties
* Colour – some metals have distinctive colour, such as copper.

* Hardness – how solid or soft the metal is and how resistant it is to damage.

* Porosity – how much air and water can pass through, which can weaken a metal.

* Toxicity – whether it is poisonous, such as lead.

Physical properties
* Thermal expansion and contraction – whether the metal gets larger (expands) or smaller (contracts) in heat.

KEY TERMS

Opacity
– the more opaque a paint is, the harder it is to see through. Opacity is the opposite of translucency.

Corrosion
– a chemical action that damages and destroys metals.

REED
TIP

To reduce wastage, use only the amount of paper that you need to fit on your rubbing block, either by cutting it or folding it.

* Electrical conductivity – the ability to carry electricity.
* Tensile strength – the maximum amount of stress that a metal can withstand when being stretched without breaking.
* Compressive strength – the maximum stress that a material can withstand when it is compressed.

Type of metal	Applications	Surface properties	Physical properties
Ferrous metals			
Cast iron (Fig 5.42)	Used for decorative and complex shapes such as stairs, handrails, fireplaces	Non toxic Hard on the surface Softer underneath the skin More porous than wrought iron	Brittle Strong Corrodes/rusts
Wrought iron (Fig 5.43)	Used for delicate patterns and ornamental ironwork, some pipework	Relatively soft Carries coatings well Non-porous	Tough, though brittle when cold Malleable, will bend rather than break Does not corrode/rust as easily as steel because low in carbon
Mild sheet steel (Fig 5.44)	Used for general engineering purposes such as steel girders, screws, nuts and bolts; also used for garage doors	Porous	Corrodes/rusts easily, due to high carbon content Tough (not brittle) Malleable, bends easily Will resist extreme conditions without twisting, warping etc
Steel (Fig 5.45)	Used in construction, cars, appliances, shipping containers	Hard and tough	Very strong Withstands high stress High carbon content means corrodes/rusts easily Brittle – the higher the carbon content, the harder and more brittle
Non-ferrous metals			
Copper (Fig 5.46)	Used for pipes, electrical wire, and decorative purposes	Distinctive red colour Can be easily damaged Will tarnish easily	Malleable and flexible Tough and **ductile** Conducts heat and electricity well Resistant to corrosion
Aluminium (Fig 5.47)	Used for window frames and kitchen items	Soft and light White/grey in colour, but can be tinted Non toxic	Conducts heat and electricity well Highly malleable Resists corrosion very well
Lead (Fig 5.48)	Used in some paints (less so today), flashings and roof covering	Blue-grey in colour Very soft Shiny, but dulls with oxidisation Toxic	Dense and very heavy Highly malleable and ductile when cold Becomes more brittle when heated Resistant to corrosion
Galvanised steel (Fig 5.49)	Used for flashing, gutters, roofing, pipes, girders and frames	Hard	Coated in zinc to protect against corrosion Light Tough and strong *Note that galvanised steel is sometimes referred to as a ferrous metal due to its original state

Table 5.8 Types of metal, their uses and properties

Figure 5.42 Cast iron

Figure 5.43 Wrought iron

Figure 5.44 Mild sheet steel

Figure 5.45 Steel

Figure 5.46 Copper

Figure 5.47 Aluminium

Figure 5.48 Lead

Figure 5.49 Galvanised steel

Corrosion

Corrosion is a chemical reaction in which oxidation or electrolysis takes place between a metal and the environment. It slowly destroys metals.

The level of corrosion depends on the amount of oxygen, hydrogen, moisture and atmospheric pollution/exposure to acids/alkalis.

Oxidation

Oxidation is the most common type of corrosion. When the metal and oxygen mix, they form oxides. Iron rust is the result of oxidation, and it is often seen on iron and steel products. The higher the iron content, the more easily a metal will rust.

Pitting is a type of corrosion causing small holes in metal that are concentrated in a small area.

Electrolysis is an **electrochemical** process that drives corrosion. If a metal is exposed to an **electrolyte** (water in the atmosphere containing **acids, alkalis** or salts) it can cause an electrical current to flow. This current flows from the anode (an area of higher energy) through the electrolyte (which can conduct electricity) to the cathode (an area of lower energy) and returns to the anode through the metal. As a result of this current flow, destructive corrosion and rust occur at the anode. The cathode is protected from corrosion. The stronger the electrolyte the greater the corrosion, so in areas of high salt content (e.g. in coastal areas) and atmospheric pollution (e.g. sulphur from industry), more corrosion is likely to occur.

The effects of surface corrosion can be managed by **cathodic protection** (see below) along with the application of special coatings (e.g. rust-inhibitive primers (see page 151)) to prevent the formation of rust and weather-resistant paints. Some types of surface corrosion are seen as aesthetically pleasing, e.g. natural rusting on steel roofs caused by the breakdown of protective coatings.

Ferrous metals can also suffer from **millscale** (or just 'scale'), a bluish/black layer of metal oxides, that occurs during production of metal sheets. It sticks to the surface of the metal, which protects the surface underneath, but can flake off later even when finished with paint. Once the millscale has come away, the metal underneath becomes vulnerable to corrosion until it is primed.

KEY TERMS

Oxidation
– a reaction between metal and oxygen causing a change in the surface, such as iron rust.

Pitting
– small holes in a metallic surface caused by localised corrosion.

Electrolysis
– the separation of substances through an electric current.

Electrochemical
– a chemical reaction brought about by electricity.

Electrolyte
– a liquid through which electricity can pass.

Acid
– a solution with a pH of less than 7.0.

Alkali
– a solution with a pH of more than 7.0.

Cathodic protection
– a protective way of stopping a metal corroding by using another metal to attract the corrosion.

Millscale
– a flaky surface formed during production of ferrous metal in a rolling mill when the metal comes into contact with air. It is also known as scale.

Figure 5.50 Oxidation

Figure 5.51 Pitting corrosion

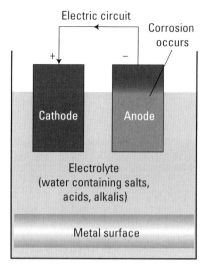

Figure 5.52 Electrolysis in a metal

Figure 5.53 Millscale

Figure 5.54 A rotary wire brush

Galvanic corrosion occurs when two different metals come into electrical contact with one another in a corrosive electrolyte. One metal becomes the anode and the other becomes the cathode. The metal that is the anode corrodes faster than it would on its own while the metal that is the cathode corrodes more slowly than it would on its own. In some cases this reaction can be beneficial, e.g. cathodic protection (see below). In other cases, where mixed metals are used together (e.g. in piping) it can speed up corrosion.

Cathodic protection

In cathodic protection the metal that is the anode (and which corrodes faster) is sacrificed to prevent the other metal from corroding. For example, some metal coatings can have a metal added to them that will corrode first and therefore protect the substrate underneath it. This is also the reason why steel is galvanised with zinc – the zinc will rust before the iron.

Preparation of metal surfaces

Degreasing and solvent wiping

Non-ferrous metals need degreasing prior to painting. (See page 140 about degreasing.)

Priming

Metals, like timber, need priming prior to painting. Primers specifically designed for use on metals will give the best result (see page 151).

Rust removal and descaling

On metal surfaces, you may notice rust or corrosion. This needs to be removed or deoxidised before any paint can be applied. Where there is only light rusting, it can be removed with an abrasive paper or cloth, it can be scraped away by hand using a chisel knife or 1-inch scraper, it can be scrubbed away using a wire brush, or a solvent may be used.

Power tools for removing more heavy rust, loose paint and millscale include rotary wire brushes, powered chipping hammers, chisel guns and needle guns. Needle guns are particularly useful for removing rust around nuts and bolts.

Rust or corrosion on metal surfaces can also be removed using temperature.

Flame cleaning

A hot oxy-acetylene torch is passed over the surface of the metal. The heat causes the rust to flake off as it expands at a different rate to the metal surface. Any loose rust can be removed by abrading with a wire brush, leaving the surface ready for painting or further treatment. Flame cleaning is not very effective at removing all the rust/corrosion and may leave burn marks on the surface.

Blast cleaning

This is a very effective method or removing corrosion/rust. Abrasive particles, such as grit, steel shot or aluminium oxide, are thrown onto the metal surface at high speed. The force of the impact loosens the rust and corrosion from the surface leaving it ready for painting or further treatment.

In wet abrasive blast cleaning, water is introduced to the abrasive particle stream resulting in less dust. This method is therefore often used when removing rust from hazardous materials, such as old lead-based paints.

Classification of standards for prepared ferrous metals

The Swedish Standard is the main standard used to classify the preparation of ferrous metals. The Standards are outlined in Table 5.9.

Figure 5.55 Blast cleaning

PRACTICAL TIP

It may be noisy when working with power tools, so make sure you wear ear defenders to protect your hearing.

Standard	Description	Explanation
St 2	Hand tool cleaning	• Manual wire brushing • Manual abrading • Manual chipping • Other handheld tools
St 3	Power tool cleaning	• Powered wire brushing • Powered chipping hammer • Powered sanders
Sa 1	Light blast cleaning	• Loose millscale, rust and foreign matter should be removed • Slight discoloration from foreign matter
Sa 2	Commercial blast cleaning	• Virtually all millscale, rust and foreign matter should be removed • Greyish in colour • 66% of each square inch of surface should be free of visible residue
Sa 2½	Near-white blast cleaning	• A surface from which all millscale, rust and foreign matter has been completely removed except for very slight discoloration in the form of spots or stripes • Approximately 95% of each square inch of surface should be free of visible residue
Sa 3	White metal blast cleaning	• A surface from which millscale, rust and foreign matter have been completely removed • Has a grey/white uniform metallic colour • Slight roughness to allow paints to adhere • Completely free of any contamination

Table 5.9 The Swedish Standard for the preparation of ferrous metal surfaces

Preparing ferrous and non-ferrous metal surfaces for the required finish

Table 5.10 shows the preparation processes for ferrous and non-ferrous metal.

Surface type	Preparation processes used
Ferrous metal	• Solvent wipe with white spirit • Remove rust, millscale and paint with a power tool (e.g. chipping hammer or rotary sander) and wire brush • Prime with metal primer or zinc phosphate
Non-ferrous metal	• Degrease/solvent wipe • Use mordant solution on galvanised metals • Abrade with rough aluminium oxide paper • Prime with metal primer

Table 5.10 Preparation processes for metals

Abrasives for use on metals

Abrasive	Material	Properties	Uses
Emery	Natural emery (carborundum) attached to cloth	Used by hand, sold in sheets or narrow rolls. Used less often now due to increased use of power tools. It is also expensive.	Dry or with a spirit lubricant for abrading metals by hand. If non-ferrous metal, just use a mordant solution to take shininess away.
Silicon carbide	Mixture of silica, coke and sand, stuck to a waterproof paper or cloth	The crystals are very sharp and long lasting, so long as they are rinsed and unclogged regularly. Comes in grades of 120 to 600. There is a self-lubricating type that does not clog.	Wet or dry, for abrading all surfaces to a very smooth finish (though not for plaster or for bare surfaces); used with mineral oil for polishing metals.
Steel wool	Threads or strands of wire made from steel or stainless steel which are twisted together	Comes in different grades: coarse, medium and fine. The strands tend to be very sharp and can easily appear as splinters in your hands, so you must wear gloves.	Etching and degreasing metal and plastic surfaces; scrubbing timber after it has been taken back with paint stripper; achieving a flatter finish on a gloss surface.

Table 5.11 Abrasives for use on metals

Abrasive papers can be used wet or dry on metals depending on the type.

When dry abrading by hand:

* use a dry abrasive paper
* do not add any water or lubricant
* sand in the direction of the grain of the timber
* use the right grade of paper
* change paper if it becomes clogged or is not working.

When wet abrading by hand:

* wet both the surface and the abrasive
* sand in a circular motion
* rinse out the paper to avoid the grain clogging up
* sand in the longest direction of the surface.

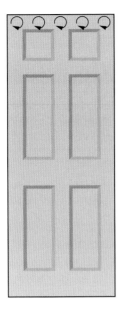

Figure 5.56 Wet sanding direction 1

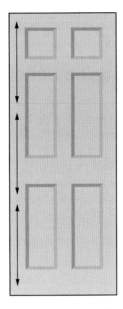

Figure 5.57 Wet sanding direction 2

The use of wet abrading for surface defects

Wet abrading can be used, for example, if gloss has bloomed in damp conditions. The wet abrading process will bring the gloss back to a shiny surface. It could also be used to give the surface a key when applying gloss over gloss.

Primers for use on metals

Primer	Description and use
Solvent-borne	
Zinc phosphate	Used for both ferrous and non-ferrous metals. It contains a rust inhibitor. Thin and clean with white spirit. Overcoat in 6–16 hours, depending on conditions.
Etch	Used for preparing non-ferrous metals to create a key for better adhesion, especially shiny metals such as zinc, aluminium etc. Thinned with butanol. Cleaned as per manufacturer's instructions. Can be used as an alternative to mordant solution. Must be applied in dry conditions. Drying time 1–4 hours; overcoat within 12–16 hours. Once mixed, it must be used straightaway.
Red lead	A bright red primer used to protect ferrous metals, such as iron and steel, from corrosion. It has good **rust-inhibition** properties. It is toxic, because it contains lead, so it is restricted to use in industrial plants and marine steel work. It is being replaced by other lead-free primers, such as zinc phosphate. It may take longer to dry in damp/cold conditions. Drying time 2–4 hours. Overcoat 24 hours after application.
Calcium plumbate	An oil-based primer used on **galvanised** iron and steel. It can be used internally or externally and has excellent rust-inhibition properties. It can also be used on ferrous metals (iron and steel) and some hardwoods. It is also lead-based meaning that it is toxic. Drying time 6–8 hours. Overcoat 24 hours after application.
Zinc chromate	A yellow-coloured rust-inhibiting primer for use on aluminium and ferrous metal. It has its own thinner. As well as inhibiting rust it is resistant to solvents and lubricating oils. Highly toxic. Drying time 16–24 hours. Overcoat should be applied as soon as the primer is dry (approximately 24 hours).
Water-borne	
Metal	Used for non-ferrous metal. Cleaned and thinned with water. Overcoat in 4–6 hours.

Table 5.12 Primers for use on metals

See practical tasks on pages 169–71 for step-by-step instructions on preparing ferrous and non-ferrous metal surfaces.

Preparing trowelled finishes and plasterboard

Trowelled finishes refers to surfaces that are created using a trowel such as plaster, render, bricks and blocks. Plaster and plasterboard are used on ceilings, internal and dividing walls.

Plasterboard

Plasterboard is usually made from **gypsum** which is then sandwiched between two layers of paper. Gypsum is made of fine crystals which look like white rock when water is added and then dried.

Plasterboard is used for **dry lining** – where the boards are attached to internal stud walls, forming part of the basic structure. It can either have coatings applied to it directly, or can receive a final skim of plaster to smooth the surface.

Plasterboard is not generally used in wet areas, such as kitchens and bathrooms, although there are some types of moisture-resistant plasterboard which will help it last longer when exposed to water. There are other types of plasterboard that have extra fire protection and sound-proofing qualities.

Feather edge (or tapered edge) plasterboard is used for taping/jointing or skimming with plaster, whereas square edge plasterboard is used for directly applied finishes, i.e. skimmed plaster. The gaps created by the tapered edges allow for easier filling of the joints, which leaves a smoother finish when painting over it.

Plaster

Plaster is used to finish internal surfaces, leaving them ready to receive paint or paper coverings. There are two main types of plaster: lime plaster and gypsum plaster.

Gypsum plaster is the most commonly used type for internal construction. It is a dry powder that forms a paste when water is added, and then hardens when it dries.

Sometimes plasterboard is finished with a thin skim of plaster. They are both very absorbent materials and must be completely dry before coverings are applied. You can tell if it is dry by its light pink colour.

Bricks and blocks

Bricks and blocks can be used as structural elements on both the outside and inside of buildings.

Blocks are made of concrete, and because of their weight and strength, are used on surfaces that will take a lot of weight. They are held together by mortar. Lightweight versions are available, which are much easier to lift, in line with health and safety guidance on manual handling. Internal blockwork is usually covered with plasterboard then painted.

Bricks are used for similar purposes to blocks, but are smaller. Like blocks they are held together by mortar. External brickwork is sometimes covered in render, but may also be painted on directly. However, is difficult to apply paint directly to bricks because they are very porous and absorbent.

KEY TERMS

Galvanised

– iron and steel that has been coated with a protective layer of zinc to prevent it from rusting.

Gypsum

– a naturally occurring white, chalky rock used in plaster and plasterboard.

Dry lining

– another term for the fixing of plasterboard on an internal background surface.

DID YOU KNOW?

Etch primer is also known as mordant solution.

PRACTICAL TIP

It is important to prime a metal surface immediately after preparation to prevent it from absorbing any moisture or other substances and rusting. A layer of oil could be applied to the clean metal to stop oxidation.

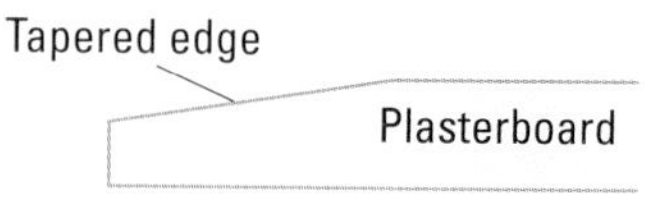

Figure 5.58 Feather edge plasterboard

PRACTICAL TIP

It is the ivory side that should be skimmed, painted or papered, not the brown side.

Figure 5.59 An example of painted brickwork

DID YOU KNOW?

There are three main types of brick: common/fletton bricks which should not be painted because of their high salt content and efflorescence; engineering bricks which cannot be painted because of their non-porous surface; and rustic/sand-faced clay bricks which will accept paint well.

PRACTICAL TIP

To prepare brick surfaces for painting:
- Clean the brick surface with fungicidal wash to remove any mould and/or efflorescence and allow it to dry as per the manufacturer's instruction.
- Fill any cracks or holes and abrade the filler.
- Prime the surface prior to applying an exterior masonry paint.

Wet dash (roughcast) and dry dash

Wet dash (or roughcast) and dry dash are coarse plaster surfaces used on exterior walls that consists of sand, small gravel, pebbles or shells.

In wet dash the rough **aggregate** is mixed into the **butter coat** before applying to the wall, whereas with dry dash (or pebble dash) the aggregate is thrown onto the wall after the butter coat has been applied.

Prior to painting any loose gravel, sand and pebbles should be removed using a nylon brush. The surface should then be washed and primed using a stabilising primer.

Concrete

Concrete is a building material made up of a mixture of sand, coarse aggregate, cement and water. When it has been mixed and hardened it forms a stone-like material.

Concrete is used both on the outside and inside of buildings. Internal concrete walls may be covered in plasterboard prior to painting. Both internal and external concrete walls can be painted on directly. However, it is difficult to apply paint directly to concrete because it is very porous and absorbent.

Cement render

Render is made of cement, sand and lime (i.e. plaster). It is mainly used on external walls but can also be used as a feature on internal walls. It is often painted after application. However, it needs to be completely dry (which takes about six weeks) before it is painted.

KEY TERMS

Aggregate
– granular material, such as sand, gravel, pebbles or shells.

Butter coat
– a soft, wet coat of render.

DID YOU KNOW?

Roughcast is also known as harling, especially in Scotland.

PRACTICAL TIP

As wet dash/dry dash isn't fixed to the wall securely, some of it will come loose during the painting process.

KEY TERMS

Saponification

– a process that creates a type of soap that attacks painted surfaces.

Stonework

Natural stone has been used for construction for thousands of years. Stone is very porous and therefore very absorbent. It should be prepared in the same way as brickwork.

Properties and characteristics of trowelled finishes

Plaster, brick and blockwork have their own properties and characteristics, which can be described as follows:

* tactility – how easy it is to work

* porosity – how much air or water can pass through

* capillarity – how much water is absorbed, which can weaken the surface

* adhesion – how well it will stick to another surface

* cohesion – how strong a surface is in terms of resisting fracture.

* surface tension – the property of the surface of a liquid to resist an external force. This property is caused by cohesion.

* acidity – how much acid is found in the surface

* alkalinity – the opposite of acidity; how much the acids are neutralised

* inertness – how little reaction is caused when plaster is adhered to a surface

* soluble salt content – how much salt is found in bricks.

The characteristics that apply to plaster include tactility, porosity, adhesion and capillarity. Brickwork characteristics include porosity and capillarity.

Alkalinity in finishes, such as plaster, can cause a chemical reaction with paint coverings called **saponification**. This can be avoided by preparing and priming the surface properly to create a barrier between the paint and the surface.

PRACTICAL TIP

The alkalinity of a surface can be tested using either litmus paper or the universal indicator. Litmus paper that comes into contact with an alkaline surface will turn blue. The universal indicator or pH scale will tell you the degree of alkalinity in the surface. Alkaline surfaces will have a value of more than 7.

PRACTICAL TIP

If you are working on buildings that over 100 years old, remember that they need to breathe. The materials you use should allow for this. The mortar in more modern buildings would be made from Portland cement and sand.

Defects associated with trowelled finishes and plasterboard

The defects that you are most likely to come across when preparing trowelled finishes and plasterboard are listed in the table below, along with the basic steps required to repair them and make good the surface.

Defect	Description	How to make good
Settlement cracks	These occur as a result of gradual building movement, e.g. as a result of the soil compacting beneath the foundation. It is often noticed on ceilings and walls, around door frames and window lintels, either on the outside or inside. Most houses will have some movement in their first few years and small cracks are not usually a cause for concern.	• Rake out and undercut the crack • Wet in crack • Fill with gypsum plaster stopper (internal) or cement plaster stopper (external) • Dry abade
Shrinkage cracks	These tend to affect concrete floors and walls, and often start near the corner of windows. They are a result of the ageing and drying process and can easily be spotted because they are not continuous cracks. Depending on the seriousness of the crack, they may need extra sealant so the wall stays waterproof.	• Rake out crack • Wet in crack • Apply sealant or stabilising solution • Fill with gypsum plaster stopper (internal) or cement plaster stopper (external) • Abrade wet or dry
Defective pointing on brickwork	Mortar joints between bricks or blocks can dry out and become crumbly. The surface needs to be sound before primer and paint can be applied.	• Rake out joint • Wet in joint • Apply sealant or stabilising solution • Fill with cement mortar • Apply second coat of stabilising solution
Defective pointing on stonework	Mortar joints between stonework can dry out and become crumbly.	• Rake out joint • Wet in joint • Fill in with cement mortar • Leave mortar slightly proud of the joint and finish off pointing to match existing pointing
Efflorescence	Surfaces that contain lime or cement are known as 'chemically active'. The water-soluble salts come to the surface as it dries and ages, leaving white deposits. These patches of white can appear on bricks, render and plaster. Efflorescence needs to be removed before primer and paint are applied.	• Use a dry brush to remove or scrub with a wire brush • Do not wash as the salts will disappear in the water and be reabsorbed into the surface
Powdering	A fine white powder forms on the surface coating. It is caused by weathering. It can be the result of poor surface preparation. As powdering leaves a porous surface, it must be removed otherwise new coats will not adhere.	• Remove chalk residue with a stiff brush and water containing a mild detergent • Rinse and allow to dry • Apply a stabilising solution to the surface • Once dry test the surface for any further chalk residue by rubbing your hand over it

Defect	Description	How to make good
Spalling	Spalling is when a surface, such as concrete or brickwork, has broken up, flaked or become pitted. It is often caused by water entering the surface and causing it to peel or flake.	*Bricks* • Drill holes around the affected brick and remove the brick • Replace the brick with a new one *Concrete* • Remove damaged area using a scotch hammer • Mix replacement concrete with water according to manufacturers' instructions and apply to the cut out area • Fill level with the surrounding surface
Blown plaster and render	Plaster and render can break away from the wall as it loses adhesion to the surface behind. It can be caused by drying out, ageing, moisture, heat, or as a result of poor application. Patches of blown plaster and render can be repaired, but if the entire surface is affected, then it may be best to strip it back to the brickwork and have the whole surface reskimmed.	• Remove loose render or plaster using a hammer and chisel • Remove any dust with a masonry brush • Rake out cracks • Undercut crack • Apply a layer of sealing coat • Fill with cement mortar (for render) or gypsum plaster (interior) and level off • Blend new cement with the surrounding area using a wet sponge
Defective plasterboard joints	If plasterboard has not been jointed correctly when installed, over time the joins between the boards may start to show. Scrim tape can work itself loose, or could even crack over time due to poor application.	• Rake out joints • Apply joint filler to larger gaps • Apply jointing tape to reinforce the joint • Abrade • Prime
Dry out	Plaster that has dried out due to rapid moisture loss (e.g. from evaporation). The top coat can flake off when applying paint.	• Rake out any cracks to remove loose material • Apply a fine surface filler using a filling knife • If surface has crumbled, apply a new layer of plaster
Defective roughcast	Roughcast that has come loose and fallen off a wall.	• Apply a coat of fungicidal wash if fungus or mould is present • Rake out cracks with stripping knife to remove loose material • Use a masonry brush to remove any dust • Mix and apply exterior filler using a filling knife • Scrape back filler to leave flush with the surface • Blend in filler while wet to match the surface • When filler has dried, apply stabilising primer ready for painting with exterior masonry paint

Table 5.13 Common defects found in trowelled finishes and plasterboard

Common reasons for cracks in plaster

Cracks in plaster and render occur for various reasons.

* Settlement cracks – most buildings will move to some degree as a result of soil movement. If the foundations have not been prepared as well as they should, then this movement can be more severe and can cause settlement cracks.

* Weathering – if a building is repeatedly exposed to severe weather such as storms, the cracks may begin to show.

- Shrinkage – as plaster and render is applied when it is wet, when it dries, the evaporation of the water can cause the surface to shrink and then crack. This can occur due to incorrect preparation of the plaster or cement.

- Dry out – plaster that has dried out due to rapid moisture loss (e.g. from evaporation). The top coat can flake off when applying paint.

- Age – eventually plaster will naturally start to crack at points of stress in a structure, e.g. from the weight of roofs and upper floors.

- Moisture damage – plaster that is exposed to water again and again will weaken. This might be the result of a leak inside or from water entering the building from outside. Even moist areas such as bathrooms and kitchens can be the cause of damaged and cracked plaster over time. This can be made worse if the wrong paints are used on a surface, e.g. a non-breathable paint on lime plaster. Efflorescence can also result from moisture on plaster and would need to be removed before decoration.

- Heat – if plaster has been forced to dry more quickly than it naturally would, e.g. so that coatings can be applied sooner, then it can result in cracking. If heat is applied too close to the plaster, such as from a steam stripper or hot air gun, this can blow out the plaster.

Rectifying trowelled finish and plasterboard defects

Some of the processes for rectifying trowelled finish and plasterboard defects have been covered earlier in this chapter: abrading (page 141); proud filling (page 142); flush filling (page 142) and degreasing (page 140).

Scraping

Scrapers can be used for a variety of tasks, such as removing nibs from a fresh plaster surface, pulling out staples, and removing wallpaper and flaking paint. There are different types of scraper available – some are sharper than others.

When working with a fresh plaster surface, be careful not to damage the flat finish. When removing wallpaper, take care to scrape evenly, otherwise you may damage the plaster underneath which would then need further repairs.

Raking out

Raking out is a way of preparing old render that is cracked, loose or rotting. The brick joints are raked out to ensure the same depth of all the joints before filler or new render is applied. This will also remove any loose debris. You will need a hammer and pointing chisel to rake out the joints. Once the raking out is complete, it will need to be cleaned out with water.

Where there is loose or crumbling debris left in and around a crack or hole, it should be raked out so that you are left with a solid surface to fill and a good key for the filler to attach to. You can use a filling knife for this.

Undercutting

Where the surface is damaged and needs filling, after brushing down and raking out the crack or hole, dig into the underside of the surface that is already there. This provides something for the filler to hold onto. Use a knife and cut the inside surface back at an angle, leaving a wedge shape. See Fig 5.60.

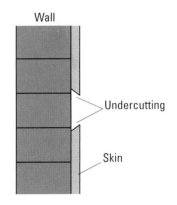

Figure 5.60 Undercutting

KEY TERMS

Scoring

– cutting or scratching through a surface, leaving small holes.

PRACTICAL TIP

When wetting in, adding a bit of detergent to the warm water will help speed up the softening of the wallpaper.

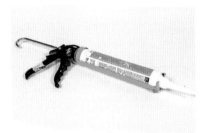

Figure 5.61 Caulk in a silicone gun

DID YOU KNOW?

When a produce is labelled acrylic, it means that it is water-based, not solvent-based.

Wetting in

When removing old wallpaper coverings, after the paper has been **scored**, water must be applied to the wall using a bucket and sponge. It will seep through the holes created by the scoring, and soak into the paper and backing, softening it and making it easier to scrape off. If working with plasterboard under the wallpaper, take care not to add too much water as it will damage the board.

Wetting in is used before applying fillers and stoppers. After you have raked out and undercut a surface that is to be filled, take a wet paintbrush and use it to apply water to the crack or hole to be repaired. Lightly brush – there is no need to soak it through. This stops the filler from drying out too quickly, which can lead to shrinking and, worse, falling out of the surface completely.

Applying caulk and sealants

Caulk is a flexible, acrylic filler. It tends not to shrink and is quick drying so it is particularly useful for filling gaps and sealing joints in skirting boards, architraves, dado rails and similar. Caulk can be applied using a silicone gun and then moulded into the right shape. It should not be sanded down afterwards. There are both paintable and non-paintable types of caulk. The acrylic type is water-based and paintable.

Silicone is also a useful sealant, often used in kitchens and bathrooms because it is waterproof, but also around external window frames. Silicone-based caulk is not paintable. It is more flexible and durable than decorator's caulk and holds up well in direct sunlight and extremes of temperature.

Refer back to pages 142 and 144 for more on the different types of filling and abrading.

Stoppers

Plaster-based stoppers, or plaster itself, are used on internal surfaces only. These stoppers should only be mixed with clean water.

Cement or vinyl-based stoppers are used for exterior surfaces because they are waterproof and harder wearing. Exterior surfaces, because they are porous, must be primed with a sealant or stabilising solution.

Taping

When applying plasterboard, or **dry lining**, board filler is applied to feather edge plaster board, and covered with scrim tape over the top of the joint. Board filler is then added in layers to provide a smooth finish that can be primed and painted onto directly.

Preparing trowelled finishes and plasterboard to receive the required finish

Table 5.14 below shows the preparation needed for different types of trowelled finishes and plasterboard.

Surface type	Preparation processes used
Fresh plaster/plasterboard	• Scrape to remove nibs from plaster • Coat first with primer or thinned emulsion • Fill cracks and holes if necessary, then rub down filler • Dust down (do not abrade) • Prime alkali-resisting primer if using an oil-based top coat
Previously painted plaster/plasterboard	• Degrease with sugar soap and rinse • Abrade • Fill cracks and holes • Sand back filler • Spot prime only if bare plaster revealed • Dust down
Covered plaster/plasterboard	• Score wallpaper • Wet in • Scrape or steam off wallpaper • Wash off glue residue • Fill cracks and holes • Sand back filler • Dust down
Brickwork, blockwork, masonry and render	• Scrape and brush off dirt • Scrub if there is efflorescence • Apply fungicide if mould present • Wash down • Rake out if loose render • Dust down • Primer with solvent-based primer sealer or stabilising solution

Table 5.14 Preparation processes for trowelled finishes and plasterboard

Primers for use on trowelled finishes and plasterboard

Primer	Description and use
Solvent-borne primers	
Alkali resisting (ARP)	Designed for alkaline surfaces containing lime or cement (e.g. render, plaster, concrete). Thinned and cleaned using white spirit. On very porous surfaces, two coats may be needed. Drying time is 8–12 hours, but overcoating must wait 16–24 hours.
Water-borne primers	
Size	A sticky glaze used to seal in porous surfaces. Traditional size is made of crushed animal bones and is usually used when paperhanging. It can also be a weak solution of wallpaper paste. Mainly water-based but can also be solvent-based.
Emulsion	Can be used as a primer when watered down, mainly for use on fresh plaster where emulsion will be used as the topcoat.

Table 5.15 Primers for use on trowelled finishes and plasterboard

Figure 5.62 Blistering

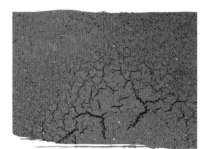

Figure 5.63 Cracking or crazing

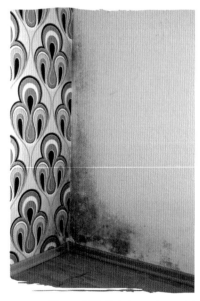

Figure 5.64 Patches of mould

REMOVING PREVIOUSLY APPLIED PAINT AND PAPER

Defects in surfaces

Redecoration may be necessary to replace the existing paint colour or to change the type of wall covering. It may also necessary to remove the paint or wallpaper that is already on a surface to rectify the following defects:

- blistering – where small pockets of trapped air appear, caused by moisture behind the surface, exposing surface to heat or sunlight, or resin seeping out of knots
- cracking or crazing – where paint splits because it is unable to expand to the same degree as the coats underneath it, caused by using incorrect paint systems
- flaking or peeling – where paint splits and lifts away from the surface because of poor adhesion, shrinking or expanding surfaces, corrosion, efflorescence and poor preparation of a surface
- excessive film thickness – where too much paint has been applied to the surface, leaving a poor finish
- mould – where fungus has grown on organic matter in the finish, due to damp, poorly ventilated areas.

Removal processes

Removal of paint systems

If the existing paint is sound, then preparation might only be degreasing and **spot-priming** where necessary.

Where paint surfaces are peeling, flaking, blistering or cracking, then the existing paint will need to be stripped back to the bare surface. Paint can be removed with chemicals (paint stripper), with heat or light.

Liquid paint removal

Paint stripper contains chemicals that soften the old paint ready for it to be scraped off with a stripping knife (scraper) or shave hook.

Regardless of the type of liquid paint stripper used, you must make sure the surface is left free of chemicals before continuing your preparation work. If you don't remove the contamination then your coatings may react with the chemicals and affect your finish. There is also the risk of nasty chemical burns from paint strippers. You must always wash the surface down with warm water and detergent, then rinse.

Solvent-based strippers will remove thick layers of paint from various surfaces, including metals, timbers and masonry. The thicker the paint, the longer the stripper will need to stay on the surface – anywhere from a couple of hours to five days. To stop the stripper from evaporating, it will need to be covered with cling wrap.

There are also water-based paint removers which are much safer to use and more environmentally friendly. However, they can take a long time to work and are not cheap to buy. They can be used on most paints, but take care when using on plastic or timber as they can damage the surface. Always follow the manufacturer's instructions.

Peelable stripper

Peelable paint strippers remove many layers of paint from timber, metal, masonry and trowelled finishes. A layer of paste is applied to the surface, followed by a paper to stop the paste evaporating while it acts on the paint. The paint then adheres to both the paste and the paper and can be peeled off in one go. They are considered more environmentally friendly than solvent-based liquid strippers.

Removing paint with heat

Paint can also be burned off. The extreme heat softens the paint so it can be scraped off with a stripping knife. When working with a large, flat surface, using heat is faster and cheaper. On more delicate surfaces, such as carved woodwork, glass, and flammable surfaces, it is better to use paint stripper.

LPG torch

Liquefied Petroleum Gas (LPG) torches are commonly used for large surfaces with thick layers of paint, but they do present a high fire risk because they use a naked flame. When using your LPG torch, it is essential to have a fire extinguisher nearby. The equipment should be checked for any leaks before starting work. If carrying out any burning off work, you must make sure you are on site for at least an hour after finishing to make sure that there is no smouldering timber. Always keep the area well ventilated and wear appropriate PPE as paint fumes are highly toxic when being burned off.

Hot air stripper

A hot air stripper uses hot air, like a hairdryer, and is safer to use than an LPG torch or gun because it uses electricity rather than a naked flame. It can also be better controlled when working on more intricate surfaces.

Removing paint with light

Infra-red lamps use light to remove several layers of paint very quickly and efficiently. It can be used to remove lead-based paints, unlike heat, because the light heats to a much lower temperature than a flame or hot air gun. Once the paint has softened, it is scraped off as usual. The lamp can then be moved to the next location to work while you are scraping off the first section.

Removal of wall coverings

To remove wallpaper by hand:

1. Score the paper using a scorer or nail block to make small holes in the top coating.

2. Soak the paper with warm water to soften the paper and glue (you may need to do this more than once).

3. Allow the water to soak into the paper. You can tell when it is ready to be peeled when it is mostly dark.

4. Remove the paper using a scraper, applying even pressure so as not to damage the plaster underneath.

If the wallpaper has been painted over, you may find it difficult to remove. In this case, use a long-handled stripper with a blade to cut through the paper, in conjunction with a steam stripper or hot soapy water. It will also take a longer time to soak while the water makes its way through the scoring, behind the paint, and to the paper behind.

Steam strippers

You may also need to use a steam stripper for difficult surfaces, such as those with many layers of wallpaper. Steam strippers boil water and send the steam through a hose to a plate that is placed flat on the wall. Once the plate has been applied, the paper is soft enough to be removed from that section.

PRACTICAL TIP

Be aware that solvent-based paint strippers are highly flammable and toxic. Use away from any sparks or naked flames, wear gloves and make sure there is good ventilation.

DID YOU KNOW?

In some local authorities, burning off with LPG has been banned due to its high fire risks. This would be considered 'hot work' and would require a permit to work.

PRACTICAL TIP

Important: If you are removing lead-based paint, you must not use heat. When working on older or heritage buildings, it is possible that you will come into contact with lead based paints.

Figure 5.65 An infra-red paint remover

Be careful not to hold the plate on the wall for too long as it can damage the plaster behind. Also remember that steam is very hot – take extra care when using a steam stripper on ceilings: do not hold the plate directly above your head, and always tip the plate *away* from your body.

Starting point and soaking time

When soaking off wallpaper, you should identify a starting point for applying the water. This is so that when you get to the scraping off process, all of the papers will have had the same amount of time to soak all the way around the room. Note that the paper may need to be soaked more than once. The longer you leave the papers to soak, the more easily the paper should come away from the wall when you scrape it.

Storing tools and equipment

Your tools and equipment should always be stored in a clean, secure and dry environment. This will protect metallic tools such as scrapers, knives, shavehooks and metal containers from rust.

Preparation tools are expensive, particularly the power tools, keep them locked away when not in use. The damp and wet can damage power tools and electrical cables, making them hazardous to use. If your power tools came with a carry case, it is best to store them inside that.

Heavy equipment should not be stored too high as this could cause a health and safety risk when lifting items in or out.

LPG equipment should be stored outside in a well-ventilated area. LPG is highly flammable and if it gets too hot inside or leaks there can be serious health and safety risks from combustion.

Fire extinguishers should be stored safely, kept in good repair and working order, and must also be clearly identified. When using a naked flame for burning paint off timber, make sure you have the correct red band or black band fire extinguisher (refer to Chapter 1) nearby.

Dust sheets should be shaken out regularly when decorating, then washed at the end of a job. Once clean they should be folded and stored somewhere clean and dry to prevent mildew. Polythene sheets are unlikely to last beyond one or two uses, but if you are storing them, ensure they are clean and dry first.

For the correct storage of brushes see Chapter 6.

RECTIFYING SURFACE CONDITIONS

Surface conditions and rectification processes

Surface condition, defect or cause of unsound paint	Description	How to avoid and/or rectify
Efflorescence (Fig 5.66)	Surfaces that contain lime or cement are known as 'chemically active'. The water-soluble salts come to the surface as they dry out and age, leaving white deposits. These white patches can appear on bricks, render and plaster.	Often it will go away without any treatment, but if the surface needs to be coated, the efflorescence must be removed by dry brushing or scrubbing with a wire brush. It should not be washed because the salts will disappear in the water and only be reabsorbed into the surface.
Moss and lichen (Fig 5.67)	Excessive damp or shade can cause algae (moss or lichen) to grow on exterior walls and roofs. It can not only spoil the appearance of a building but also damage the substrate.	To reduce the chance of algae returning, remove the source of damp or shade if possible. Apply a moss and algae killer and leave it as according to the manufacturer's instructions (usually up to an hour). Then wash down surface to remove both the chemicals and the algae. Treat with a moss prevention solution or primer.
Moulds and fungi	Mould is a type of fungus caused by spores in the atmosphere which grow and feed on organic matter found in various surface finishes. It often grows in damp, poorly ventilated areas, especially in old buildings, in corners, behind furniture or curtains where air flow is poor.	Mould must be properly removed before paint can be applied. Sterilise and wash down the area with a fungicidal solution. Scrape mould off. Wait a week and re-treat if necessary. Coat with paints that contain a fungicide.
Contamination (dirt, grease, silicone, wax polish, carbon/smoke) (Fig 5.68)	Some contaminants will create a thick coating on the surface that would resist the application of new coatings. Smoke can stain and yellow a surface and leave a residue which should be removed.	Moderate dirt and grease is easily removed with a mild degreaser such as detergent or sugar soap. Heavy duty grease may require solvent wiping. Silicone can be cut and scraped or peeled away, and any leftover dissolved with a solvent-based silicone remover. Use acetone products with care as they can also dissolve plastic such as uPVC. Wax polish will come off with either warm water and sugar soap or a solvent. Smoke stains can be removed with warm water and detergent or sugar soap. Discoloration/stains can be removed using a wood bleach.
Friable surface (Fig 5.69)	A friable surface crumbles easily when touched. This condition can be found on old brickwork or rendering. If the surface is not properly prepared, any paint you apply will also crumble.	Brush down with a stiff brush to get rid of any loose debris. Then apply a stabilising solution.

Surface condition, defect or cause of unsound paint	Description	How to avoid and/or rectify
Chalking or powdering (Fig 5.70)	This is where a fine white powder forms on the surface coating. It is caused by weathering, and looks as though the colour is fading, but is actually sitting on top of the coating. It can be the result of poor surface preparation, e.g. not priming using a sealant before painting. The chalking leaves a porous surface, so it must be removed or new coats will not adhere.	Ensure that the right type of paint is applied to the surface, e.g. interior vs exterior paints, and that it is not overthinned or overspread. Remove chalk residue with a stiff brush and water with a mild detergent. Rinse, allow to dry, then test the surface for any further chalk residue.
Saponification	Saponification is a process that produces a foamy soap due to a chemical reaction. It occurs on areas with concrete, brickwork or cement rendering. The alkalinity of the surface will act on the paint coating and essentially strip it back.	Treat with an acid wash and apply an alkali resisting primer (ARP).
Cissing	Cissing is where paint is not continuously joined on a surface. The paint rolls back towards itself and forms beads. Cissing occurs on smooth and shiny surfaces.	To avoid this patchy effect, surfaces must be free of grease, oil, polish or wax. They should be abraded to create a rougher texture. Before coating again, the surface must be degreased and completely dry.
Slow or non-drying surface coating	This might include a bituminous paint or a speciality coating such as anti-climb paint.	Add a barrier by using a different type of solvent on surface, e.g. if using white spirit based paint, use methylated spirit to create barrier. This will put a skin over the top and prevent the stain block bleeding through so you can overcoat with a water- or solvent-based paint.
Bleeding (resin, nicotine, bitumen) and discoloration	The tannins in wood can cause a yellow/brown discoloration when they bleed through the paint film. In particular, if a knot has not been treated properly with knotting solution (shellac), then it may bleed through. Nicotine stains from heavy cigarette smoke can bleed through paint. Bitumen (tar) paints are used for protecting exterior surfaces from weathering. If they are overcoated with another paint, however, the bitumen will bleed through and stain the surface.	Ensure all knots have either been treated with knotting solution, or large resinous knots have been cut out. For nicotine staining, wash down the surface with sugar soap, rinse, dry and prime with an oil-based stain blocking or shellac primer (water-based will allow stains to bleed through) To avoid bitumen bleeding, use a bitumen sealer before applying other paints. Note: you cannot remove bitumen bleeding, only cover it up.
Loss of gloss (Fig 5.71)	Loss of gloss occurs with high gloss paint and varnishes when the paint has been exposed to condensation or cold before it has dried properly. The finish will appear dull and matt. It tends to occur in damp areas such as bathrooms, or in very cold weather.	Avoid painting with oil-based gloss coatings in cold or damp weather. Once the surface has dried completely, abrade with wet and dry paper then re-coat when the conditions are better.

Surface condition, defect or cause of unsound paint	Description	How to avoid and/or rectify
Wrinkling or shrivelling (Fig 5.72)	When the surface of a paint dries too quickly or has been exposed to wet weather before drying, and the layer of paint underneath is still wet, the top layer forms a skin.	Avoid coating too thickly and make sure previous coats are thoroughly dry before overcoating. The paint needs to dry and harden first, then abrade with wet and dry paper. Wash down and rinse the surface before reapplying paint.
Cracking or crazing	Where paint splits because it is unable to expand to the same degree as the coats underneath it. Cracking or crazing is caused by using incorrect paint systems and by painting outdoors in excessive heat or sun.	To avoid, ensure previous layers of paint are completely dry, use the correct paint system (e.g. don't use water-based paints on top of gloss paint), and prime the surface correctly.
Flaking	Where paint splits and lifts away from the surface because of poor adhesion, shrinking or expanding surfaces, corrosion, efflorescence and poor preparation of a surface	Flaking can be avoided by ensuring the surface has been properly prepared, made free of dirt or dust, and the correct paint used. To remove flaking, use a scraper to remove as much a possible, then abrade by hand or if a large surface then use a power tool.
Blistering	Where small pockets of trapped air appear, caused by moisture behind the surface, exposing surface to heat or sunlight, or resin seeping out of knots.	Can be avoided by not applying too much paint at once, making sure the substrate is dry before painting and preparing the surface properly. To make good, abrade surface and reapply paint.
Bittiness (Fig 5.73)	Bittiness is the appearance of dust or grit on or below the surface of the paint. This can happen before painting, where a surface has not been cleaned and dusted off properly, or during painting where debris has attached itself to the wet surface.	Can be avoided by proper preparation, i.e. ensuring the surface is clean before painting, and also by clearing the area of dust and debris before you start painting. Make good by lightly abrading the surface and thoroughly dusting it down before repainting.
Wet rot (Fig 5.74)	Wet rot grows in wood that has been exposed to moisture. It is a brown fungus that eats away at the timber.	The source of moisture must first be removed. Rotten timber should be raked out, dried, then coated with wood preservative and spot primed with wood primer. Insert wood screws into the timber, fill with a two-pack filler (more than once if necessary), then abrade.
Dry rot (Fig 5.75)	Dry rot is a serious form of timber decay. It is a fungus that eats away at the timber for food and causes the wood to lose its strength. It has the appearance of cotton wool. It is also caused by exposure to a high moisture content (of between 18 and 22%) and poor ventilation. These conditions create high humidity which can speed up the growth of the fungus. It is more damaging than wet rot because it spreads further to other timbers.	The source of the moisture must first be removed. An area 300 mm around the rot should be treated. Rotten timber should be cut out and replaced. Wood preservative should be applied to replacement timber and surrounding timber. Adjacent plastered areas should be treated with dry-rot killer before replastering.

Surface condition, defect or cause of unsound paint	Description	How to avoid and/or rectify
Runs, sags or curtains (Fig 5.76)	When working with too much paint on your brush or roller, the force of gravity can take over and leave drips of paint running down the surface. Curtains refer to a line of drips, which give the ragged effect of a curtain edge.	Easily avoided by applying paint carefully and spreading it evenly across the surface. Where runs have already occurred, wash and dry the problem area, abrade it to a smooth surface, then repaint.
Sunlight (denaturing)	Wood that has been exposed to the UV rays in sunlight can become grey and friable.	Brush down with a stiff brush to get rid of any loose debris. Then apply a stabilising solution and primer.

Table 5.16 Surface conditions and defects and how they can be avoided

Figure 5.66 Efflorescence

Figure 5.67 Moss and lichen

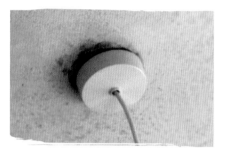

Figure 5.68 Nicotine staining

Figure 5.69 A friable surface

Figure 5.70 Chalking or powdering

Figure 5.71 Loss of gloss

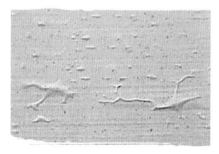

Figure 5.72 Wrinkling or shrivelling

Figure 5.73 Bittiness

Figure 5.74 Wet rot

Figure 5.75 Dry rot

Figure 5.76 Paint runs showing the curtain effect

Unsound paint can also appear on metallic surfaces, both ferrous and non-ferrous, as a result of exposure to:

* grease
* corrosion
* moisture
* pollution
* oxidation
* millscale.

See pages 145–7 of this chapter for more information on metals, corrosion and its causes.

Figure 5.77 Denaturing

Stoppers

See pages 142–3 on the use of stoppers. Table 5.17 below provides a summary of the different types of stoppers available and their uses.

Cleaning agents

Solvent-based cleaners

It is very important to note that where possible you should avoid using pure solvents for tasks such as cleaning. There are other specially designed products available that contain the necessary solvents and are less risky to use. See pages 134–5 for information on methylated spirit, white spirit and acetone, detergents and sugar soap.

PRACTICAL TIP

When washing down you should actually work from the bottom of the surface upwards to prevent streaks and staining.

PRACTICAL TIP

Remember the correct health and safety precautions when working with solvents or solvent-based products.

CASE STUDY

South Tyneside Homes

South Tyneside Council's Housing Company

Preparation is key

Jenny Sibley recently completed her Level 3 apprenticeship in painting and decorating and is now working for the National Trust.

'I am currently working as part of a team refurbishing over 90 Grade II listed cottages on the National Trust's Cotehele Estate in Cornwall.

As part of my job I have to prepare a variety of surfaces, including plaster, woodwork and stonework, for painting. As the cottages we are working on are old, you just don't know what you are going to come across.

So far most of the defects that I have had to make right have been due to damp problems from defective damp proof courses – for example peeling wallpaper, mould, efflorescence and blistering.

Within the team we have people with a good range of experience and skills. For example, we get advice from the stonemason on dealing with damp in stonework to sort out the structure that is letting water in. Once we have removed the cause of the damp we take back the surface until it is sound, apply filler and plaster and use a breathable paint so that the damp can come through rather than get trapped and flake the paint off.'

Products	Uses/properties											Special comments
	Interior	Exterior	Solvent-based	Water-based	Timber	Plaster	Plaster-board	Concrete	Ferrous metal	Non-ferrous metal	Brick/block-work	
Solvent-based wood fillers	x	x	x		x							
Water-based wood fillers	x	x		x	x							Easy to clean up
Linseed oil putty	x	x	x		x					x		Mainly used as glazing putty.
Water-based putty	x			x	x	x					x	
Expanding foam	x	x	x	x			x	x	x	x	x	Good for awkward gaps, e.g. pipe entries. Also for insulating and stopping draughts.
Decorator's caulk	x	x		x	x	x	x		x	x		Use along tops of skirting board, around door/window frames
Silicone	x	x	x						x	x		A good adhesive for non-porous materials such as around sinks and baths.
Lightweight filler	x	x		x	x	x	x					
PVA primer/sealer	x	x		x		x	x				x	
Gyspum plaster	x			x		x	x				x	
Cement plaster	x	x		x		x	x	x			x	
Cement mortar	x	x		x				x			x	
All purpose filler, e.g. Polyfilla	x			x		x	x					Needs to be mixed, therefore harder to use.
Ready mixed filler	x			x		x	x					Easy to use

Table 5.17 Fillers and stoppers, their characteristics and uses

PRACTICAL TASK

1. PREPARE FERROUS METAL SURFACES

OBJECTIVE

To prepare a ferrous metal exterior (e.g. a balcony or girder) ready to receive paint using a rotary sander.

PPE

Ensure you select PPE appropriate to the job and site conditions where you are working. Refer to the PPE section of Chapter 1.

> **PRACTICAL TIP**
>
> This task brings a higher than normal health and safety risk due to dust, noise, surface contamination, trip and slip hazards, therefore signs and barriers should be used. Remember also to place a dust sheet around work area to catch any falling paint and rust.

TOOLS AND EQUIPMENT

Rotary sander

Wire brush (hand tool)

Old paintbrush

Degreaser

Dusting brush and dust pan

Dust sheet

Rubbish bag

Work area signage and barrier system

STEP 1 Remove any surface contamination, such as grease or oil. Brush a coat of grease remover onto the surface, following manufacturer's instructions.

The degreaser you choose will depend on how much contamination there is – for light grease, sugar soap will be fine; for heavy-duty grease you may need a solvent-based proprietary degreaser.

Figure 5.78 Degreasing the surface

STEP 2 Wash the surface with clean water and allow it to dry. If using a water-based product, allow to dry prior to rust removal or abrading.

Figure 5.79 Washing the surface

STEP 3 Check that the rotary sander is fit for use:

- check it is the correct type, i.e. size and voltage
- check the power supply type, i.e. how many volts
- check for any damage evident, e.g. electrical supply cable with cuts through the outer covering
- check for any poor repairs to the sander.

STEP 4 Connect the rotary sander to the mains electricity supply and check that it has the right kind of paper attached.

PRACTICAL TIP

Remember to wear eye, ear and hand protection as well as a dust mask.

Before you turn the sander on, make sure the cable is well out of the way. Power tools can be dangerous if not used correctly – be aware of the risk of cutting through the electrical cable.

STEP 5 Standing on firm ground with your feet apart, hold the sander using both hands: one on the handle, the other on the rest found usually on the top towards the front.

PRACTICAL TIP

If working on a surface that isn't fixed, it is important to make sure it will not move.

Figure 5.80 Using the rotary sander

STEP 6 Pass the sander over the surface, allowing the abrasive paper to make contact with the surface being abraded. Pass over the surface until all corrosion is removed.

PRACTICAL TIP

When working in small or awkward areas, you can use a stripping knife to loosen the rust and then abrade with narrower wire brush.

Figure 5.81 Abrading metal with a wire brush

STEP 7 Use a dust pan and brush to remove any dust, paint and rust particles from the metal surface. Carefully empty the dust sheet of any dust, paint or rust particles from the metal surfaces and place them in a sturdy rubbish bag and dispose of in a responsible manner (e.g. local council waste refuse site).

Re-lay the dust sheet to provide further protection around work area when applying primer, undercoat and topcoat.

PRACTICAL TASK

2. PREPARE NON-FERROUS METAL SURFACES

OBJECTIVE

To prepare a ferrous metal exterior balcony ready to receive paint.

PPE

Ensure you select PPE appropriate to the job and site conditions where you are working. Refer to the PPE section of Chapter 1.

TOOLS AND EQUIPMENT

Old paintbrush

Degreaser

Acid etch primer

STEP 1 Remove any surface contamination, such as grease or oil. Brush a coat of grease remover onto the surface, following manufacturer's instructions.

The degreaser you choose will depend on how much contamination there is – for light grease, sugar soap will be fine; for heavy-duty grease you may need a solvent-based proprietary degreaser. Abrade the surface with aluminium oxide paper.

PRACTICAL TIP

If the material is weathered, it will need to be solvent wiped (see page 140). Use a mordant solution on galvanised metals.

Figure 5.82 Degreasing the surface

STEP 2 Apply an acid etch primer (mordant solution) according to manufacturers' instructions using a paintbrush.

Figure 5.83 Applying an acid etch primer

TEST YOURSELF

1. Which of the following primers can be used on non-ferrous metals?

   a. Red lead

   b. Calcium plumbate

   c. Etch

   d. Zinc chromate

2. Which of the following methods would be used to rectify small cracks or dents in a surface?

   a. Back filling

   b. Flush filling

   c. Stopping

   d. Priming

3. Galvanic corrosion:

   a. is rust stimulative

   b. is rust inhibitive

   c. occurs during the production of metal sheets

   d. can be both rust inhibitive and rust stimulative

4. Which of the following defects is caused if plaster is not left to dry out completely?

   a. Saponification

   b. Efflorescence

   c. Shrinkage cracks

   d. Spalling

5. Which of the following statements about silicon carbide as an abrasive is true?

   a. It is made from strands of silicon that are twisted together

   b. The strands can splinter so you need to wear gloves when using it

   c. It cannot be used on plaster surfaces or bare surfaces

   d. It comes in grades of 200–600

6. Which of the following is a sign of wet rot?

   a. Small holes in the timber and bore dust

   b. Cracks in the timber

   c. A brown or white fungus on the timber

   d. A fungus with the appearance of cotton wool

7. Timber should not be painted if the percentage moisture content of the timber is above:

   a. 5%

   b. 8%

   c. 10%

   d. 20%

8. Blast cleaning is:

   a. a method of removing rust using abrasive particles

   b. a method of removing rust using an oxy-acetylene torch

   c. a method of removing paint using an oxy-acetylene torch

   d. a method of power washing surfaces to remove defects

9. Which of the following are all softwoods?

   a. Oak, Beech, Mahogany

   b. Cedar, Spruce, Beech

   c. Pine, Cedar, Oak

   d. Cedar, Spruce, Pine

10. Which of the following is the correct description of Sa 2 in the Swedish Standard of ferrous metal preparation?

   a. Hand tool cleaning

   b. Near-white blast cleaning

   c. Commercial blast cleaning

   d. Light blast cleaning

Chapter 6

Unit CSA L3Occ 119
APPLY COATINGS BY
BRUSH AND ROLLER

LEARNING OUTCOMES

LO1/2: Know how to and be able to prepare the work area to apply coatings by brush and roller

LO3/4: Know how to and be able to prepare for the application of coatings by brush and roller

LO5/6: Know how to and be able to apply water-borne and solvent-borne coatings by brush and roller

LO7/8: Know how to and be able to clean, maintain and store brushes and rollers

LO9/10: Know how to and be able to store materials

INTRODUCTION

The aims of this chapter are to:

* teach you how to prepare the work area

* help you to apply water-borne and solvent-borne coatings by brush and roller.

PREPARING THE WORK AREA

Whether you are working indoors, in someone's house, or outdoors, on a building site, the area must be prepared for painting. Any items that cannot be removed and stored should be carefully covered to protect them from paint or other damage. The area should also be cleaned and cleared of debris and dust before you begin painting to avoid any contamination to your materials.

Domestic and commercial factors

Working in a domestic setting

Painters and decorators do not always work in new or empty buildings. Often, paintwork is needed in a domestic setting, such as a home, that is still in use. Or you may work in a business setting that still has items in it, such as furniture and carpeting that need to be protected from damage. Look at Fig 6.1 and think about what you should do about each item. You will need to consider: doors and windows; fixtures and fittings; furniture and flooring.

Protecting furnishings and fittings

Furnishings and fittings can be very expensive or even impossible to replace.

Door furniture

Door furniture (handles, finger plates, locks, letterboxes, numbers, knockers, etc.) should be removed. Wrap the parts individually in a protective cover, e.g. newspaper or bubble wrap, then place them in a container or box along with the screws that attached them. Another option is to cover them up with masking tape, but removal is the safest and easiest way to protect them from scratches and paint spatter.

Window furniture

Window furniture (curtains, blinds, pelmets and poles) should also be removed before any work takes place. With curtains, you should carefully remove them from their pole, lay them flat and gently fold them so as to avoid odd-looking creases. Place them in a plastic bag. They should then be stored in a safe, dry place, away from the work area. When removing blinds, retract them (pull them up) first. When removing poles, rails and brackets, be sure to keep all of the small parts together.

Fixtures and fittings

Light switches and power points on walls should be covered with masking tape to avoid dust and paint spatter. Light coverings, such as lamp shades, should be put in a safe, dry place, wrapped or covered up. Large light coverings, that cannot be removed, should be covered with plastic sheeting then taped down. If light fittings have to be removed, then an electrician should do this.

REED TIP

Think about customer service. Taking care of people's belongings will make a good impression on your clients and reflect well on you and your company.

PRACTICAL TIP

If you're removing lots of small pieces of furniture and fittings, it is useful to label the containers or boxes you put them in. This will make it quicker and easier for you to replace them all once the painting is finished.

Ventilation ducts and smoke alarms should be unscrewed and removed, then covered and stored in a safe place. If this is not possible, use masking tape to protect them.

Items on walls, such as mirrors, pictures, shelves and ornaments, should all be carefully removed, individually wrapped with a protective cover, e.g. bubble wrap, then stored somewhere safe, clean and dry. Shelves that cannot be removed (along with their screws and brackets) can be covered with plastic sheeting.

Flooring

If the carpets have not been removed, do not attempt to do this yourself. Instead cover them fully using dust sheets, polythene sheeting (for more waterproof protection), and carefully tape them down. Other surfaces such as boards, laminate or tiling should be covered with polythene sheeting. Rugs that can be rolled up, should be removed and placed somewhere safe, clean and dry.

Furniture

Furniture, such as chairs, tables, sofas and electrical equipment, should be removed from the working area. If this is not possible, then bigger items can be moved into the centre of the room and carefully covered with sheeting. Plastic sheeting will stop any paint seeping through.

Commercial factors

While you are less likely to be dealing with people's personal belongings in a commercial setting, there are still items of considerable value that need to be protected including office furniture, machinery and equipment.

Workstations, equipment and furniture

The workstation area includes all furniture (e.g. chairs, desks, storage cupboards), machinery and equipment (e.g. computers, printers, phones). These items either need to be removed or covered properly with protective sheeting.

Lighting

Use masking tape and polythene sheeting to protect any lights from paint spatter, roller or brush marks.

Climate, weather and temperature

Depending on the environment you are working in, certain precautions may be required to protect your work and your materials. Painting in the wrong conditions can result in a cracked, blistered or flaky finish.

Painting in cold

Painting in cold conditions can slow down the process. First of all, a surface such as plaster might not dry fully. Each coat can take much longer to dry, which in turn means you may not be able to apply the surface coat when you are scheduled to. The wall temperature should be above 10°C. Cold conditions can make the paint itself harder to apply.

Painting in heat or direct sunlight

In hot weather or direct sunlight paint may dry too quickly, which can result in uneven patches, or cracking and peeling, that have to be re-prepared and repainted. Even if it is not a particularly warm day, direct sunlight can heat a surface so that its temperature is much higher than the air temperature. You may need to plan your work around the path of the sun throughout the day, i.e. following the shade by painting the western side of the house in the morning and the eastern side in the afternoon.

Figure 6.1 A typical workstation

Painting in wet and humid conditions

Painting outside in the rain or snow is should be avoided. In other humid conditions, and even indoors such as in a bathroom, the paint may still not dry properly.

Painting outdoors

It is not always possible to avoid painting outside in poor weather conditions. In order to continue work in poor weather conditions, such as wind, rain, hail and snow, you will need to protect the area. This can be achieved by putting up a large tent-like structure, using a frame and some plastic sheeting. This has the added benefit of stopping others from coming into contact with the work.

Public access to premises

You will sometimes be working with products that give off toxic fumes. While is it important to protect yourself from these with your own PPE, you should also remember to protect members of the general public. You could be working on commercial premises that are still being used for business by employees and even customers. Ensure all access points are clear of obstacles and hazards.

Ventilation

Similarly, to protect yourself, your clients and the general public, when applying paint or using solvents, e.g. for thinning or cleaning, there must be somewhere for the fumes to escape. Keep windows open or use extractor fans to help reduce this hazard.

Dust and debris

Before you begin to paint, or even open the tin, it is essential that you clean and tidy the area. Preparation of the surface may have left a lot of dust and debris behind, especially if removing wallpaper or abrading the surface. Dust and debris left around can not only be a trip hazard, but it may also get into the atmosphere and leave you with a poor surface finish, such as bittiness, which you would then have to fix.

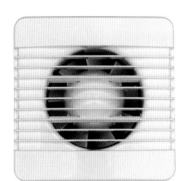

Figure 6.2 An extractor fan

PRACTICAL TIP

Be sure to use decorator's tape (often blue) rather than household masking tape (beige), which tends to tear and pull off the paint below. Each tape has a time limit for the amount of time it should stay on a surface before being removed. If you leave it for longer than the stated time, it can be harder to remove without taking paint away or leaving residue behind.

PRACTICAL TIP

It is best to remove masking tape straight after you have finished painting the surface around it. If you wait until the paint is dry, then you risk peeling back part of the new paint. If you are doing two coats, remove it after the first coat and re-tape before the second coat.

Masking tape

Masking tape comes in a variety of widths and strengths and levels of adhesion for different purposes. It is used to protect surfaces by 'masking' them from the paint, as well as to attach protective sheeting to surfaces.

* Exterior masking tape is designed for use in cold, hot or wet weather. It can be used on more uneven or rough surfaces that you find outdoors, as well as for fixing sheets for protective tenting.

* Interior masking tape is designed for use indoor use only. It tears very easily and has low adhesion, which makes it easy to remove.

* Low-tack tape is used for delicate surfaces, e.g. those that have been recently painted.

* Crepe masking tape is a more stretchy tape, which makes it good for surfaces that aren't straight.

* Seven-day masking tape can remain on a surface for 7 days before the adhesive would potentially cause damage when removed and 14-day tape is also available.

Protective sheeting and masking

Type of protective sheeting	Description and use	Maintenance and storage
Dust sheets	Made of cotton twill, they should be folded over once to make them thicker. Used for protecting furniture and flooring. They come in various sizes, often 4 m × 6 m. They are not waterproof, so will not protect from heavy spills and can become a fire hazard if soaked with flammable materials and not cleaned. Dust sheets come in different thicknesses, from lightweight to heavy duty. They are more costly than polythene sheeting, but they are reusable. For a more water-resistant dust sheet, you can buy cotton sheets with a protective backing.	• Shake out regularly • Wash at the end of the job • Once clean, fold and store somewhere clean and dry so they don't suffer from mildew.
Polythene sheets	They can be used on their own or under dustsheets for extra protection from wet preparation. They are usually cut to fit and taped down. They can also be used outdoors to protect gardens and pathways and come in rolls of various sizes. They are very cheap. Any paint that drips on the sheet will take longer to dry because it is not absorbed and can be hazardous. As they do not hold in place well there is an increased risk of slipping. A less professional option than cotton twill.	• Usually thrown away at the end of a job – a professional decorator might invest in cotton twill and canvas sheeting with protective backing, or double/triple up their dust sheets when working with very messy materials.
Tarpaulin	These are waterproof covers used to protect members of the public from splashes. They can be rubber-coated fabric, canvas and coated nylon. They offer good protection from the weather and are good for use in areas where there are lots of people moving about.	• Once dry, roll it up, otherwise it will get mouldy • Store in a clean, dry place.
Drop sheets	Drop sheets come as both fabric and plastic. They are best used outdoors to protect from rain, as well as paint and dust.	• As for dust sheets.
Corrugated sheeting	A good option for very messy work, they are light and tough waterproof sheets that can be bent or cut to suit the shape of the area. They are more expensive than polythene sheeting, but can be reused. The sheeting is waterproof, but try to keep it dry when in use otherwise it could be a slip hazard.	• Wipe down, roll up and stack the sheets in a clean dry place so they can be reused.
Masking paper	A big roll of plain paper that is used, with masking tape, to cover up large areas when painting. It is particularly useful when using spray equipment because it is more absorbent. It can be applied with a masking machine. Self-adhering types that do not need to be taped are also available.	• Dispose of responsibly after use • Store rolls of paper in a dry place
Masking machines	Hand-held machines that can be loaded with masking film or paper. The dispenser allows you to cover large areas in one long stroke, so they are quicker and easier to use than masking tape and sheeting. There are models of different sizes for light- to heavy-duty use. Replacement dispensers of paper or film are available, so the purchase of the machine is a one-off cost.	• The cutting blade will need to be replaced as it will become blunt through repeated use. • Store in a clean, dry place off the ground as water will affect the adhesion of the tape.

Type of protective sheeting	Description and use	Maintenance and storage
Masking shield or spray shield	A board made of aluminium or plastic, this can be held by its handle, or even with an extension pole, to protect a specific area from overspray when using spray equipment. It is portable and can save a lot of time applying masking around doors and windows.	• The shield should be wiped down before moving to the next area to make sure wet paint is not transferred.
Self-adhesive masking film	Similar to cling film, this self-adhering polythene film will stick to most surfaces, or will need minimal taping. The surface also allows dust and paint overspray to stick to it so that it doesn't float around in the atmosphere and stick to wet paint. It can be applied with a masking machine.	• Dispose of responsibly after use.

Table 6.1 Types of protective sheeting

DID YOU KNOW?

Decorator's masking tape is often blue in colour to make it easier to see where it has been applied, especially when painting light-coloured surfaces. Decorator's tape now comes in other bright colours too.

PRACTICAL TIP

A number of new types of protective sheeting are available, including protection for different surface types (e.g. breathable materials for timber), anti-slip polythene sheeting, chemical resistant sheeting, sheeting made from recycled materials, and flame retardant plastic sheeting. Depending on the type of site you are working on, some of these might be worth further investigation.

Figure 6.3 A masking machine

DID YOU KNOW?

The terms drop sheets and dust sheets are often used to mean the same thing.

Protecting your work and the surrounding area

In addition to the methods of protection already outlined in this chapter on pages 174–8, there are some additional tools and equipment that can be used to protect your work and surrounding area:

Signs

When working in a public place, you should put up a sign to show that you are working. You can buy brightly coloured warning tape for this, or put up a simple 'wet paint' sign.

Barriers

You can use barriers to seal off your working area from other trades who are working around you, or the general public. There are lightweight, foldable workgate barriers that have reflective panels so they can be easily seen.

Other tools and equipment needed for protecting the work area

* Pliers – for removing nails.
* Screwdrivers (including slotted, cross-head and posidriv) – for erecting signs.
* Claw hammer – for removing nails or putting up signs.
* Brushes – for removing dust and debris before you paint.
* Brooms – for clearing up the work area before painting.
* Shovels – use along with brooms to clear up rubbish.

PREPARE FOR THE APPLICATION OF COATINGS BY BRUSH AND ROLLER

Brushes and rollers are the main tools for applying paint. Brush bristles can be made from natural bristle or synthetic fibres (see Chapter 5 on page 134).

Brushes

Brushes are made of different parts: the handle, ferrule, setting, and filling (see the diagram in Fig 6.5).

* Handles can be made of timber and plastic. Certain timber handles are treated and sealed so that water, solvent and paint are not absorbed into them.

* The ferrule joins the handle and the filling. It is often made from plated metals, but also from plastic.

* The setting is an adhesive that sticks the bristles together at the base of the brush under the ferrule.

* The filling is simply the bristles.

Rollers

Rollers are used for applying paint to large, flat areas. They are faster to use than a brush, unless painting small areas, corners or irregular shapes. Rollers are made up of a frame or yoke, which can be single arm or double arm (see Figs 6.5 and 6.6), and a sleeve which slips onto the frame and is removed for easier cleaning. The handle can be attached to an extension pole so the roller can extend your reach so you don't need to move access equipment around as often.

Sleeves can be of the following types:

* Woven fabric – resistant to shedding, needed on smooth surfaces, can be used with all paints, less pressure needed to apply paint

* Mohair – goat's hair, useful for fine finishes on smooth surfaces, for use with both solvent and water based paints

* Sheepskin – higher in density than man-made coverings, they pick up the most paint and so do not need to be loaded as often as other rollers

* Lambswool – the wool is attached to a man-made backing; it is not as good at picking up paint as sheepskin and is often used with solvent-based paints

* Knitted – can be used on medium or rough surfaces, has high capacity for holding paint, gives fastest coverage; use with matt or satin paints, more pressure needed to apply paint

* Foam – do not shed fibres, good for solvent-based paints but not water-based paints as it can leave an uneven finish.

Selecting the right application tools

In general, you can use most tools with most coatings, with just a few exceptions:

* Short pile rollers are good for smooth surfaces and long pile is better for rough or uneven surfaces, as the fibres can get into the small holes.

PRACTICAL TIP

Revisit Chapters 4 and 5, pages 109 and 129, to remind yourself about the need to protect your work and the surrounding area from damage caused by general work activities, contact with the public, and poor weather conditions.

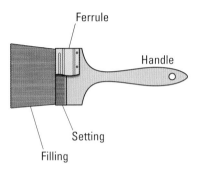

Figure 6.4 Parts of a brush

Figure 6.5 Single arm or cage roller frame

Figure 6.6 Double arm roller frame

Medium pile rollers are a good all-rounder if you are dealing with a variety of surfaces.

* When working on exterior surfaces, such as when painting brick or render, you will be working with both a porous surface and thicker paints. Therefore you will need to use a textured roller with long pile and a coarse wall brush so that you get into all the small holes and cracks.

* When working on interior surfaces, particularly plaster, you should use a synthetic brush or a universal roller sleeve with your water-based paints, stains or varnishes.

* When using textured coatings, you will also need some specialist tools, such as a rubber stippler, a lacer or a texturing comb.

* When painting with oil-based materials inside, it is better to use a mohair roller or a pure natural bristle brush.

* When priming metals you will need a very coarse brush that will carry thick primers such as zinc phosphate.

* Wood treatments, stains, preservatives, varnishes (matt, eggshell and gloss) can be applied using a full-bodied, tightly packed brush with a 100 per cent natural bristle filling. A slightly oval-shaped head of the brush will help when getting into slightly uneven timber surfaces. There are brushes specifically designed for applying stains and varnishes.

Main types of surface coating

The main types of surface coating are:

* water-borne coatings – both for interior and exterior surfaces

* solvent-borne coatings – both for interior and exterior surfaces

* interior and exterior systems for timber, metal, trowelled finishes and plasterboard, e.g. primers

* treatments for wood, such as stains, preservatives and varnishes

* specialist coatings, e.g. flame-retardant coatings and anti-fungus coatings.

These different coatings together make up a **paint system**. The parts of a paint system may include primers, sealants, undercoats, coloured top coats (pigmented), clear (non-pigmented) coatings (such as glaze), and stains. For each substrate, e.g. timber, metal, plaster, there is a particular combination of these that should be applied for a quality finish. Each part of the system has its own characteristics and role to play in preparing and finishing a surface, e.g. sealant is designed to bond with the surface and create key for the other coats to stick to.

Water-borne coatings

Water-borne coatings have improved greatly to provide appropriate alternatives to oil-based paints. In water-borne paints water is the thinner (see below). At one time only oil-based paints could be used on exterior surfaces, but there are now a number of water-based products that are as durable and weather resistant. They are also much faster drying and better for the environment.

Acrylic paints have long been used on interior surfaces, particularly trowelled finishes, as primers, undercoats and top coats. There is a very wide range of products, pigments and finishes (matt, mid sheen, silk, eggshell and even gloss). Acrylic exterior paints now exist for use on masonry (e.g. render, roughcast) and on timber surfaces, including primers, undercoats and topcoats, with finishes from a flat matt to a shiny gloss (see below).

KEY TERMS

Paint system

– a number of different coatings applied to a surface to decorate and protect it.

PRACTICAL TIP

The manufacturer's instructions on the tin will tell you whether the paint is suitable for the surface you're coating and conditions you're working in.

DID YOU KNOW?

Water-borne paints are also known as acrylic paints or latex paints. These three terms are often used interchangeably.

Solvent-borne paint systems

In solvent-borne paints organic solvents are the thinner (see pages 185–6). Traditionally, oil-based paints have been used for outdoor purposes and for surfaces needing better durability and high gloss finishes, such as timber. Sealants, primers, undercoats and top coats can all be solvent-based, and there are different finishes available from matt to gloss (see below).

High and low solid coatings

High solid coatings were introduced to meet the legislation around volatile organic compounds (VOCs) (see page 190). Traditionally, coatings were **low solid** and contained a high percentage of solvents, which are harmful to the environment. In high solid coatings the percentage of solvents has been reduced by about 40 to 50 per cent and replaced by solids.

Interior and exterior systems for timber, metal and trowelled finishes

Interior and exterior paint systems can be both solvent-borne and water-borne and include:

* masonry, brick and stone paints

* wood and metal primers (e.g. galvanising primer, knotting solution, stain sealers)

* specialist finishes e.g. anti-corrosion paint, damp proofing paint. (See pages 182–4 for more information on specialist finishes.)

Wood treatments

Wood treatments were once only available with a solvent base, but there are now more water-based alternatives. Stains, preservative and varnishes are available for both interior and exterior timbers.

Stains

A stain contains natural wood coloured dyes and adds colour to a wood to improve and bring out the appearance of the grain. Stains can be semi **translucent**, which will not cover up any defects, so the wood should be well prepared first. They can also be **opaque.**

Stains are usually designed only to give colour and they do not offer any protection for the wood. However, there are now stains available that also include protective ingredients such as fungicide, so that you can avoid having to apply two separate coatings.

Preservatives

Wood preservatives are coatings used on external timber, e.g. decking, fences and garden furniture. The coating protects the surface from the effects of weather (sunlight and moisture), rust, mould and fungicides.

Note also that some timbers, e.g. those designed for outdoor use such as decking, are already treated with preservatives, so they may only need an oil or stain coating.

Varnishes

A wood varnish is a translucent non-pigmented coating that is a combination of a film-former and thinner (see below). Varnishes come in different finishes (matt, eggshell and gloss). The purpose of varnish is to protect the wood from water and heat. You will normally need to apply more than one coat for it to be effective. Varnishes can be oil or water based. Water-based varnish is very hard wearing and doesn't yellow like oil-based or yacht varnish. Some varnishes contain a stain (colour or tint); others are clear.

Figure 6.7 Applying a stain

Figure 6.8 Varnished wood

Specialist coatings
Flame retardant

Description:	Most paints are flammable if exposed to flames. Flame-retardant paints are designed to slow down the spread of the flames. There are two types of flame retardant paints: Flame-retardant paint – water-borne paint that prevents the spread of flames. It can be used on internal wall surfaces, ceilings or timber. The surface carbonises slightly when it comes into contact with the flame and forms a charred layer. This charred layer is harder to burn and prevents further burning. However, if the source of the flame is continuous it may burn through the paint and burn the substrate. Intumescent paints – water-borne, **high build**, matt coatings generally used on timber and steel but they can also be used on walls and ceilings. When exposed to heat the chemical compounds in the paint react causing the charred layer to expand which insulates the substrate, helping to keep it cool. As a fire progresses, the charred layer gets thicker and provides a better layer of insulation.
Preparation:	Surfaces should be prepared so that they are clean and free from oil and grease. Steel and cast iron surfaces should be prepared so that they are free from rust and millscale and should be primed with a suitable etch primer.
Application:	Flame-retardant paints can be applied by brush, roller or airless spray. Drying time: Flame-retardant paint: 4 hours; Intumescent paint: Approximately 4–6 hours.
Health and safety:	Intumescent paints may contain hazardous materials that cause skin irritation and are harmful if swallowed. Refer to the manufacturer's health and safety data sheet and wear appropriate PPE for both types of paint.

See the practical task on page 207 for step-by-step instructions on how to *Apply flame retardant coatings.*

Anti-fungal paint

Description:	Anti-fungal paints are water-based matt paints that can be used internally (plaster and plasterboard) and externally (cement rendering, roughcast, brick and concrete) to protect surfaces from the growth of fungi, algae and mould that can cause the surface to degrade. Anti-fungus paints contain a biocide – a chemical substance that can kill or reduce the impact of living organisms.
Preparation:	The surface should be prepared so that it is clean and free from oil, grease and dust.
Application:	Apply using a brush, a roller or by spraying. Drying time is about 1 hour. Overcoat after 2–4 hours.
Health and safety:	As the product is semi toxic, PPE (overalls, gloves, goggles and masks) should be worn when applying it, particularly if spraying, to prevent irritation and inhalation and the area should be well ventilated. If using in rooms with low ventilation or in rooms below ground level (e.g. cellars), make sure you wear the correct respiratory protective equipment (RPE).

Floor paint

Description:	One-pack or **two-pack** solvent-based paints, these are used on internal concrete, timber and metal floors (e.g. in entrance halls, schools and warehouses) to give a hard wearing surface.
Preparation:	Abrade previously painted surfaces to remove any loose surface material and to improve adhesion.
Application:	Apply using a brush or long pile roller depending on the area to be covered. Drying time is anywhere from 2 hours if using a two-pack paint to 8 hours if using a solvent-based paint. Overcoat after 24 hours.
Health and safety:	Floor paints contain various chemicals that are flammable and can irritate the skin (e.g. de-aromatised white spirit, xylene, cobalt carboxylate and ethyl methyl ketoxime). They are highly flammable so keep them away from heat and flames. Wear overalls, goggles and gloves to prevent irritation and a mask to prevent inhalation. Ensure the area is well ventilated. If the area is not well ventilated, ensure you wear the correct respiratory protective equipment (RPE).

See the practical task on pages 208–9 for step-by-step instructions on how to *Apply a specialist two-pack floor coating.*

Figure 6.9 Floor paint gives a hard-wearing surface

KEY TERMS

High build

– paints that can be applied more thickly to hide imperfections.

Two-pack

– a paint that is supplied in two parts which need to be mixed together in the correct amounts before application.

PRACTICAL TIP

Anti-fungal additives are sometimes added to paints for use in damp conditions such as bathrooms or outdoor areas.

DID YOU KNOW?

Anti-climb paint is also known as anti-vandal paint.

Micaceous iron oxide (MIO)

Description:	A one-pack or two-pack high build coating for use on iron (e.g. street lights) and steel. The micaceous iron oxide provides a barrier to moisture so the paint has good anti-corrosion properties.
Preparation:	All surfaces should be clean, dry and free from grease, rust and millscale prior to painting. The surface does not need to be primed prior to application.
Application:	It can be applied by roller, brush and conventional spraying and airless spraying. Drying time: 2 hours until touch dry; 16 hours until fully dry. Allow 24 hours before overcoating.
Health and safety:	MIO paint is flammable and may cause irritation so be sure to follow correct health and safety and wear appropriate PPE.

See the practical task on page 209 for step-by-step instructions on how to *Apply micaceous iron oxide paint*.

Anti-climb paint

Description:	A single pack, solvent-based, non-drying paint with a semi-gloss finish that is applied to exterior surfaces, such as drains, railings and fences. It remains slippery after application to stop intruders from being able to climb up the structure it has been applied to.
Preparation:	The surface on which the paint is to be applied should be dry and free from loose material or flaking paint work. High gloss surfaces should be abraded lightly to ensure the paint adheres to the surface.
Application:	The paint can be applied with a short-bristled brush, a trowel or by hand using a protective glove.
Health and safety:	Although the paint is non-toxic, it may cause irritation to the skin and is flammable, so refer to the manufacturer's data sheet and follow the correct health and safety procedures.

DID YOU KNOW?

Pliolite® is another commonly used solvent-borne exterior specialist paint for use on cement rendering, roughcast, brick and concrete. Due to the nature of Pliolite® resin, it can be applied all year round – i.e. at really low temperature (down to –20°C) and in wet conditions as it is showerproof 20 minutes after application.

Self-cleaning paint

Description:	A water-based paint for use on exterior surfaces i.e. cement rendering, roughcast, brick and concrete, it works by providing a really smooth matt finish that dust and dirt can't adhere to. Microscopic particles in the paint also attract rainwater to the surface of the paint so that any dirt is washed away.
Preparation:	The surface should be prepared so that it is clean, dry and free from loose or flaking material and fungal growth prior to application.
Application:	It can be applied using a brush or long-pile roller. Drying time: Around 1–2 hours. Overcoat can be applied after 12 hours.
Health and safety:	Although the paint is non-toxic, it may cause irritation to skin and should not be inhaled, so refer to the manufacturer's data sheet and follow the correct health and safety procedures.

Figure 6.10 Self-cleaning paint reduces the rate at which dirt collects on exterior surfaces

> **DID YOU KNOW?**
>
> In the early to mid 1900s, a great deal of pigment came from lead. Its general use has been banned since 1992.

> **DID YOU KNOW?**
>
> Traditional paints used drying oils but to reduce the long drying time synthetic resins, such as alkyd resins, were introduced.

> **DID YOU KNOW?**
>
> As film formers containing synthetic resins need oxygen to be able to get right through them in order to dry, the film thickness is low to prevent **skinning** and wrinkling.

> **KEY TERMS**
>
> **Skinning**
>
> – when a layer of paint dries out and forms a skin.

Paint finishes

The finish refers to the final coat of paint that is visible on the surface. On interior surfaces, such as walls and ceilings, the finish would usually be matt or silk.

* A matt finish will leave a non-reflective, even surface.

* A silk finish has more of a sheen and can be wiped down, so it tends to be used in wetter areas such as kitchens and bathrooms.

* Eggshell finish is designed to be more durable and is also used in wet areas.

* Gloss finish is most commonly used on woodwork, such as mouldings and doors, because it is particularly durable and resistant to chipping and weathering.

Components of paint

Paint is made up of various parts, depending on the type of paint and its uses. The main ingredients of all paints are:

* pigment (the solid, colour part)
* film former (or binder) (resin which makes paint stick to the surface)
* thinner (water or solvent which disappears as the paint dries).

Pigment
This is the solid part of paint that gives it its colour. Its opacity prevents the colour of the previous paint layers showing through. Pigments can be synthetic (man-made) or natural. Natural pigments include clay, calcium carbonate, talc, mica and silica.

Pigments can be crystalline or non-crystalline (amorphous). In crystalline pigments the atoms in each molecule of pigment are arranged in a structured pattern. In amorphous pigments the atoms are randomly organised. The structure of the pigment can affect colour.

Film former or binder

The film former is also known as the medium, vehicle or binder. It is a liquid **polymer** that hardens to form a film as the paint dries. The purpose of the binder is to:

* make the paint adhere to the surface

* bind the pigment together

* improve the gloss potential, durability, elasticity and strength of the Film formers are categorised by the way in which they dry or are cured to form a film, i.e.:

* oxidation

* solvent evaporation

* chemical reaction.

Further information on the ways in which paint dries (can be found on pages 189–90).

Oxidation

Natural drying oils and **oleoresinous** binders dry in this way by combining with oxygen to form a hard film.

* Natural drying oils commonly used are linseed oil, soya been oil and tung oil as they provide a consistent dry film.

* Oleoresinous binders are a mixture of oil and resins and a solvent. Resins are added to make the film more durable and improve the gloss. If the binder contains more oil than resin it is called a '**long oil**' and if it has less oil than resin it is a '**short oil**'. Oleoresinous binders have been replaced by alkyds and other synthetic resins (to reduce the drying time) but they are still used in lacquers, primers and undercoats.

Solvent evaporation and chemical reaction

Solvent evaporation – resins are dissolved in solvents which evaporate to leave a dry film. Paints that dry in this way include vinyls, rubbers and bitumens. These binders are **non-convertible.**

Chemical reaction – a chemical reaction (oxidation, polymerisation or coalescence – see page 189) occurs within the binder. Resins are dissolved in the solvent, which evaporates as the paint dries, and the resin particles join together to form a protective film. These binders can be **convertible**.

Resins

The resins used in binders can be synthetic (man-made) or natural and the type of resin or polymer used determines the strength of the film:

* synthetic resins – alkyd resin, polyurethane resin, epoxy resin and silicone resin

* natural resins – shellac resin and copal resin

* water-based resins – acrylic, vinyl and polyvinyl acetate (PVA).

Solvent/thinner

Solvents are also called thinners because they dilute or thin the paint. The purpose of the **solvent**/thinner is to dissolve the resin (**solute**) so that the paint (**solution**) is at the right **viscosity** for application. It is the solvent part of paint that evaporates into the air when paint dries. Solvents are highly flammable and have a **volatile flash point** of about 40°C.

Emulsion paints

Emulsion paints are water based so their solvent is water. The paint dries as the water evaporates and leaves behind a layer of pigment.

KEY TERMS

Polymer

– a molecule made from joining together many small molecules called monomers i.e. a film former/binder is made up of polymers.

Oleoresinous

– binders that are made up of a combination of a natural or synthetic **resin**, a drying oil and a solvent to reduce the viscosity of the paint.

Long oil

– where an oleoresinous binder has a ratio of 3–5 parts of oil to 1 part of resin.

Short oil

– where an oleoresinous binder has a ratio of 0.5–1.5 parts of oil to 1 part of resin.

Non-convertible

– binders that do not react chemically and can be re-dissolved once dried using the original solvent (i.e. reversible).

Convertible

– binders that change chemically during drying and cannot be re-dissolved in the original solvent (i.e. non-reversible).

Solvent

– chemical term to describe the liquid in which a solute dissolves.

Solute

– chemical term to describe the substance that dissolves in the solvent.

Solution

– chemical term for the mixture that is formed when a solute has dissolved in a solvent.

Viscosity

– the thickness of a liquid.

Volatile flash point

– the temperature at which a solvent gives off enough vapour to ignite.

KEY TERMS

Biocide

– a chemical substance that can kill or reduce the impact of living organisms. Examples include pesticides, insecticides, fungicides and algicides.

Terebine drier

– a chemical substance that can be added to oil-based paints and varnishes to speed up the drying process by absorbing oxygen.

Solvent-based paints

The pigments in solvent-based paints are dispersed in oil which is dissolved in a solvent. The paint dries as the solvent evaporates, leaving behind the pigment and oil. The oil reacts with oxygen in the air by a process known as oxidation to leave a hard film. Solvents used in solvent-based paints include methylated spirit, alcohols and white spirit.

Extenders

Extenders are used for giving the paint more body or bulk. Adding extenders to paint can make them cheaper to manufacture, easier to apply and slows down the settling of pigment to the bottom of the tin. The disadvantage of extenders is that they reduce the opacity of the paint. Some extenders are added during the manufacturing process; others need to be added to the paint before application.

Dispersant/emulsifier

Dispersant or emulsifier (also known as stabiliser and plasticiser) is added to paint to keep the particles separate and prevent them from clumping together or settling to the bottom. They are also known as plasticisers because they make the paint more elastic and help give a smooth finish.

Driers

Driers are also known as hardeners or catalysts. They are added to some paints to speed up the drying process. In solvent-based paints, driers or hardeners can be added to create a harder or shinier finish, such as in enamel paint.

Additives

Additives include driers, emulsifiers and extenders as described above. Additives are extra ingredients added to the basic components of paint that change the properties of the paint such as the finish, 'spreadability' and drying time. Some paints, such as those for use in bathrooms or outdoors, also contain **biocides** to prevent mould growth and the growth of algae.

Anti-frothing or anti-foaming agent can be added to emulsion paints when they are being applied with a foam roller to stop the paint from getting too much air in it and frothing up when it is rolled onto the surface.

Oil-based paints can contain **terebine driers** – an additive that helps paint to dry in cold, damp or exposed areas.

Activators

Activators (or hardeners) are chemicals added to paint that mix with the paint to complete a chemical reaction, such as in two-pack floor paints.

Preparing surface coatings

Stirring the paint

Most types of paint must be stirred before use, but always check the instructions on the tin. They will tell you whether the paint should be stirred or not, but never shake a tin before opening it. Some paints, such as non-drip gloss, should not be stirred, otherwise their special properties will no longer work. Use a paint stirrer, rather than any old stick, because the small holes will allow the paint to pass through and mix it more quickly.

Decanting the paint

It is rarely a good idea to work directly from the main paint tin. Often they are too big and heavy to move around while you work. They may also get contaminated with debris, or if they tip over, a lot of paint would be lost.

Instead, it is better to decant your paints into a smaller container, such as a paint kettle.

Search and strain

When opening a tin of paint that has already been used, you will need to check the condition of the paint. If a tin of oil paint has been opened before then a skin may have formed. This skin will have caught any dust and debris and can be removed.

If there is any dust or debris in the paint itself, or old dried flakes of paint, then it will need to be strained. You can simply add a strainer to the top of your paint kettle so that the paint passes through when decanting.

Note that not all paints should be strained. Certain primers would be affected by the straining process so that they would not be effective any more. Their special characteristics (e.g. heat resistance or fast drying time) would disappear and they would no longer work as they are designed to. Once again, as with stirring paints, always check the manufacturers' instructions on the tin.

Adjust viscosity

Viscosity means the thickness of the paint. The viscosity of paint can be changed by adding thinner to dilute it: either solvent or water depending on the type of paint. It is this part of the paint that evaporates in the drying process.

The viscosity needs to be at the right level to allow the paint to be applied easily. If a paint has been opened before, it may have changed viscosity since its first use, i.e. some of the solvent or water has evaporated, and will need to be thinned again before use.

Thinning may also be needed to help a porous surface (bare timber, fresh plaster) to better absorb the paint, e.g. adding water to emulsion as a primer for fresh plaster. This helps to seal the surface and prolong the life of the coatings.

Some emulsion paints can be coarse in texture and therefore harder to apply. Paint conditioners can also be added to make the paint go further, to slow down drying times (keeping wet edges open) and to give a better finish. The role of a conditioner is to make the paint easier to apply, to help reduce brush marks and give a smoother finish. They can be added to paints in certain conditions where the workability of the paint might be a problem, e.g. in warm weather or when producing decorative finishes. Conditioners can be expensive, so remember that using a thinner may be enough to achieve a similar effect.

Proprietary conditioners are ready-made conditioners that you would buy off shelf and add to your paints. Non-proprietary conditioners are raw materials that you add to paint to extend drying time and flexibility, e.g. adding linseed oil into a gloss paint.

Coatings and conditions

Thixotropic coatings

Non-drip gloss paints are thixotropic paints. Thixotropic paints have a jelly-like consistency that becomes less viscous when they are stirred or shaken or when they are applied with a brush or roller. They are free-flowing during application but become more viscous once they have been applied. The advantage of thixotropic paints is that they are easy to apply and result in a smooth even finish with a reduced risk of sags or runs. Thixotropic agents are also used in primers, decorative paints and paints for corrosion protection.

REED TIP

Health and safety is important. Make sure everyone in your team is wearing appropriate PPE as necessary. Remind them if they've forgotten.

APPLY WATER-BORNE AND SOLVENT-BORNE COATINGS BY BRUSH AND ROLLER

PPE

See Chapter 1 for more details about PPE. The main items of PPE you will need to protect yourself when applying paint by brush and roller are:

* overalls (to cover your skin)
* safety glasses/goggles (if in contact with solvent-based products)
* gloves (to protect from solvent-based products that may cause dermatitis).

As always, you should check the policy of your own college or workplace and follow the guidelines set on the use of PPE.

Hazards, health and safety and risk assessment

Refer to pages 104–5 of Chapter 4 for information on health and safety including the Work at Height Regulations 2005, COSHH and the importance of following manufacturer's instructions.

The main hazard when working with surface coatings is contact with solvents and solvent-based products. Always keep in mind that they are a fire hazard, as well as being risks to your respiratory and skin health.

Primers

Priming is the very first coat applied to an untreated substrate. Priming helps to seal the substrate and prevent staining or corrosion due to atmospheric conditions. The surface must be properly prepared before the primer is applied, otherwise it will not stick properly. Primers are designed for different types of surface, such as for metal or wood, to give the best result. However, a universal primer is suitable for most jobs. Some primers can also double as the undercoat, i.e. you can apply it once as the primer, then apply it again as the undercoat. Primers should be applied by brush, or with a roller for large flat surfaces such as fresh plaster. For more detail on types of primers, see pages 144–5, 151 and 159.

Other terminology associated with coatings

Platelet pigments	These are found in leafing primers. The pigments are flat instead of round.
Breathability	This refers to the ability of a paint to dry out completely. Contract emulsion allows the surface to dry out completely.
Flexibility	Paints that are 'flexible' have a high resin content that allows movement as surfaces expand and contract.
Latex water-based	They have better water resistance and adhesion. Latex paints don't contain 'latex'. They are paints that use synthetic polymers, such as acrylic and PVA, as binders. The higher the acrylic content the better the quality of the paint. Latex paints are also known as 'emulsions'.
Dispersion	The ability of the paint to allow pigments to flow across the surface.

Drying processes

The drying process of paints, stains and varnishes is influenced by air, light, temperature and moisture.

Atmospheric conditions, such as hot or cold air, air flow (draughts), direct sunlight, darkness, and humidity will affect how quickly and thoroughly a paint will dry. They can also affect the final finish and even cause paint defects such as loss of gloss, skinning and wrinkling. The impact of atmospheric conditions will also depend on the type of coating being used, e.g. some coatings will respond well to heat.

Usually, warm, light and dry conditions will speed up the drying process; cold, damp and dark will slow it down.

There are three ways in which paint dries:

1. **Evaporation:** the thinner (solvent or water) evaporates into the air.

2. Oxidation: a solvent-based paint reacts with the air (oxidation).

3. Chemical reaction: **coalescence/polymerisation**: two paint ingredients are mixed and then solidify on the surface, forming a film.

Water-borne coatings dry by evaporation and coalescence:

*   Evaporation: the water turns into a gas and enters into the air, leaving the remaining parts of the paint to dry on the surface. Drying starts as soon as the paint is exposed to the air and applied to the surface; the thinner the coating, the faster it will dry.

*   Coalescence: after initial evaporation of the water and solvent, additives in the paint fuse together to form a film. This process is sometimes known as 'curing'.

Solvent-borne coatings dry by evaporation, oxidation and polymerisation:

*   Evaporation – the solvent turns into a gas and enters into the air leaving the remaining parts of the paint to dry on the surface.

*   Oxidation – when the film former comes into contact with oxygen in the air, it turns into a solid to form a hard film.

*   Polymerisation – the entire coating becomes a solid as a result of the resin particles joining together to form a film.

Stages of drying

You can tell when paint has dried properly by using these indicators:

*   Flow – if a paint still has 'flow' then it can still be moved by brush or roller and is still wet.

*   Set – once a paint cannot be brushed or rolled any longer, it is said to have 'settled'.

*   Tack – if the coating is still sticky or tacky when you touch it, then it is not completely dry; it cannot be brushed or rolled without ruining the finish.

*   Touch dry – if the paint is smooth and no longer tacky when you touch it, it may be only dry on the very surface; you mustn't overcoat touch dry paint.

*   Hard dry – once the paint has dried all the way through to the previous layer, then overcoating is possible; check the timing recommended on the tin.

*   Through dry – this means that the coating has dried all the way through and is securely stuck to the surface.

KEY TERMS

Evaporation
– where liquid turns to a vapour.

Coalescence
– where particles come together to make a solid.

Polymerisation
– where particles combine to form a polymer (a single film, like plastic or resin).

PRACTICAL TIP

Paints that dry by chemical reaction are non-reversible. Paints that dry by evaporation are reversible (see page 185 above).

PRACTICAL TIP

As paints that dry by chemical reaction are non-reversible, the pot life of the product may only be a matter of seconds or a few days.

PRACTICAL TIP

Solvent-based paints that dry by evaporation are unsuitable for use where chemical resistance is important because the film can re-dissolve in the solvent (i.e. the process is reversible).

Volatile organic compounds (VOCs)

Solvent-based paints are high in VOCs, although these levels have been reduced significantly. VOCs contribute to air pollution and global warming when they evaporate into the atmosphere, so VOCs are now limited by law.

Categories of coatings

There is a classification system for coatings that contain VOCs – these are used on all product labels:

Sub category	Type of coating	Examples of product
A	Interior matt walls and ceilings	Water-borne eggshell, matt, satin, textured and ceiling paints
B	Interior glossy walls and ceilings	Water-borne, silk and satin paints, kitchen and bathroom paints
C	Exterior walls of mineral substrate	Masonry, brick and stone paints
D	Interior/exterior trim and cladding paints for wood and metal	Gloss paint (including non-drip), undercoats, primers, metal paints
E	Interior/exterior trim varnishes and wood stains including opaque wood stains	Preservative wood stains, clear and coloured varnishes and glazes
F	Interior and exterior minimal build wood stains	Wood stain, decking and fence stains or dyes
G	Primers	Wood primers, knotting solution, stain sealers
H	Binding primers	Masonry stabilisers, decking sealer
I	One-pack performance coatings	Anti-corrosion paint, galvanising primers, damp proofing paints
J	Two-pack reactive performance coatings for specific end use such as floors	Floorpaints, etch primers, flame retardant coatings
K	Multi-coloured coatings	Pearlescent effect paints
L	Decorative effect coatings	Scumbles and glazes

Table 6.2 Classification system for coatings

Product labelling

A product labelling system has been created by the coatings industry to show at a glance how high the VOC content is in a particular product. There are five bands from minimal to very high:

* Minimal VOC (0% – 0.29%)
* Low VOC (0.30% – 7.100%)
* Medium VOC (8% – 24.100%)
* High VOC (25% – 50%)
* Very high VOC (more than 50%)

Sequence for painting room areas and components

Ceilings

* Ceilings should be painted first, followed by walls, and then any woodwork or decorative features.

* Cut in around the edges and the light fittings using a brush, then fill in using a roller with an extension pole.

* Ceilings may not be a perfectly flat surface and, because they are large and reflect light, it is better to use a matt paint so that imperfections are not emphasised.

* Do not lay off ceilings in one direction only – use the criss-cross method – otherwise light reflection will show up any unevenness in the surface.

Broad walls

* Cut in along the top of the wall at the ceiling line, into the corners, around window and door frames, along the top of skirting boards, and around any obstacles such as light switches and power points.

* Ensure your cutting in line is wide enough so that you can paint using a roller without touching the adjacent surfaces.

* Be sure to overlap between sections so that the two wet edges of paint join together. The overlap should be about 7–15 cm.

Painting doors

* When painting a panelled door, each section is painted in a specific order so that the wet edges of the paint can be blended together. See Fig 6.11 for the correct order to use.

* Flush doors, i.e. those with a flat, plain surface, should be painted in sections of around 30 cm square. Start painting from the top-right or top-left and then work in stages to keep a wet edge. Lay off vertically with the grain (Fig 6.12).

* Remember with all doors that if the door opens away from you, the hinge edge of the door must be painted too. If it opens towards you, then paint the edge with the latch.

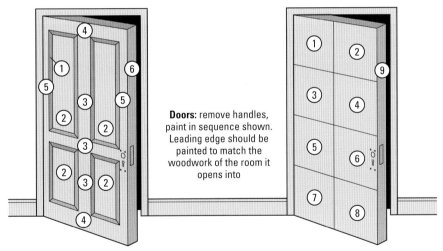

Doors: remove handles, paint in sequence shown. Leading edge should be painted to match the woodwork of the room it opens into

Figure 6.11 The correct order for painting a panelled door

Figure 6.12 The correct order for painting a flush door

Painting casement and sash windows

* Similar to door furniture, remove any fixtures and fittings such as handles and blinds before starting.

* When painting windows, you would usually work from top to bottom, then side to side.

* Always paint the sash bottom rail second to last, finishing with the uprights.

* Choosing the right size brush will depend on the size of the window frame – the bigger the frame, the wider the brush.

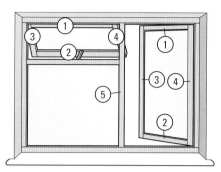

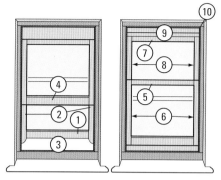

Figure 6.13 The correct order for painting **casement window** frames: paint opening parts before frame and interior sill.

Figure 6.14 The correct order for painting **sash window** frames: from inside open sashes as far as they will go, paint all accessible surfaces, reverse sashes and complete painting

Painting sash windows

* Remove any soft furnishings, such as blinds, and any window fittings, such as catches, before starting.

* Raise the inner sash and lower the outer sash to show the bottom half of the outer sash (Fig 6.14).

* Start by painting the glazing bars/muntins then the bottom rail and finally the side rails (on both sides) on the bottom part of the outer sash.

* Lower the inner sash and raise the outer sash to show the top half of the outer sash (Figure 6.15).

* Finish painting the top part of the outer sash – muntins, top rail and side rails.

* Finally, paint the inner sash (which is fully exposed (Fig 6.16) starting with the muntins, then the top rail, the lower rail and finally the side rails (on both sides).

* Leave the window in the open position until the paint has dried and then paint the frame, leaving the sill until last.

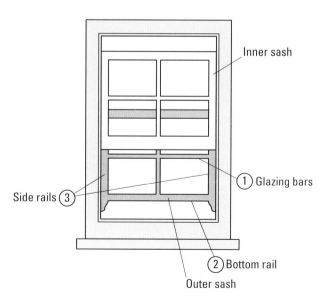

Figure 6.15 The correct order for painting a sash window – painting the top of the outer sash

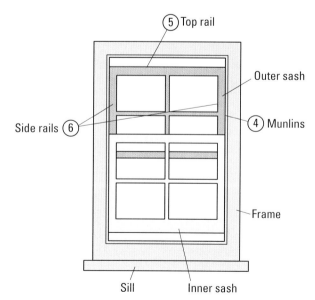

Figure 6.16 The correct order for painting a sash window – painting the bottom of the outer sash and the inner sash

Linear work and decorative mouldings

● Linear work is a term for the painting of long features such as skirting boards, dado rails, architraves and cornices. Decorative mouldings include these, but also plaster mouldings such as ceiling roses.

● It is very important that the paint on these features does not overlap with the surrounding wall, otherwise the eye will be drawn to it.

● These features are usually painted after the main ceiling and wall paint is applied. Painting in clean, straight lines is a skill you will acquire with practice and patience.

Cutting in to features

Where walls join up with ceilings, door and window frames, skirting boards, and around light switches and sockets, you will need to 'cut in' with a brush. This is because it is difficult to reach to the edge of these areas with a roller, or because one of the surfaces will be painted a different colour.

When cutting in, hold your paintbrush like a pencil at the base of the brush, rather than the end of the handle. This will give you more control and help you to stop your muscles getting tired.

Be careful to avoid 'framing' around sockets and doorframes when using a roller – get your roller as close to the cut-in area as possible without touching the feature itself, i.e. when filling in with the roller, there should be some overlap with the area you have painted with the brush.

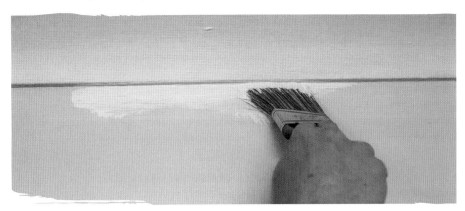

Figure 6.17 Cutting in with an angled brush using a pencil grip

Staircases

● Spindles are the smaller supporting posts under the banister or handrail on a staircase. The balusters are the large supporting posts at the top and bottom of the staircase.

● Use a smaller brush to apply any coatings, particularly if there are small decorative features.

● Begin at the top of the staircase and paint the spindles first, followed by the banisters then the balusters last.

Possible defects after applying paint systems

Defects in your paintwork can occur if:

● paint is not applied carefully and correctly,

● paint is applied in the wrong conditions (e.g. wet weather)

● a surface has not been prepared correctly.

Visible defects

The table below shows some of the visible defects that can occur when you are painting, the cause of each defect and the ways to avoid or fix it.

Defects	Causes	Ways to avoid defect or fix it
Misses – areas where the paint has not been applied (Fig 6.18)	• Not taking enough care • Poor visibility/bad lighting • Undercoat is similar colour to topcoat • Wrong method of application, e.g. wrong roller type used on textured surface	• Paint more carefully, using a method to ensure you don't miss sections • Use enough, proper lighting • Use a different coloured undercoat • Choose the correct tools, e.g. long pile roller • Once dry, apply an extra coat to the whole area
Grinning – the colour underneath the topcoat shows through (Fig 6.19)	• Trying to change colour too drastically, e.g. bright red to yellow • Applying the paint too thinly, e.g. overbrushing or overthinning the paint itself • Using the wrong colour undercoat	• Use more undercoats to mask the original colour • Use more paint and apply it more evenly, and follow instructions for thinning on the tin • Use an undercoat slightly lighter in colour than the topcoat • Apply extra coats as necessary
Runs/sags – paint is running or dripping down the surface (Fig 6.20)	• Paint has been applied too thickly or unevenly • Paint has run or dripped from mouldings, e.g. picture rails • The wet edge has started to dry and new paint applied is not blended in	• Apply evenly and not too thickly • Avoid coating mouldings too heavily • Plan carefully so that wet edges are minimised and not allowed to dry • If the run is still wet, stipple with a brush to remove the run • If it is dry, wait a few days before rubbing it down and repainting
Excessive brushmarks/ropiness – where brush marks can be seen in the paint finish (Fig 6.21)	• Applying paint carelessly, e.g. without laying off • Applying coat too heavily • Applying topcoat to a sloppy undercoat • Applying to undercoat that is not yet dry	• Make sure you apply paint carefully and always lay off • Undercoats should also be applied carefully • Make sure previous layers of paint are dry first • If paintwork has dried, then abrade the surface first, then recoat
Paint on adjacent surfaces (Fig 6.22)	• Careless application of paint • Not using masking tape where needed	• Take extra care when painting near corners and mouldings – don't paint too close • Use masking tape for tricky areas • Scrape off excess, rub down and repaint
Fat edges/wet edge build up – where an extra thick layer of paint forms along edge of painted surface (Fig 6.23)	• Accidental overpainting of a right-angled surface, e.g. a door frame	• Take extra care when painting in corners that will receive more than one coating • Let the fat edge dry, then rub down and repaint
Excessive bits and nibs – where small bits of dust and plaster are trapped under the paint (Fig 6.24)	• Poor preparation – plaster has not been denibbed before painting • An unclean surface, area or tools	• Always prepare the surface fully before painting • Remove all dust and rubbish before painting • Wait until the paint has dried, then de-nib or sand the area and repaint

Defects	Causes	Ways to avoid defect or fix it
Irregular cutting in – an uneven appearance at edges where cutting in meets the main surface (Fig 6.25)	• Careless application when cutting in • Leaving brush marks • Cutting in not in a straight line • Overloading brush • Overreaching • Having a shaky hand	• Take care when cutting in • Use the right amount of paint • Cut in towards the line • If paint touches another surface when cutting in, remove it while wet with a damp rag (water or solvent) • Overcoat the area to hide the problem
Orange peel – where spray or roller coating dries with a textured finish (Fig 6.26)	• Applying paint that is too thick, i.e. needs thinning • Applying paint too thickly • Holding the spray gun too close • Setting the pressure incorrectly	• Thin paint according to instructions on the tin • Reduce amount of paint used • If using spray gun, set it to the right pressure • Give the surface a complete rub down and repaint
Roller edge marks ('tram lines') and roller skid marks – an uneven surface with small, thicker patches of paint or spatter (Fig 6.27)	• Use of wrong roller cover • Paint build-up at ends of roller • Skid marks are caused by applying too much pressure to the roller, making it slide across the surface	• Choose the correct roller cover for the type of surface and paint • Occasionally wipe off the edges of the roller into the tray • Hold the handle firmly, but don't press too hard or force it • Sand down and repaint affected patches

Table 6.3 Visible paint defects, causes and remedies

Figure 6.18 Misses

Figure 6.19 Grinning

Figure 6.20 Runs or sags

Figure 6.21 Ropiness

Figure 6.22 Paint on adjacent surfaces

Figure 6.23 Fat edges

Figure 6.24 Bits and nibs

Figure 6.25 Irregular cutting in

Figure 6.26 Orange peel

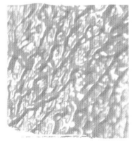

Figure 6.27 Roller edge marks

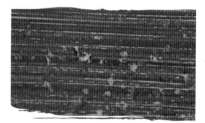

Figure 6.28 Cratering

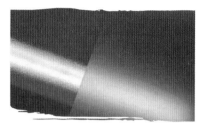

Figure 6.29 Blooming

Figure 6.30 Yellowing

Figure 6.31 Flotation or flooding

Post-application defects

The above paint defects occur while the paint is being applied. You may also come across a range of defects that can occur after the coatings have been applied and have dried. These include the following, covered on pages 163–5 of Chapter 5:

* discoloration
* retarded drying
* bleeding
* loss of gloss
* cracking or crazing
* peeling or flaking
* chalking.

Some further post-application defects are as follows:

* Cratering – this can occur when spots of rain fall on the wet paint, or when condensation or dew forms on the paint while it's still drying.
* Blooming – this is often coupled with loss of gloss and is a white appearance on a glossy surface, caused by applying paint in cold or humid conditions.
* Fading – this is a loss of pigment or colour, caused by exposure to sunlight or weather.
* Yellowing – this affects white paint and occurs when linseed oil or resin based paints are not exposed to light; there are non-yellowing white paints now available.
* Flotation or flooding – this occurs when a paint has been incorrectly mixed or an unsuitable thinner has been used and one of the paint's coloured pigments separates out on the surface. It can be seen as a mottled effect on the surface.

Film thickness

The thickness of coatings is measured using the terms wet film thickness (WFT) and dry film thickness (DFT). Once each layer of coating has been applied to the surface, it is measured using a gauge while still wet to make sure it is the correct thickness. This thickness is measured using microns – one micron is just one millionth of a meter. By measuring the WFT throughout the process, and adjusting the coatings accordingly, the DFT is more likely to be correct.

Once all coats are dry, the DFT is also measured, and if it is not even then additional coatings can be applied.

To measure the thickness, the WFT gauge is pressed into the wet paint. The two outer edges of the gauge will have paint on them as well as some of the other teeth on the gauge. If, for example, the gauge has paint on the teeth marked 4 and 5, but not 6, then the WFT is somewhere between 5 and 6 microns.

Colour systems

Working with colours is part of every painter and decorator's job. The use of colour can affect our experience of a space. Colours can help express someone's personality, bring more light into a room, or even help to calm or stimulate the people in it.

Colour wheel

The colour wheel is a basic system to help us understand, describe, identify, put in order, and mix colours. There are three primary colours: red, yellow and blue. Primary colours are pure colours that cannot be made from any other colour. When each of these is mixed together, they make three more secondary colours:

* red + yellow = orange

* yellow + blue = green

* blue + red = violet.

When secondary colours are mixed, they create a tertiary colour.

* red + violet = red violet

* red + orange = red orange

* and so on.

In total, this makes 12 basic colours or hues (see Fig 6.32).

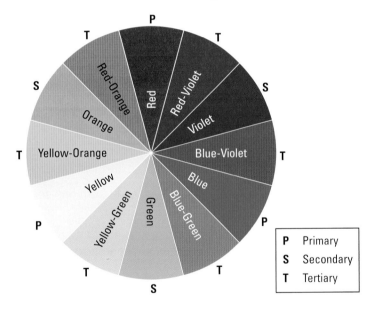

Figure 6.32 The colour wheel

Using colour

The colours that you see in the basic colour wheel above are at their full saturation. This means that they do not have any white or black added to them; they appear at their strongest. When a hue has black added to it, this is called a 'shade' of that colour. When white is added, that hue is called a 'tint'. Look at Fig 6.33 to see the outer shades and the inner tints. The lightness or darkness of a hue is called its 'value'.

Figure 6.34 Monochromatic colours

Figure 6.35 Achromatic colours

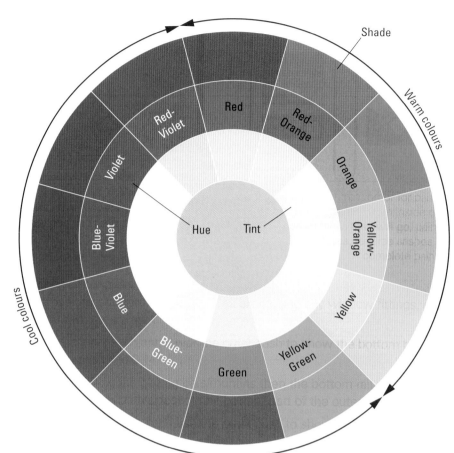

Figure 6.33 Tints and shades in the colour wheel

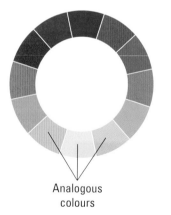

Analogous
colours

Figure 6.36 Analogous colours

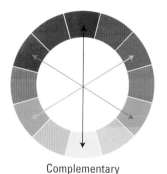

Complementary
colours

Figure 6.37 Complementary colours

The colours from red to yellow are often described as 'warm' colours. These tend to seem closer (advancing) and more stimulating. Whereas colours from violet to yellow-green are thought to be 'cool' colours. These seem further away (receding or retiring) and more relaxing.

Monochromatic colours

Monochromatic colours are where any shade from one section or hue of the colour wheel is used. See Fig 6.34.

Achromatic colours

Achromatic colours are not truly colours but are variations of black and white, i.e. greys. See Fig 6.35.

Fig 6.38 Achromatic colour scheme

Analogous colours

Analogous colours are colours that are next to one another in the colour wheel. One colour is the dominant colour and the others enhance it (see Fig 6.36).

Complementary colours

Complementary colours are those which are opposite to one another on the colour wheel. When mixed in the right proportions they will produce either white or grey (see Fig 6.37).

Hues

Hues or colours, when seen in their purest natural state form a natural scale or order: some are naturally light in hue (e.g. yellow), some are naturally dark (e.g. violet), and others are in between (e.g. green). This natural order of colours has been the basis for many colour theories, such as in Munsell's colour theory, and is clearly seen in the colour wheel. Natural order can be seen in things that occur in nature, such as a sunset or the change of leaves in autumn.

Figure 6.39 Analogous colour scheme

Figure 6.40 Complementary colour scheme

BS 4800: Paint Colours for Building Purposes

The British Standards Framework is a range of 122 paint colours that are specified for use for building and construction work. The system was introduced to bring consistency to the colours used in building materials and finishes.

The Framework categorises colours based on hue, greyness and weight.

Hue

The hue represents colour. The Framework has 12 horizontal hue rows numbered from 02 to 24 and one hue row for neutral colours (i.e. ones without hue) number 00.

Greyness

This is the amount of grey content in the colour. This is divided into five groups labelled A to E with A being having the most greyness and E having no greyness. The vertical rows show the greyness in each colour.

Weight

Colours in each column of the Framework have the same character within each of the hues. The weight indicates the variations in lightness and greyness within the hue (i.e. how strong the colour is) and is given as numbers ranging from 01 to 53. The higher the number the brighter and more intense the colour is.

Figure 6.41 BS4800 colour 04 E 53

Figure 6.42 BS4800 colour 12 B 15

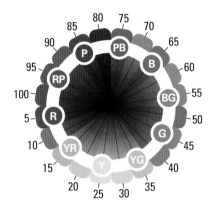

Figure 6.43 The Munsell hue circle

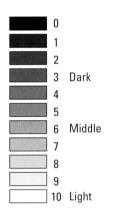

Figure 6.44 The Munsell value scale

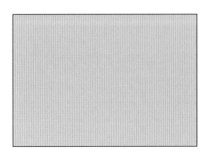

Figure 6.45 Munsell colour 5Y8/10

Colour identification codes

Each colour produced according to BS 4800 has an identification code, for example:

a) 04   E   53

    (Hue)   (Greyness)   (Weight)

Here, the 04 means it is part of the red group of colours; E means it is a very strong colour; 53 makes it a heavy colour.

b) 12   B   15

    (Hue)   (Greyness)   (Weight)

Here, the 12 means it is a green/yellow colour, B means it is quite weak, and 15 makes it very light. Figures 6.41 and 6.42 show what these two colours look like.

Munsell colour system

In the early 1900s, Albert H. Munsell created a system that has since been used on an international scale. Munsell's system formed the basis of the British Standards. His system divides colour into three parts:

1. Hue: the basic colour, i.e. red, yellow, green, blue and purple, and the colours halfway in between each of these, e.g. yellow/red, green/yellow etc.

2. Value: the lightness or darkness of a colour

3. Chroma: the greyness or purity of a colour.

On the Munsell scale, each colour appears evenly spaced to the eye, i.e. each colour is equally different from the next. Hue, value and chroma are measured by a letter or number so that a colour can be more accurately described. Colours appear differently to different people, and using a code to describe the colour means that everyone is referring to the same one.

Hue

The hues are coded as follows:

* yellow = Y
* yellow–red = YR
* red = R
* red–purple = RP
* purple = P
* purple–blue = PB
* blue = B
* blue–green = BG
* green = G
* green–yellow = GY

Each of these hues is divided into 10 sections from 1 to 10 where 5 is the purest version of the hue.

Value

Value is shown on a vertical scale of 0 to 10 where black is 0 and white is 10. Each division on the scale is a shade of grey. See Fig 6.50 below. The value number appears after the hue, e.g. 2R3. This would be a red that is closer to purple–red and is quite dark.

Chroma

Chroma, or a colour's purity, is shown on a horizontal scale where 0 is a neutral grey up to the purest version of the hue which can be as high as 14. Not all hues will have a chroma range this long, e.g. yellows have a greater chroma range than purples because of what the eye can physically see. The chroma number appears after the value, e.g. 5Y8/10 (see Fig 6.45).

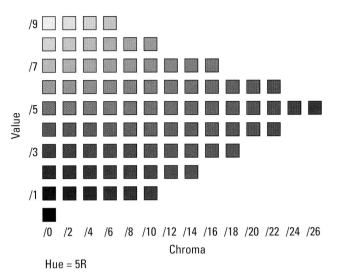

Hue = 5R

Figure 6.46 A branch of the Munsell colour tree

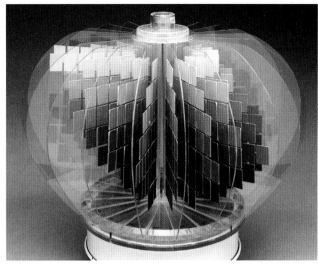

Figure 6.47 The Munsell colour tree

RAL

RAL is another standardised colour notation system used in Europe. It consists of 210 colours, available in matt and gloss, all with their own 4-digit code and name. The colours are often used for regulation warning and traffic signs, government agencies and public services. The first digit refers to the hue:

DID YOU KNOW?

You can visit the Munsell website for more information on the Munsell colour system: *www.munsell.com*

First number	Hue	Number of colours in hue range	Colour sample	
1	yellow	40	Maize yellow – RAL 1006	
2	orange	14	Traffic orange – RAL 2009	
3	red	34	Wine red – RAL 3005	
4	violet	12	Heather violet – RAL 4003	
5	blue	25	Signal blue – RAL 5005	
6	green	36	Pale green – RAL 6021	
7	grey	38	Blue grey – RAL 7031	
8	brown	20	Ochre brown – RAL 8001	
9	white/black	14	Graphite black – RAL 9011	

Table 6.4 RAL colour system

The effects of artificial light on colour

Colours vary when viewed in sunlight and in artificial light because of the way that different types of light and colours interact. As most internal painting will be carried out in artificial light, it is important to take into consideration the source of the light when choosing colours.

Types of artificial light

* Tungsten is a type of **incandescent** light used in normal domestic light bulbs. It contains a filament made of metal tungsten and gives off a warm yellow/amber light that enhance reds, oranges and yellows but make blues and greens more muted.

* Halogen bulbs are a type of incandescent bulb with improved efficiency over tungsten bulbs.

* Compact fluorescent light (CFL) is used in energy-saving domestic bulbs and contain a gas (argon or mercury) that gives off a vapour to produce ultraviolet light. They give off warm white, neutral or blue/violet rays which tend to enhance blues and greens.

* Light emitting diodes (LEDs) – an LED is an electronic device that gives out light when an electric current is passed through it. LED lamps can give out either warm or cool light depending on the type used and can therefore have a range of effects on colour.

* HPS (high-pressure sodium) is used mainly in outdoor lighting (i.e. street lighting) and in industrial premises, they give off a yellow/orange light. They are used when it is not particularly important how a colour looks.

Metameric effect

Metamerism is an effect where two colours that appear to match in one light source don't match in another. This can be the result of how the human eye perceives the colour. All objects absorb some light and reflect the rest. The reflected light is what is seen by the human eye. If the light source changes then our perception of the colour can change. It is therefore important to match a colour in the lighting conditions in which it will be used.

CLEANING, MAINTAINING AND STORING BRUSHES AND ROLLERS

As a painter and decorator, your tools are your livelihood. Treat them with care and respect so they will last a long time and ensure you can provide good quality work.

Methods of cleaning

Cleaning brushes

Brushes should be cleaned or stored in a brush keep immediately at the end of a job. If water-based paint is left to dry, it is no longer water soluble. Don't leave your brushes just sitting in water or solvent at the end of a job. The metal part (ferrule) can rust and the bristles can be pushed out of shape if they're resting on the bottom. It's best to use different brushes for water-based and solvent-based paints.

To clean a brush:

1. Using the back of a knife, carefully scrape any excess paint onto a sheet of paper.

2. Use a sheet or two of newspaper and use long brushstrokes to get rid of as much paint as you can.

3. If using **water-based paint**, wash it under clean, warm running water.

4. Rub a little washing detergent or soap into the bristles and rinse until the water runs clear.

5. Once clean, flick the brush to remove excess water.

6. Gently reshape the bristles and leave the brush to dry.

7. If using **oil-based paint**, use white spirit or a brush cleaning solvent with lower emissions.

8. Pour solvent up to the top of the bristles into a small container, only just bigger than the brush, to reduce the amount of solvent you need.

9. Dip the brush into the solvent, and swirl it around to work the solvent into the bristles.

10. Once the brush is clean, remove excess solvent by flicking the brush.

11. Put a lid on the container to allow the solids to settle.

Workshop vs on-site cleaning of brushes

Most brushes used for oil-based paints are kept in a brush keep (see below). However, brushes used for water-based finishes are cleaned on site, but are taken back to the compound or workshop where there is warm water to remove any dried paint.

Natural bristle brushes

Be gentle when dealing with natural bristle as they can come out of the brush more easily. When rinsing a natural bristle brush, don't use hot water, only lukewarm, as the heat can cause the ferrule to expand and the bristles to fall out. It is better to use natural bristle for oil-based paints. Synthetic brushes are best used for water-based paints.

Cleaning roller sleeves

Never leave rollers or sleeves sitting in water or solvent at the end of a job. The frame can rust and the sleeve can be pushed out of shape.

To clean a roller:

v1. Scrape excess paint into your roller tray using the back of a knife or scraper.

2. Move the roller over the ribbed part of the roller tray to remove excess paint.

3. Run the roller over some sheets of newspaper, cardboard or a rag to get rid of as much paint as you can.

4. Remove the sleeve from the frame.

5. If using **water-based paint**, wash the sleeve under running water and work the paint out with your hands.

PRACTICAL TIP

Remember the environment! Don't dispose of solvents and oil-based paints down the drain, and reduce your water and solvent use by removing as much paint from the brush as possible before rinsing.

PRACTICAL TIP

Don't forget to wear the right PPE when cleaning brushes and rollers in solvent.

DID YOU KNOW?

Cleaning off solvent-based paint releases more VOCs into the atmosphere, and the solvent itself is hard to dispose of safely. To reduce your impact on the environment, choose water-based paints where possible.

PRACTICAL TIP

You can also use a paint roller cleaning tool, which is a small, round or semi-circular plastic device which fits over the roller sleeve, and helps you squeeze the excess paint from your roller back into the roller tray or stock pot. You can then use the tool to squeeze the water and paint from the roller when washing it out.

DID YOU KNOW?

It is better for the environment to use a brush keep instead of cleaning out your brushes with solvent each time.

PRACTICAL TIP

Before storing your brush, try looping a rubber band loosely around the end of your brush to keep the bristles together.

6. Rub some detergent or soap into the pile, making a good lather, then rinse until the water runs clear.

7. Squeeze as much water out of the sleeve as you can and leave it to drip dry.

8. If using oil-based paint, sit the sleeve in some clean white spirit for a short while.

9. Run the roller along a clean roller tray or scuttle.

10. Once all paint is removed, use detergent and warm water to rinse off the white spirit.

If you have used a foam roller for applying oil-based paint, then you may wish to dispose of it rather than clean it.

Storing brushes and rollers

Short-term storage
When taking short breaks from painting, e.g. overnight, it is not necessary to clean your brush or roller. However, you must take care that the paint does not dry up and harden them. Brushes and rollers can be steeped in water or solvent (depending on the paint you're using), but this is not ideal as it can bend the bristles out of shape.

Instead, wrap up your rollers in a plastic bag when leaving them overnight. Similarly, a paintbrush can be wrapped in cling film or tin foil to keep the bristles and paint wet. But remember that the solvent acetone can dissolve some plastics.

For shorter breaks, e.g. lunch breaks, you can leave the roller sleeve fully submerged in the tin of paint that you are using. This is called 'suspension'.

Long-term storage
When storing brushes and rollers between jobs, you can use a storage tub. This will suspend your brushes in water or white spirit without having the bristles resting on the bottom and getting out of shape. The disadvantage of this method is that the water or solvent can drip out when starting your new paintwork, affecting your finish.

There is also a wet storage system, known as a brush keep, so you can avoid having to clean your brushes after each job. The brushes are suspended not in liquid but in vapour so they can be reused straight away for the same colour. The fumes come from a pad or wick that can be topped up with fluid.

Otherwise, if you have cleaned and air-dried your brushes and rollers, they can be gently moulded back into shape and then wrapped in brown paper, lint-free cloth, or the jacket or container they came in. Beware that a pure bristle brush can come under attack from insects and moths if stored for a long time in a moist or damp area.

Environmental and safety considerations

Disposing of contaminated materials
Paints and solvents are often quite toxic and bad for the environment. You should never just tip solvent or paint down a drain. Even water-based paints can damage a drainage system. Instead, you can pour paint-contaminated liquids into a large bucket and leave it out to evaporate. Once the water has disappeared, you can peel off the solids and place them in the bin.

Figure 6.48 A brush keep

When working with paint thinners, i.e. solvents, use a small container to reduce the amount of solvent you need to use. Once it is used and dirty, you can pour it into a larger plastic container, seal it, and allow the solids to settle to the bottom. When this has happened, you can drain off the clean solvent and use it again. Once you have filled a container with paint residue, you can take it to a licensed waste disposal business or local council recycling centre who will charge to dispose of the waste for you.

If you have any solvent-contaminated rags, these should be disposed of in the same way, however, it is very important that you don't leave a solvent-soaked rag bunched up as it can spontaneously combust. Open it out to let it dry before deposing of it responsibly with your other solvent-contaminated waste.

PRACTICAL TIP

The best way of managing contaminated and toxic materials is to avoid using them in the first place. Solvent-based paints can mostly be avoided now that water-based paint technology has greatly improved.

STORING PAINT MATERIALS

Storage conditions

Water based coatings should be stored:

* with clear labels
* on a shelf or rack off the ground
* in a storage area that is free of frost
* in a storage area with a constant temperature
* out of direct sunlight.

Solvent-based coatings should be stored:

* with clear labels
* with their lids tightly sealed
* in a storage area with a constant temperature below 15°C.

Dry powder products, such as sugar soap and powder fillers, should be stored in an environment that is well-ventilated, low humidity, and frost-free.

Ready-mixed fillers need to be kept away from strong light and heat. If exposed they would dry up in the pot, or expand the pot and let moisture content in, and potentially thin the material down. When working with a two-pack product, make sure that the activator is sealed and stored correctly away from the filler.

PRACTICAL TIP

Make sure the rims of your paint tins are clean, otherwise you may not be able to seal them properly.

PRACTICAL TIP

It is always best to use the oldest materials first. Water-based paints in particular have a limited shelf-life. Check the use-by date on the tin and when storing your paints, place the oldest tins at the front of the rack.

Hazards of storing materials

Many of the materials you are working with can be hazardous when stored. Solvents and solvent-based coatings in particular contain volatile organic compounds (VOCs), which means that there is a high risk of fire or explosion, especially if they are stored at too high a temperature.

These materials fall under the COSHH Regulations (see Chapter 1, page 3). Your employer should have assessed the risks, put in place precautions and procedures, and made sure you are aware of these risks and know how to store materials safely.

You will also be working with materials that can be heavy or bulky, e.g. sacks of powdered products or large tubs of paint. Store heavy items close to the ground and avoid awkward or heavy lifting. Remind yourself of the manual handling techniques in Chapter 1, pages 20–1.

DID YOU KNOW?

All hazardous waste must be stored safely and securely. Consider whether it can be reused. Follow the government guidance on disposal of hazardous waste.

Effects of incorrect storage

If water-based paints are stored in a very cold, frosty environment, the water part of them can freeze. This will affect the finish, even once it has defrosted.

If solvent-based paints or varnishes are stored without the lid tightly on, they may suffer from skinning. The paint will form a skin on the top, though this can usually be removed.

Solvent-based paints should also be inverted (turned upside-down) from time to time to stop the contents from separating or **settling**.

Storing paints incorrectly can also cause **fattening** or **livering**. This is where the paint has become so thick that adding a thinner to it does not work. It can sometimes happen when a paint is used past its use-by date. When paint has been stored for too long, the paint ingredients can stick together in lumps. This is called **flocculation**. If you try to use paint that has already flocculated in the tin, then you will get a streaky or patchy finish. Good stock rotation, should help to avoid this.

If dry powders are stored in an environment that has too much moisture, they can become hard or 'set', which would leave them unusable.

CASE STUDY

The customer is king

Kieran St Julian is a self-employed painter and decorator and was a World Skills UK finalist in 2013

'After leaving school I went into landscaping but decided to come back to college to retrain as a painter and decorator. As a landscaper I dealt with many different customers and developed good customer skills.

Good customer service is about:

* Sticking to schedule: I like to give customers a realistic schedule and am flexible to make sure that I stick to it. Sometimes unexpected things happen and you need to go the "extra mile" to get things done on time.

* Customer satisfaction. It is really rewarding when a customer is happy. I get loads of referrals and word of mouth is my biggest form of advertising!

* Good communication: I have an initial meeting with the customer to discuss their requirements. I give them a colour chart to look at, I ask them about the purpose of the room, and I assess the other colour schemes throughout the house. We then discuss and agree a colour scheme and I provide advice if needed. It is important to make sure the client is happy to ensure I meet their expectations.

* Respecting a customer's house is important: I treat their house as I'd expect mine to be treated. I clean and tidy up afterwards and leave it as I found it. I also make sure I protect it really well with sheeting as I don't want to risk paint getting on anything.'

PRACTICAL TASK

1. APPLY FLAME RETARDANT COATINGS

OBJECTIVE

To apply a water-based flame retardant coating to a plasterboard wall and measure the wet film thickness

TOOLS AND EQUIPMENT

Flame retardant coating (water-based)

Paint kettle

Paint roller

Paintbrush (65–100 mm)

Paint roller extension handle

Paint stirrer

Paint tin opener

Low level working platform

Dust sheets

Wet film thickness gauge/micrometer

STEP 1 Prepare the surface ready to receive coatings in the same way as if applying standard water-based coatings.

STEP 2 Decant the flame retardant coating into a paint kettle and cut around the ceiling, skirting and architrave using a paintbrush.

Figure 6.49 Using a paintbrush to cut in around the skirting

STEP 3 Paint the remaining surface area with a roller, ensuring a generous application.

STEP 4 Measure the wet film thickness (WFT) using a micrometer/WFT gauge. Place the micrometer onto the surface and read off the level of the thickness in microns. Check against the manufacturer's instructions to make sure the correct level of film application has been achieved.

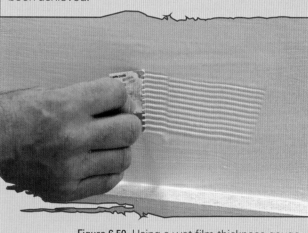

Figure 6.50 Using a wet film thickness gauge

STEP 5 Clean, dry and store your brush and roller in the way that you would for water-based coatings.

PRACTICAL TASK

2. APPLY A SPECIALIST TWO-PACK FLOOR COATING

OBJECTIVE

To apply a two-pack floor coating to a 1 m² floor area.

PPE AND RPE

Ensure you select PPE and RPE appropriate to the job and site conditions where you are working. Refer to the PPE section of Chapter 1.

PRACTICAL TIP

Ensure the room is well ventilated. If the level of ventilation is not sufficient, make sure you wear the correct RPE (mask and ventilator) as the paint is high in VOCs. Also make sure that someone else knows that you are applying a two-pack coating. Even in a well-ventilated room the solvents can still affect you.

TOOLS AND EQUIPMENT

Two-pack coating

Stirring stick

Paintbrush (65–100 mm)

Paint roller

Masking tape

Paint tin opener

Vacuum cleaner, brush and mop

RPE equipment – mask and ventilator

STEP 1 Clean, vacuum and mop the floor to ensure any dust and other surface contaminants are removed.

STEP 2 Protect and prepare the surrounding area using masking tape as necessary.

STEP 3 Open the container for the base material and add the activator to the base material. Stir until both parts are mixed. Leave to stand according to manufacturer's instructions.

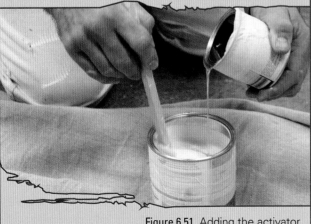

Figure 6.51 Adding the activator

STEP 5 Using a paintbrush, cut in around the skirting – starting at the end furthest from the door.

Figure 6.52 Cutting in to the area being painted

PRACTICAL TIP

Use a good quality but inexpensive brush that can be disposed of after the job.

STEP 6 Using a roller, apply the paint to the remaining surface area.

Figure 6.53 Using a roller to apply paint to the remaining area

STEP 7 Once you have finished, dispose of the paintbrush and roller sleeve correctly according to environmental legislation.

PRACTICAL TIP

If the floor is smooth concrete, the floor needs to be scarified using a mechanical scarifier prior to application to key the surface and avoid lifting.

PRACTICAL TASK

3. APPLY MICACEOUS IRON OXIDE PAINT

OBJECTIVE

To apply micaceous iron oxide (MIO) paint to a ferrous metal surface (e.g. an iron girder).

TOOLS AND EQUIPMENT

MIO coating (water-based)	Paint stirrer
Paint kettle	Paint tin opener
Paint roller	Dust sheets
Paintbrush (65–100 mm)	Wet film thickness gauge/ micrometer

STEP 1 Prepare surface thoroughly and degrease it.

STEP 2 Stir the paint thoroughly as it is highly pigmented.

STEP 3 Apply the paint using a brush and roller ensuring a generous application.

Figure 6.54 Using a brush to apply the paint

STEP 4 Test the wet film thickness using a wet film thickness gauge/micrometer (see page 196).

Figure 6.55 Testing wet film thickness

TEST YOURSELF

1. How do humid conditions affect the application of paint?

   a. They can make the paint more viscous and harder to apply

   b. They can reduce the drying time

   c. They can increase the drying time

   d. They can cause cracking and peeling of the paint's surface

2. If working in an area with little or no ventilation, for which of the following specialist paints would you need to wear RPE?

   a. Micaceous iron oxide

   b. Two-pack floor paint

   c. Intumescent paint

   d. Fire retardant paint

3. Which of the following are all painting defects that would occur after painting?

   a. Grinning; sagging; bits and nibs; fat edges

   b. Blooming; cratering; chalking; orange peel

   c. Ropiness; misses; fading; flotation

   d. Loss of gloss; bleeding; peeling; cratering

4. Which of the following specialist coatings is a flame retardant?

   a. Self-cleaning paint

   b. Bitumen paint

   c. Intumescent paint

   d. Micaceous iron oxide paint

5. Which of the following is the definition of non-convertible binders?

   a. Binders that change chemically during drying and can't be re-dissolved in the original solvent

   b. Binders that don't react chemically and can be re-dissolved in the original solvent once dried

   c. Binders that change chemically and can be re-dissolved in the original solvent once dried

   d. Binders that don't react chemically and can't be re-dissolved in the original solvent

6. Which of the following is the correct sequence for painting a sash window?

   a. Top part of inner sash, bottom part of inner sash, outer sash, window frame

   b. Bottom part of outer sash, top part of outer sash, inner sash, window frame

   c. Window frame, bottom part of inner sash, top part of inner sash, outer sash.

   d. Window frame, bottom part of inner sash, top part of outer sash, window frame

7. What is the name of the drying process in which a solvent turns into a gas and enters the air leaving the film to dry on the surface?

   a. Coalescence

   b. Polymerisation

   c. Oxidation

   d. Evaporation

8. What are analogous colours?

   a. Shade that uses one hue of the colour wheels

   b. Variations of black and white

   c. Complementary colours that lie opposite one another on the colour wheel

   d. Colours that are next to one another on the colour wheel

9. What is loss of gloss?

   a. A mottled effect resulting from poorly mixed paint

   b. Loss of pigment or colour

   c. A white appearance on glossy paint that occurs in cold or humid conditions

   d. A dull matt finish caused by condensation or cold temperatures

10. What does the letter E relate to in the following BS 4800 colour identification code 04 E 53?

    a. Greyness

    b. Weight

    c. Hue

    d. Chroma

Chapter 7

Unit CSA L30cc120
APPLY WALL COVERINGS TO CEILINGS AND WALLS

LEARNING OUTCOMES

LO1/2: Know how to and be able to prepare the work area to apply wall coverings to ceilings and walls

LO3: Know the characteristics of specialised wall coverings and how they are produced

LO4/5: Know how to and be able to select and prepare adhesives for applying wall coverings

LO6/7: Know how to and be able to apply papers and wall coverings to walls and complex surfaces

LO8/9: Know how to and be able to store materials

INTRODUCTION

The aims of this chapter are to:

- help you prepare specialised wall coverings for ceilings and walls
- show you how to apply specialised papers to walls and complex surfaces.

PREPARING THE WORK AREA

Look again at Chapters 4 and 5, pages 109 and 129, on the need to prepare
the work area so that you protect your work and the surrounding area from
damage caused by general work activities, contact with the public and poor
weather conditions. Look also at Chapter 6, pages 174–8, which covers
protecting different types of environments and methods of protection.

Figure 7.1 A Supaglypta® paper

WALL COVERINGS

Wallpapers have been used for centuries as a way of decorating rooms and
also to cover up uneven surfaces, protect surfaces, and even to provide
a degree of insulation. Today, there is a very large range of standard and
specialised papers for different purposes.

Production and printing methods for wallpapers

Standard papers are made from two basic materials: wood pulp or vinyl. There is
more than one process for manufacturing textured wallpaper.

Production methods

Wet embossing

A thick and heavy paper is rolled and moulded into shape while wet between
a steel roller and a roller with a printed design that is coated in rubber. This
produces a heavy wall covering that is sold in panels, designed to look
like brick or stonework, and also sold in rolls of textured paper known as
Supaglypta® (made by Anaglypta®).

Dry embossing

Similar to the wet **embossing** process, paper made from wood pulp is
inserted between two rollers while dry. This creates white, textured papers
which can be painted. The textured finish means that it can cover up small
imperfections in the surface. This technique was patented by the Anaglypta®
company in the late 1800s.

Heat expansion

While heat expansion papers look similar to embossed papers, the process
is different again. Instead of wood pulp, vinyl wallpaper has a layer of PVC
(polyvinyl chloride) added. The PVC layer is heated (to about 190°C) to soften
and expand the PVC. It is then impressed with the design and cooled quickly
to fix the shape. The paper produced is called relief vinyl or blown vinyl.

Laminating

This is the process of bonding a solid vinyl decorative surface to a paper, fabric
or fibre-glass backing. This process is used in vinyl paper papers.

Figure 7.2 An Anaglypta® paper

Printing methods

Printing technology has developed considerably from the time, 500 years ago, when wallpapers were hand-painted after being applied to the wall.

Block printing

Block printing was the first wallpaper printing method used, possibly as far back as the early 1500s. Blocks of wood from fruit trees were carved with a design, covered in ink and then hand-pressed onto paper. More recently, the same process has been developed by using plastic, metal and linoleum 'blocks' to achieve the same effect.

Pins in the corner of the blocks would pierce the paper so that the printer could find the exact end point of the previous section. Papers were then left to hang and dry. Once each layer of colour dried, a new set of block designs and a second colour were printed over the first.

Figure 7.3 A wallpaper printing block

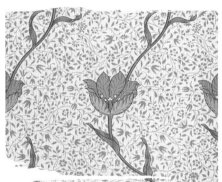

Figure 7.4 An example of block printed wallpaper by designer William Morris

Screen printing

Screen printing began to be used on wallpapers in the late 1940s. A silk or nylon screen of mesh is coated completely in an emulsion or polymer. A stencil of the design is applied to the wet polymer and the uncovered section dries and hardens under a special light. The design is then removed which uncovers still wet polymer which is then rinsed away, leaving the design on the open mesh. To print the design, ink is applied to the screen and then 'squeegeed' through the tiny open holes in the mesh. The screen printing process is time-consuming and requires a lot of labour, therefore these types of wallpapers are expensive.

Figure 7.5 The screen printing process

Machine printing

Most papers today are produced by machine. The first wallpaper printing machine was invented in the late 1700s, and wallpapers became more affordable as they were produced faster. There are different types of machine printing that have been developed including surface printing, gravure and flexographic printing.

Wall covering types

Standard

* Pulps – wood pulp is used to make foundation papers (e.g. lining paper), preparatory papers that receive paint (e.g. wood ingrain, Anaglypta®), or finish papers that can be embossed, washable, and patterned. Lining and preparatory papers are often used to cover up small surface defects before paint or a finish paper is applied. Pulp-based finish papers are commonly used in living areas.

PRACTICAL TIP

Wet printing refers to the more traditional methods of printing on paper such as block printing or roller printing which use wet ink.

PRACTICAL TIP

Dry printing is a more modern process of digital printing, used when photographic images are being transferred to the paper.

213

KEY TERMS

Wide-width vinyl

– vinyl for use in commercial premises that comes in rolls that are 130 cm wide.

Jute

– a natural, rough fibre made from a tropical plant.

Selvedge

– a narrow strip left on the edges of rolls of wall coverings to protect the edges of the roll from damage during production and transportation.

Solid relief

– a flat-backed paper with a raised pattern.

- Embossed – these are decorative papers that have a raised pattern such as Anaglypta® and Supaglypta®. Embossed paper may also have a coloured print, or can be painted after application.

- Simplex – this type of paper has only a single layer. It is a foundation paper used for covering up small defects in the surface or to even up any surface porosity. As a foundation paper, it should either receive a coat of paint or another layer of decorative wallpaper.

- Duplex – this type of paper has two layers which are glued together, then embossed, such as Anaglypta® papers. These can be used on walls or ceilings and may be painted over.

- Ready-pasted – also known as pre-pasted paper, this is made of vinyl or washable paper and has a coating on the back that will become adhesive once soaked in water. The advantage is that they can save pasting time.

- Paste the wall – these papers are applied directly to the wall, which has been pasted a section at a time. It can reduce pasting time and the mess of working on a pasting table. Traditional papers cannot be used in this way because they need to expand during the soaking process. Paste-the-wall paper has a special backing that does not expand. It is also lighter to handle so reduces the chance of tearing.

- Borders – these are narrower strips of wallpaper that are hung horizontally to create a decorative effect around a room. Border papers can be applied on their own on a painted surface, or over the top of other wallpaper using an overlap or border adhesive. You can find self-adhesive border papers, but they are harder to position.

Vinyls

- Fabric and paper-backed vinyl – these papers can be smooth or embossed (by heat) and are made by laminating polyvinyl chloride (PVC) on a fabric or paper backing. Fabric-backed vinyls are good for high traffic areas as they tend to be washable and can cope with ongoing rubbing and handling, and are also good for higher moisture areas such as kitchens and bathrooms. They have two layers so that when redecorating, the top layer can be peeled away from the backing, leaving the bottom layer to be repapered (similar to a lining paper), removed if unsound, or painted over. Vinyls come in a variety of widths from narrow width (53–70 cm) to medium width (70–100 cm) and **wide width** (130 cm).

- Blown vinyl – this is a decorative PVC paper that has been given a raised or relief surface using the heat expansion method. It can look similar to Anaglypta® and be used for the same purpose, or it may be printed with a pattern. The pattern is easily flattened so blown vinyl should not be used in high traffic areas.

Specialist wallcoverings

Type of paper	Description
Hand printed (Fig 7.6)	This was the original way in which wallpapers were printed by block printing. These specialist papers are now mostly produced by silkscreen printing. They have untrimmed edges and need to be ordered specially, which makes them expensive. They are still used in period houses/properties.
Weftless (Fig 7.7)	Natural fibres, such as wool, **jute**, hemp and cotton are dyed and laminated to a paper backing. The product is printed using standard methods after lamination. They give a natural, textured finish. As these wall coverings are made of natural fibres, there will be some shading (see page 229). They are used in low-traffic areas, such as living rooms and dining rooms. Adhesive is applied to the wall rather than the paper.

Type of paper	Description
Lincrusta® (Fig 7.7)	A raised pattern/textured wallcovering made from a paste of linseed oil, cork, resins and pigment on a thick paper backing. It is used for dado panels and wallpaper friezes and requires painting after hanging. Lincrusta® is moisture resistant and is used where a good hard-wearing textured decorative finish is required (e.g. Victorian buildings) or to imitate stone. It comes in rolls (with a **selvedge**), panels or as borders. Lincrusta® is a **solid relief** paper.
Supaglypta® (See Fig 7.1)	A textured **hollow relief** wallpaper made from cotton fibre, rosin size, china clay and alum. It is produced by wet embossing. It is often used in houses, restaurants and pubs to give a decorative finish. Cotton fibres are used instead of wood pulp which produces a more deeply embossed pattern. Its textured finish can hide imperfections in surfaces.
Flocked (Fig 7.8)	One of the oldest types of wallpapers, flocked wallpapers have a velvet effect created by the process of flocking where fibres of wool, silk or synthetic material are applied to an adhesive surface to give the paper a raised surface or pile. Flocked papers can be handmade or machine made. They provide a luxurious, velvet-like textured finish and can be found in restaurants or large houses. Flocked wallcoverings are very delicate and the pile can be crushed easily if stored lying down so they need to be stored on their end. They need to be cleaned with great care when hung.
Hessian (Fig 7.9)	A type of paper-backed fabric wallcovering, it is made from a woven fabric produced from jute. The woven fabric is laminated on to paper. Hessian paper is hard wearing and is good for use in areas of higher traffic, such as offices and other public buildings.
Silk (Fig 7.10)	Fine threads of silk are laminated to a paper backing. It is expensive and is often used in areas of low traffic, such as in restaurants and hotels, to provide a high-quality finish with a sheen. The surface should be prepared thoroughly prior to application as otherwise the defects will be emphasised by the silk's sheen. Avoid getting any adhesive on the paper's surface as it can't be cleaned off and avoid soaking the paper as it will delaminate.
Felt (Fig 7.11)	A type of fabric wallcovering made from pressed wool fibres. The fabric is bonded to a paper backing. Felt wallpaper provides a luxurious finish. As with other fabric wallpapers, avoid using in areas of high traffic to avoid damage.
Metallic foil (Fig 7.12)	Made by laminating a paper backing with a thin sheet of aluminium or by spraying a metallic PVC on to the paper backing. The pattern is printed onto the foil. Sometimes there is a polyester sheet between the paper backing and the foil to prevent water in the adhesive from coming into contact with the foil. Metallic papers can make a room seem bigger as they reflect light and they can also be used in display areas in offices. Care needs to be taken when hanging and storing these papers to avoid damaging the metallic print. Clay-based premixed vinyl adhesive is needed to hang these papers.
Glass fibre (Fig 7.13)	Made from spun fibreglass it has either a coarse, medium or fine woven texture. The nature of the paper makes it flame resistant, mildew resistant and ideal for reinforcing surfaces with imperfections. Glass fibre coverings are extremely durable and are resistant to abrasion. They are therefore suitable for use in high-traffic areas such as hallways. Glass fibre paper is available in wide-width roles. When hanging, the adhesive should be applied to the wall surface rather than the back of the wall covering as the structure of the paper can be affected if it gets wet. Once hung, it is hard to remove.
Mural (Fig 7.14)	A pictorial wall decoration that continues over two or more panels of wall covering and covers most of the wall without a repeat. The scene may be printed digitally or by custom, hand or machine printing. Mural wallpaper is used both in domestic premises (e.g. bedrooms) and commercial buildings (e.g. offices). Durability varies according to the materials used. Careful planning of the layout and joints is needed if hanging murals that cover more than one panel.
Liquid wallpaper (Fig 7.15)	It is made from fabrics and natural products such as silk and cotton. The wallpaper base mixture is mixed with water and then applied to a prepared wall using a trowel. Drying time is 24–48 hours. It can be used on uneven surfaces to cover defects and the finished covering has no joints. It is durable and easy to repair.

Table 7.1 Types of specialist wallcoverings

KEY TERMS

Hollow relief

– a hollow-backed paper with an embossed design. The back of the paper has hollows that match the design.

215

Figure 7.6 Hand-printed wallpaper (with untrimmed edges)

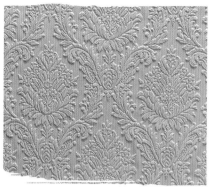

Figure 7.7 Lincrusta®

Figure 7.8 Flocked wallpaper

Figure 7.9 Hessian wallpaper

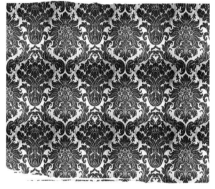

Figure 7.10 Silk wallpaper

Figure 7.11 Felt wallpaper

DID YOU KNOW?

Supaglypta® was invented by Thomas J Palmer, an employee at the Lincrusta® factory. The disadvantage of Lincrusta® was that it was quite rigid and heavy so Palmer invented Supaglypta® – a similar product made from wood pulp and cotton that is lighter and more flexible.

Figure 7.12 Metallic foil wallpaper

Figure 7.13 Glass fibre paper

Figure 7.14 Mural wallpaper

Figure 7.15 Liquid wallpaper

International symbols

Each roll of wallpaper will have a label with a series of symbols. These will give you important information such as whether the paper is washable, how it is to be pasted, how it can be removed, and the type of pattern. Sometimes the explanation of the symbol will appear beside it, but sometimes it will not, so it's helpful to remember what they each mean. Figure 7.16 below shows all the symbols and their meanings.

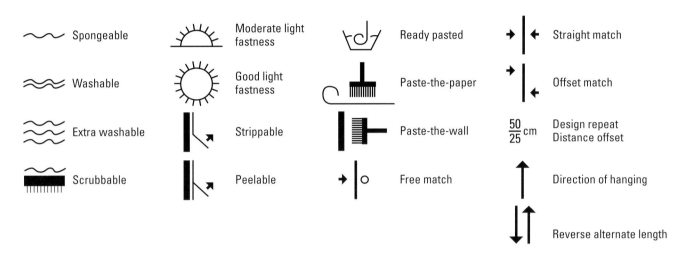

DID YOU KNOW?

Silk is a very delicate material. It will disintegrate if exposed to sunlight for long periods of time and it is also sensitive to artificial light.

Figure 7.16 International wallpaper symbols

Pattern types

* A set or **straight match** has a repeating horizontal pattern (see Fig 7.17).

* A **drop match** or offset match has a repeating pattern but it is not horizontal, i.e. the match occurs at different levels across the paper (see Fig 7.18).

* A random or **free match** has a pattern that does not repeat in a particular sequence, meaning it can be hung without aligning the pattern between drops (see Fig 7.19).

KEY TERMS

Straight match

– also called a 'set pattern', this is a pattern that repeats itself horizontally, i.e. if you place two lengths side by side, the match will be at the same level on each edge.

Drop match

– also called a 'drop pattern' or 'offset match', this is a pattern that does not repeat horizontally, i.e. if you place two lengths side by side, the match will not be at the same level on each edge.

Free match

– also called a 'random match', this is a pattern that does not need to be aligned horizontally when hung.

Figure 7.17 A straight match   **Figure 7.18** A drop match   **Figure 7.19** A free match

SELECTING AND PREPARING ADHESIVES

Types of adhesive

Adhesives (wallpaper paste) are applied to the back of papers to stick them to the surface. Each type of adhesive has advantages and disadvantages, so choose carefully. Table 7.2 compares the main types.

Type of adhesive	Usage	Advantages	Disadvantages
Cellulose paste – has the highest water content of all the adhesives. It comes as a white powder in boxes and sachets and must be mixed with water just before use.	• Tends to be used with lightweight papers. • Prime/size porous surfaces (e.g. new plaster) before application.	• Ease of application • Cheap to buy • Easy to mix • Doesn't tend to stain the wallpaper • Has a long life before application – does not rot • Contains a fungicide to stop mould growth	• Not strong enough to hold heavier papers • Not as adhesive as starch paste • Excess water content can cause paper to expand and distort • High water content can become trapped behind non-breathable papers, causing damage to plaster
Starch paste – also known as 'flour adhesive' or 'cold water paste', it is made from wheat flour. It comes in sachets to be sifted to cold water, then whisked to avoid lumps. It has less water content than cellulose paste, but more than ready-mixed.	• Can be used for lightweight to heavyweight, textured papers • Prime/size porous surfaces (e.g. new plaster) before application.	• Now contains a fungicide to stop mould growth and sometimes preservatives to prolong life	• More expensive than cellulose • Harder to mix • Shorter life – rots after a couple of days, so must be used fresh • Easily stains the wallpaper
Ready-mixed – a PVA based paste that comes pre-prepared in a tub. Particularly good for use on vinyl papers.	• Can be used with lightweight papers if thinned down with water, and heavy weight papers, and vinyls • Prime/size porous surfaces (e.g. new plaster) before application.	• The most adhesive of the three types • Can be pasted directly onto the wall surface for some papers • Contains a fungicide to stop mould growth • Has a long life before application – does not rot	• More expensive than other pastes as it is ready mixed • Comes in larger, heavier packaging than flakes or powders • Can be too thick (hard to apply) and may need to be diluted
Overlap adhesive – used for overlaps around corners etc when working with vinyl paper, and for hanging borders to painted or vinyl papered walls.	• Can be used on external or internal angles • Can also be used for pasting on border papers • Designed to bond vinyl to vinyl.	• Very strong adhesive • Adheres vinyl on vinyl • Can provide extra strength for high-condensation or hot areas	• Some overlap adhesives can stain the wallpaper
PVA – pastes containing polyvinyl acetate are used for extra strength on heavy papers and come pre-prepared in a tube or tub.	• Can be used for heavy fabric papers and vinyl • Prime/size porous surfaces (e.g. new plaster) before application.	• Ready-mixed • Strong adhesive with good long-term adhesion • Saves time as you can use it straight out of the tub	• As for ready-mixed pastes

Type of adhesive	Usage	Advantages	Disadvantages
Multi-purpose – usually starch-based pastes with added fungicide, they can be mixed to different thicknesses or strengths and can therefore be used on a variety of paper types.	• Can be used on many types of paper including heavy-weight pulps and lightweight vinyls • Prime/size porous surfaces (e.g. new plaster) before application.	• Contains fungicide to prevent mould growth	• These tend to be more expensive • Can mark the paper • Difficult to mix • Short shelf-life (1–2 days)
Proprietary (easy strip, light, medium, heavy, sealed surfaces)	• For use with specialist papers e.g. wide-width vinyl.	• Ready-mixed • Formulated for use with a specific product so should have the correct bonding properties for the wallcovering that is being hung	• As for ready-mixed pastes
Lincrusta® glue – a thick, ready-mixed paste made from clay with good adhesive properties.	• Suitable for use with Lincrusta® and other heavy wallpapers. • Prime/size porous surfaces (e.g. new plaster) before application.	• Strong adhesive with good bonding properties	• Difficult to mix and apply • Can stain surfaces if left to dry

Table 7.2 Advantages and disadvantages of adhesive types

Factors affecting the consistency of adhesives

It is always best to follow the paper manufacturer's instructions when choosing your paste. A good rule of thumb is: the heavier the paper, the stronger the paste needs to be.

Adhesives do not always produce the same results when used under different conditions. The more consistent your paste, the better your final finish will be. Here are some factors that will affect how consistent your adhesive is and how well it works:

* Incorrect preparation – e.g. adding too much or not enough water, not leaving paste on paper long enough before applying to wall, not stirring paste enough.

* Paper type – e.g. using cellulose paste for vinyl paper which may not give immediate adhesion.

* Paper weight – e.g. using cellulose paste for a heavy paper such as woodchip.

* Surface – e.g. pasting onto an uneven surface without a lining paper, or where a porous surface absorbs the paste.

* Room/air temperature – this can affect how quickly the paste dries, e.g. a cold damp room will cause paste to dry too slowly, and a hot room will dry the paste out more quickly, leading to dry edges.

* Shelf life – starch paste has a very limited life and shouldn't be used if it has perished.

Figure 7.20 Different types of adhesive

Checking the consistency of paste

If your paste is too thin, it can soak through to the face of the paper and damage it. The paste can also be messier to work with. If your paste is too thick, it will be harder to apply. The weight of the paper you are working with can also affect whether the paste is the right consistency to stick to the surface.

Always follow the manufacturer's instructions when mixing your own paste. It is better to start with too little water, and thin it gradually. But you can't take excess water out of the paste!

You may need to let some pastes stand, perhaps for 10–20 minutes, to help achieve the right consistency. Take care not to add your adhesive to the water too quickly, as this can cause lumps.

Defects caused by incorrect consistency of adhesives

If you do not achieve the correct consistency of adhesive for the paper you are using, you risk ending up with the following defects:

Defect	Description and causes
Blisters (Fig 7.21)	Raised pockets of air called blisters or bubbles can appear on the wallpaper. Blisters can occur if: • the paper has not been properly brushed after application • the wrong paste has been used • patches of wall have been missed or pasted twice • the lining paper underneath has not been applied correctly.
Delamination (Fig 7.22)	The patterned or visible side of the wallpaper has come away from its backing. This can happen if the paste has soaked through the paper too much.
Stretching (Fig 7.23)	If the wallpaper has been moved up the wall after it has been pasted, it can cause horizontal stretching. The paper can also stretch vertically from the weight of the paste if too much has been used, or if its consistency is too thick. The problem with stretching is that it can create gaps between the edges or seams. If the paper has a pattern, then stretching can mean the pattern no longer matches up.

Table 7.3 Defects resulting from incorrect consistency of adhesive

Figure 7.21 Blistering

Figure 7.22 Delamination

Figure 7.23 Stretching

Health and environmental hazards and PPE

Wallpaper adhesives should never be inhaled, ingested (eaten), touched with bare skin for long periods or allowed to get into eyes.

Most wallpaper adhesives now contain a fungicide to reduce the chances of mould forming on or behind the paper. However, these can cause allergic skin rashes or conditions such as eczema.

Like many other decorating materials, leftover wallpaper adhesive should not be released into the ground or into the drains. Even though some pastes are water based, they should still not be disposed of into the water supply, particularly if it contains a fungicide. You can wait for the paste to dry out, then dispose of it as household waste.

As starch paste is made from a natural ingredient, it is not considered harmful to the environment, however, pouring any sort of adhesive down a drain can cause it to be blocked.

APPLYING WALL COVERINGS TO CEILINGS, WALLS AND COMPLEX SURFACES

Suitability of surfaces

It is essential to check the suitability of the substrate on which you are going to hang the covering before you start. For example, a surface might need to be:

* primed/sized – porous surfaces must be sized to prevent adhesives from being absorbed into them and to give a good surface for adhesion

* lined with lining paper to smooth out any imperfections, such as cracks, and to cover any solvent-based paints that have low absorbency to create a more porous surface to which the paint can adhere.

Lining paper

Lining paper is used as a base for applying finishing papers or paint. Lining paper is available in different types and grades and the type you decide to use will be determined by the condition of the wall and type of finishing paper you are hanging:

Type of lining paper	Description and usage
Standard lining paper	• A lightweight paper for use with most standard papers. • Usually off-white in colour but can also be brown or red.
Pitch paper	• A brown paper with bitumen on one side. • Temporarily prevents moisture seeping through.
Scrim-backed paper	• A heavy-duty white paper. • Backed with a cotton scrim. • Use on walls that are likely to expand and contract.
Metal foils	• Thin non-ferrous metal sheets prevent moisture seeping through. • Can be lead or aluminium.
Expanded polystyrene	• Sheets of foam resin. • Improves insulation.

Table 7.4 Different types of lining paper

PRACTICAL TIP
PREPARING ADHESIVE

Ensure you select PPE appropriate to the job and site where you are working. Refer to the PPE section of Chapter 1.

Choose an appropriate adhesive (cellulose, starch or ready-mixed light or heavy) after considering the following:
• the weight of your paper
• whether your paper is breathable
• where the paper will be hung (e.g. in a wet or dry area)
• how long the papering job will take
• whether it needs to be applied to the paper or directly to the wall.

Mix and stir the paste according to the manufacturers' instructions. Adjust the thickness if necessary.

Grade	Usage
800–1000	• For standard walls that are in relatively good condition with minor cracks and imperfections
1200–1400	• Medium to heavyweight paper to give coverage of rough surfaces, e.g. new plaster
1700–2000	• Super heavyweight paper available to cover major imperfections. • More difficult to work with. • Extra time should be allowed for the paste to soak – 12–15 minutes.

Table 7.5 Different grades of lining paper

The type of finishing paper you use may affect whether or not you choose to use a lining paper. For example, when using a heavy vinyl or Anaglypta® there is a risk that the joints may spring open, therefore the use of lining paper helps the joints stay tight. If you were using a light embossed paper or wood chip, then you may not need a lining paper.

When applying a lining paper, you should use the **cross-lining** technique. This is where the lining paper is hung horizontally instead of vertically. It avoids leaving the two layers of paper with the same joins over the top of each other. Generally lining paper is a different width to decorative paper so this is unlikely to occur often, but cross lining is the safer option.

Manufacturers' instructions

Each roll of wallpaper contains manufacturers' instructions on how to prepare and hang the wallcovering to get the best results. For example, they contain information on:

* type of wallcovering – e.g. fabric it is made from, and type of backing e.g. fabric-backed or paper-backed

* priming/lining – i.e. whether the wall needs to be primed/lined and the products/processes to use

* adhesive – the type of adhesive you should use and how to mix and apply it

* soaking time – how long the product needs to be soaked for

* method of cutting from rolls – i.e. cutting in descending order (see page 235)

* hanging advice – e.g. whether the product should be straight or reverse hung, shading checks (see page 235)

* tools and techniques – e.g. tools that should be used for smoothing and methods for jointing

* health and safety.

It is essential to follow the manufacturers' instructions to ensure you follow best practice to get a good finish. Failure to do so may result in defects (see pages 239–40) due to incorrect preparation and hanging or accidents/injuries due to poor health and safety.

Hazards, health and safety

Many of the potential hazards of applying papers are the same as those for preparing surfaces and painting. In particular, you should remind yourself about working at height and be familiar with the Work at Height Regulations 2005 and Control of Substances Hazardous to Health, outlined in Chapter 1.

Tools and equipment

Table 7.6 below describes the tools and equipment you will need for applying wallpaper, what they are used for, and how to care for and maintain them.

Tools and equipment	Correct uses	Care and maintenance
Tape measure (Fig 7.24)	An essential tool for taking measurements. Usually retractable.	Keep clean and free of adhesive which can stop it from retracting properly.
Folding ruler (Fig 7.25)	A metre long, wooden ruler that can be folded away. Used for measuring lengths and widths of an area.	
Plumb bob (Fig 7.26)	A small metallic weight attached to a line of cord, used for checking whether a wall line is vertically straight. A spirit level can also be used.	
Chalk and line (Fig 7.27)	Chalk is used for marking along the plumb line which then guides where to hang your first length of paper. The chalked line is held taut, then plucked so that it springs back against the wall and leaves a chalk mark. It comes in a reel (often 30 m long) and a case.	If the string begins to fray, cut off the affected section and reattach the line to the metal clip. Chalk line cartridges will need to be replaced from time to time.
Spirit level (Fig 7.28)	Can be used instead of a plumb bob or chalk and line for marking lines, both horizontal and vertical.	Can be easily damaged if dropped or struck against things, which will affect the accuracy of the readings. It is possible to get replacement bubbles for some levels.
Laser level (Fig 7.29)	Comes in standalone models, attach to the wall, or as spirit levels with a laser. Projects a vertical or horizontal line for marking drops on walls.	Take good care of this equipment as it is expensive. Follow any manufacturer's guidance. Any damage to the level could affect the readings, which would mean replacing the entire laser level.
Paste brush (Fig 7.30)	A brush used for applying paste to the paper. Can also be used for washing down.	Wash after use with warm, soapy water. Rinse and allow it to dry before storing.
Paste table (Fig 7.31)	A folding table used for cutting, measuring and pasting papers. Also known as a pasteboard.	Keep surface and edges clean and free from adhesive as you go. A dirty surface will affect the paper you are working with.
Sponges (Fig 7.32)	Used for cleaning down your paste table regularly, wiping paste from surfaces such as door frames and skirting boards, and excess paste from washable papers.	Keep your sponge clean itself or it will contaminate the surfaces you are trying to clean.
Buckets (Fig 7.33)	Used for mixing and keeping your paste in.	Clean after use to make sure that adhesive doesn't dry on the inside of it.
Seam roller (Fig 7.34)	Used for rolling down the edges of papers where they join, i.e. at the seams. Should only be used for non-embossed papers.	Keep roller clean and free from adhesive. Lubricate only as needed.
Paperhanging shears (Fig 7.35)	Long-bladed scissors, used for cutting lengths and trimming paper.	Must be kept sharp and clean. When cutting pre-pasted paper, wipe clean after each use. Do not scrub with abrasive paper or it will blunt the blades.
Paperhanging brush (Fig 7.36)	A wide brush of natural or synthetic bristles, used for smoothing air bubbles from paper when applied to wall or ceiling.	Avoid getting paste on the bristles. Keep clean by washing in warm, soapy water after use. Hang to dry.

Tools and equipment	Correct uses	Care and maintenance
Trimming knife (Fig 7.37)	Used for trimming and cutting in tight spaces or at angles. Sometimes with a retractable blade.	Keep knife edge sharp, snap off blade or replace blade as needed. A blunt blade will tear rather than cut your paper.
Caulker or caulking tool (Fig 7.38)	A blade used for smoothing vinyls and some lining papers.	Keep clean of paste – wipe clean after each use.
Spatulas (Fig 7.39)	A flexible plastic tool used for smoothing out air bubbles or wrinkles from wallpaper. Particularly good for use on vinyl papers.	Keep clean of paste – wipe clean after each use.
Metal straight edge (Fig 7.40)	Used in conjunction with a knife as a guide for cutting the top and bottom of wallpapers (edging).	Keep clean of paste – wipe clean after each use.
Ridgley straight edge and trimmer	A machined stainless steel straight edge that contains a guide track. A cutting wheel mounted in a spring-loaded handle is attached to this track. When the handle is pressed the cutting wheel trims the paper. The cutting wheel is then pushed along the straight edge to cut the wall covering to the required size.	Make sure cutting wheel is kept sharp and clean.
Fabric-backed vinyl joint cutter (Fig 7.41)	Used for trimming off the selvedge or cutting joints for fabric-backed vinyls. The joint cutter has a guard to protect the wall as any damage to the substrate could result in joints opening out at a later stage.	Make sure cutting blade is kept sharp and clean.
Cotton gloves	Used when applying wallpaper adhesive containing fungicide to prevent skin irritation.	Wash thoroughly to remove any contaminants.
Rubber roller (Fig 7.42)	A solid rubber roller on a roller arm. Used to roll out air pockets when hanging vinyl papers and murals.	Keep roller clean and free from adhesive. Lubricate only as needed.
Felt roller (Fig 7.43)	A roller head made of a number of felt discs. Used instead of a paperhanging brush when applying delicate wall coverings.	Keep roller clean and free from adhesive. Lubricate only as needed.
Protective strip (plastic for paper-backed wide-width vinyls) (Fig 7.44)	For use as protection behind the overlap when cutting the joint. Usually used with paper-backed vinyl.	Keep plastic strip clean and free from adhesive.

Table 7.6 Tools and equipment for applying papers

Figure 7.24 Tape measure   Figure 7.25 Folding ruler

Figure 7.26 Plumb bob

Figure 7.27 Chalk and line

Figure 7.28 Spirit level   Figure 7.29 Laser level   Figure 7.30 Paste brush   Figure 7.31 Paste table

Figure 7.32 Sponges   Figure 7.33 Buckets   Figure 7.34 Seam roller   Figure 7.35 Paperhanging shears

Figure 7.36 Paperhanging brush   Figure 7.37 Trimming knife   Figure 7.38 Caulking tool   Figure 7.39 Spatula

Figure 7.40 Metal straight edge   Figure 7.41 Fabric-backed vinyl joint cutter   Figure 7.42 Rubber roller

Planning the work

Before you cut any paper or apply it to the surface, there are several things you need to check and consider. Unless you plan ahead, you may find yourself in the middle of a job using the wrong materials, having to buy extra supplies, or having to start over.

Open up the roll to:

* read the manufacturers' instructions
* check for any damage to the paper
* see if the batch numbers are all the same
* check for any colour variance (shading) between the rolls
* look at the pattern and note the type of match
* work out which way up the paper should go on the wall.

Figure 7.43 Felt roller

Figure 7.44 Protective strip

Starting and finishing point

You should always choose a starting and finishing point for your wallpapers so that the pattern matches up and that any mismatches are not too visible. Your starting point should be marked with a plumb bob and line (or spirit level or laser level) from the top of the wall, either marking with chalk or pencil.

When working with plain, unpatterned papers, it is usual to start papering next to the natural light source, i.e. the window. Use your marker line to line up the edge of the paper. You should then work around the room in a direction away from that light source so that you don't cast a shadow over the area you're papering and to avoid the light reflecting off any of the joints, making them more obvious.

Remember to plumb each wall when you turn a corner.

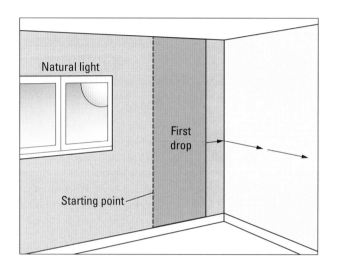

| Figure 7.45 First drop | Figure 7.46 Centralised first drop |

KEY TERMS

Focal wall

– the wall that your eye naturally goes to upon walking into a room.

PRACTICAL TIP

When cutting around obstacles with wood ingrain paper, you cannot use a craft knife, but must instead use scissors. It is very hard to cut through the chips of wood and a knife can cause the paper to tear.

Centring

When using a patterned paper on a feature such as a chimney breast (or the **focal wall**), then the paper should be aligned in the very centre of the feature or focal wall rather than to one side of your marker (see Fig 7.46). However, this may depend on the width of your focal wall: if the paper edges would end up too close to the corners, you may need to adjust your starting point. A large pattern should be centralised with a full motif appearing at the top of the wall.

Walls, features and obstacles

Not all rooms are perfectly square and flat. Some rooms will be of different shapes, or have features such as chimney breasts, staircases, window reveals, windows and doors. Attics and loft rooms will have sloping ceilings and dormer windows. Most rooms will have small obstacles such as light switches, power points and ceiling roses. These shapes, features and obstacles are not a problem, but need to be considered when planning and applying the paper.

Alcoves, niches and reveals (internal and external angles)

If you were to keep papering around a corner without trimming it first, you would end up with wrinkles and the next drop would not necessarily be straight. Also, if a corner is out of plumb this could mean the edges would not meet. You will need to cut one of your lengths in two and then rejoin them on the wall.

Borders

Some wallpaper borders can be applied at ceiling height or at dado height. In both situations, if you are applying the border to a raised pattern (such as an Anaglypta® paper), then a flat border needs to be planned because some border papers will not stick over the top of a raised pattern wallpaper.

Staircases

Staircases have awkward corners and long drops to cope with. It is therefore essential to erect a safe working platform that allows you to access all areas and suits the layout of the stairwell. There is a lot of wastage when papering stairwells, especially with patterned paper, so it is essential to calculate how much you need to make sure you have enough and to avoid any unnecessary wastage. Divide downstairs and upstairs into two areas as shown in Fig 7.47 by imagining that the skirting upstairs goes all the way across the stairwell and measure the dimensions as normal (see below). Start preparing the surface by removing handrails and any other wall-mounted obstacles. Before hanging, plan your starting point. This should be the longest drop of wall covering.

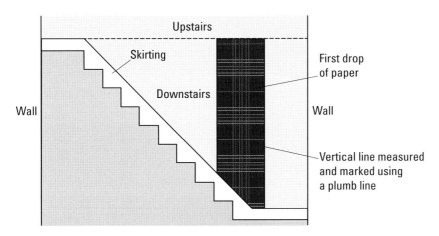

Figure 7.47 Starting points for wallpapering a staircase

Free-standing column pillar

When papering a free-standing column or pillar you need to allow for an even number of drops. The **circumference** of the column should be measured and the number of drops calculated, measured and trimmed (allowing a 50mm overlap for each drop).

Door frames

When hanging paper on the second side of a door, it is essential to use a plumb bob or spirit level to make sure the frame is vertical as you won't have an already hung drop to guide you.

Calculating the quantity of paper needed

There are two ways of calculating how much paper you will need to cover an area. It is important to work this out so that there is not too much excess or wastage or alternatively so that you do not run out of paper once you've started. Don't forget that, when working with either method, you will need to allow for more paper when using patterned papers with a drop match.

Figure 7.48 Checking alignment after a door frame

KEY TERMS

Circumference
– the distance around the outside of an object

Reverse hanging
– hanging alternate lengths in the opposite direction to mask slight variations in colour.

Shading
– variations in colour between different rolls of paper of the same pattern/colour.

Girthing method

This way of measuring out the room uses the width (girth) of the wallpaper roll itself. Using a roll of paper, mark around the room where each width ends. Then measure the height of the room to see how many drops you can cut from each roll. This is a quick and easy method but may lead to inaccuracies in calculations meaning you could run out of paper. You may wish to allow for an extra roll of paper in case of any miscalculations.

Area method

The area method is more precise as it uses the actual dimensions of the entire room, taking into account any areas that will not be papered, such as doors and windows.

Each wall must be measured separately, or otherwise taken from a drawing. The area of each wall or ceiling is calculated by multiplying the width by the height, including any doors or windows. Next, the area of each non-papered surface should have its area calculated. This amount should be subtracted from the total area of the room.

Once you have the total area that needs papering, you must find out the surface area of each roll of paper you are using. Divide this amount into the total surface area for the room.

Note that there can be a quite a bit of paper wastage in the wallpapering process. It is best to allow for this by adding an extra 15 to 20 per cent to the amount of paper you bring to complete a job.

Cutting papers

Cutting considerations

It is very important when hanging wallpaper that each length has been cut correctly. The following are factors to consider when cutting:

Pattern type

If you are working with a bold pattern that has a prominent repeat, or a small or indefinite pattern, then you will need to take extra care to match each length when cutting on the table. You will also need to choose the best point for the pattern to start at the top of the wall.

Pattern match (set/straight, offset/drop, random/free match, reverse)

- If you are working with a straight match, then you can work from the same roll of paper.

- If it is a drop match, you will need to cut lengths from two different rolls side by side. This will make less waste than cutting from just one roll.

- A random/free match design doesn't need to be matched so you can work from the same roll of paper. However, to minimise visual effects such as shading, alternate lengths should be reverse hung.

Batches

A batch is the set of rolls that are produced from one print run. Each batch will have its own number so that, when you're buying several rolls for a job, you can make sure they come from the same batch and should therefore have exactly the same colour and pattern.

Most wide-width vinyls are made in large batches. As a result of the manufacturing process, there may be subtle differences in colour between rolls at the beginning and end of a batch. Manufacturers number each roll and suggest that the wall covering is hung in that order to minimise the risk of shading between rolls.

Wastage

Some wallpapers are very expensive, so you will need to make the most of them and avoid cutting off more than you need to. Straight or set match papers will have frequently repeating patterns, so you will not find much paper wastage when cutting. However, drop match patterns may have very large gaps between the repeats. The bigger this gap, the more wastage there will be. As mentioned, working from two rolls will reduce this wastage.

Shading

Each roll of wallpaper printed will have a batch or shade number code which shows whether the rolls were part of the same print run. Sometimes the wallpaper rolls can have different shades depending on whether they were printed at the start or end of the print run. If there is a difference of shade between rolls in the same batch number, then you should not use the paper and return it. If there is a difference in shade from one edge of the paper to the other, you may need to reverse every second length before pasting so that the same shades match up edge to edge. Some types of paper will tell you to reverse alternate lengths, and in this case you should follow the manufacturer's instructions.

The same printing defect can happen with embossed papers where the level of the indentation differs from one part of the paper to another.

Even very small differences in shading or colour would be obvious on a wall, therefore you should check the shading of each roll before cutting.

Figure 7.49 A star cut

Cutting methods

* Star and half star cuts – when you come across an obstacle such as a round light fitting, a pipe or a ceiling rose, a series of small triangular cuts is made from the centre of the paper to form a star shape. These can then be trimmed and the edges smoothed under the feature. A half star cut would be used if only half the length of paper is covering the feature.

* **Mitre cuts** – when applying border papers around a room, you may need to paper down staircases or around door frames. To make sure there is no gap in the paper when you change angle or encounter an obstacle, the border paper should be cut at a 45° angle called a mitre cut.

* **Splicing** – when you need to overlap two lengths of paper, e.g. papering around a recessed window, to make a perfect edge or line between the two lengths, you use a straight edge and a sharp knife to cut through both layers, then remove the offcuts.

Figure 7.50 A mitre cut

Marking lines

Marking lines on your surface before you start is essential. Without a straight line to work from, you cannot achieve straight wallpaper. A plumb line should be made:

* at your starting point
* after an internal or external angle or corner
* over and around window reveals
* after a door frame
* after any feature or obstacle.

KEY TERMS

Mitre cuts
– cuts at a 45° angle used when applying border papers around obstacles, such as door frames, to ensure there are no gaps in the paper.

Splicing
– cutting through two layers of overlapped paper.

Remember, however, that if you are working with patterned paper, your starting point may be in the centre of a strong focal point, such as a chimney breast.

When papering vertically, you will find your marking line using a plumb bob or a spirit level. When papering horizontally, e.g. when applying lining paper or a border, you still need to find a straight line using a spirit level, laser level or a chalk line held at both ends.

When marking lines you will need to consider some of these factors:

* Access required – what access equipment will you need to safely reach the tops of walls or ceilings?

* Light source – from which direction will your light be coming? Which side of your marking line will you need to stand to see better? Have you planned to work your way around the room away from the light source?

* Room dimensions – when marking a plumb line, are the corners of a room exactly plumb? If not, will you need to adjust your first length? Are your room measurements accurate? Do you have enough paper?

* Economy – have you planned ahead to minimise paper waste and costs?

To mark behind your plumb bob line, you can use a pencil or a chalk line with a weight attached. The chalk line is pulled away from the wall, then released, which flicks the wall and leaves a chalk mark. To mark a line using a spirit level, simply use a pencil to draw along the straight edge of the level. For a more high-tech option, you can use a laser level to find your line. You can either attach it to the wall while you are working (although this can be awkward) or you can draw a mark underneath the laser line. There are also standalone laser lines that stand in the centre of a room and cast a horizontal line.

Pasting methods

There are different ways that paper is pasted to the wall. Some papers come ready with their own dried adhesive, some need the paste to be applied directly to the wall and others need the paste to be painted onto the back of the paper. Check the manufacturer's instructions before pasting your paper – they will tell you what types of paste you should use.

The method requiring paste to be applied directly to the wall is becoming more common. 'Paste the wall' paper has a special backing that will not expand when it is wet, whereas traditional papers need to expand, which is why they are left to soak with the paste on.

This type of wallpaper can be a quicker and cleaner way of applying paper. Because the paper is dry, it weighs less. This reduces the possibility of tearing. Note that the wall should only be pasted one length or section at a time. If you paste the whole wall in one go, the paste will dry too quickly and the paper won't stick.

Many border papers are self-adhesive and do not need to be pasted, however, it can be easier to position the border when using a paste. If you do need to paste a border paper, cover your pasting bench with lining paper to protect the bench from paste. You can paste more than one length at a time, depending on how wide the paper is. You may need to replace the lining paper to prevent contamination of the face of the border paper. Each length you cut will usually be the size of the entire wall, which once pasted should be concertina folded.

Figure 7.51 A pasting machine

The following methods can be used when applying paste:

* Pasting machine – designed to save time and apply paste evenly without having to use a brush and pasting table, it has a well of paste that allows several lengths to be soaked at once. It should be cleaned thoroughly at the end of a job.

* Brush – using a large pasting brush is the traditional way of applying paste to papers. You can avoid mess by using the correct technique of aligning the paper with the table edges and pasting from the centre out to the edges.

* Roller – using a paint roller can speed up the pasting process, whether you're pasting the wall or pasting the paper. Care needs to be taken when applying the paste with a roller as it can go on too thickly and cause hanging problems such as stretching.

* Spatula – used for getting rid of any air bubbles and smoothing out wall coverings. In particular it is used by contract/commercial decorators as it works well on vinyl papers.

Reasons for choosing folds

Concertina folds

A **concertina fold** is used when papering ceilings or when papering horizontally. The paper is folded in a series of small folds so that it is more manageable to hold when working above your head. It is also makes it easier to work with the very long lengths needed to paste all the way across a wall or ceiling.

End-to-end, end-to-centre or lap folds

When working with more normal lengths, such as for a vertical ceiling to floor drop, then it is best to use an **end-to-end fold**. The top fold should be about two-thirds of the fold and the bottom fold should be about one-third.

Methods of jointing and trimming

Overlap joint

Overlap joints are used on corners to compensate for irregularities (e.g. corners that aren't square or an uneven wall surface). There is a slight overlap (about 20 mm) in the drops of paper and this is then cut away to give a seamless match.

Once the wallpaper drops have been hung a sharp knife and metal straight edge should be used to cut off the overlapping material and the joint should be smoothed using a spatula to give a neat finish. When using overlap joints, start papering from the lightest side of the room so the overlap joints won't cast shadows.

Butt joint

Butt joints are used on flat surfaces. Two lengths of wallpaper are positioned so that they meet edge to edge without any gap. There is no need to cut the paper. A seam roller is used to seal and smooth the seam. The finished joint may not be as seamless as an overlap joint. Care should be taken to align the pattern and the surface of the wall covering should be smoothed away from the joint. This type of joint is suitable for use on pre-trimmed papers.

KEY TERMS

Concertina fold

– a way of folding paper that looks like a concertina (piano accordion). The small folds of about 35 cm are made paste-side to paste-side and face-side to face-side. They can be easily lifted and unfolded when applying the paper to the surface.

End-to-end fold

– also known as end-to-centre or a lap fold, this is a way of folding paper that is simply folding each end of the paper into the centre with pasted edges together.

PRACTICAL TIP

If you need to reposition a butt joint, be careful not to force the paper too hard. Forcing the paper too hard can stretch it so that when the adhesive dries the seam opens up. It might also tear the paper.

PRACTICAL TIP

Selvedge needs to be trimmed accurately to ensure a good pattern match.

KEY TERMS

Decorator's crutch

– a support made of a roll of paper, some cardboard or a straight edge, used to stop your folded paper from creasing or bending when papering a ceiling.

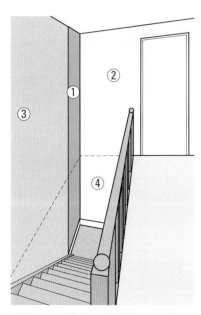

Figure 7.52 The order for papering a staircase

Double-cut joint

When both layers of paper in an overlap joint are cut through with a sharp knife it creates a double cut joint. This is essentially a butt joint on overlapped paper. This type of joint is effective on papers with unfinished edges and wide-width fabric-backed vinyl and can also be used around the head of a window.

Trimming

It is essential to make sure that wallpaper trimming is done neatly and accurately. Poor trimming will be obvious and stand out and will detract from the finished result. Most wall coverings now come pre-trimmed (i.e. the edges have been removed).

Some specialist papers, such as hand-made papers and Lincrusta®, still have a protective strip, called a selvedge, along the edges of each roll. The selvedge needs to be removed by using a metal straight edge and trimming knife before pasting.

Hanging processes for standard paper

Walls

When papering a wall vertically, the first length of paper is held up to the plumb line at the starting point. The longest fold of the paper is placed at the top of the wall, taking care not to let the rest of the folded paper drop. Once the paper is touching the wall, it is slid along the surface to meet the marked line. The paper should then be smoothed with a paperhanging brush from the centre outwards to remove any bubbles. Once smooth, the bottom section should be unfolded gently, then smoothed in the same way. The same process is followed for the next lengths of paper, matching them flush with the edge of the previous length and making sure that any pattern matches up.

Use a seam roller to press down the join between the lengths, unless you are hanging embossed paper. To trim the top and bottom edges, press the paper into the corner edges of the ceiling and the skirting boards using the back of your scissors or a straight edge. Once a crease is made, the edge is gently lifted back and cut along the crease, and finally pressed back into place on the wall. To keep the wallpaper and any surrounding areas clean, it is good practice to wipe away any extra paste using a wet sponge.

Ceilings

When applying papers to ceilings you will need to use concertina folding. Note that you will need to use a **decorator's crutch** to support your concertina folded paper as you will be working with longer lengths. It is very important that you use the correct access equipment for this task (see Chapter 4 for the most appropriate and safest options).

Staircases

Look back to page 227 to remind yourself of the planning and preparation you need to do before papering a stairwell. To find your starting point measure the height of each wall to find the longest drop. Once you have worked out where this will be, use a plumb line, chalk line or spirit level to find a true vertical line and mark it. You should always paper this length first, then up the stairwell, then the landing, and finally downstairs.

Around windows and doors

When hanging paper around a door or window frame:

1. Hang the paper as you usually would from the ceiling edge, but allowing the paper to drop over the frame.

2. Mark the corner and edges of the frame on the paper.

3. Cut the excess paper away into the corners you've marked.

4. Smooth the paper onto the wall along the frame into the corners.

5. Trim the excess paper from the edge of the frame.

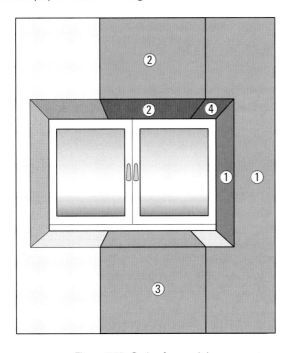

Figure 7.53 Order for applying paper to recessed window

Light switches and power points

When papering around light switches and power points, you will need to cut and trim the paper using **star cuts/half star cuts** to make room for the fitting (see Fig 7.50 for a star cut).

Internal angles

When papering into an internal corner:

1. Measure the width from the last length you hung into the corner. Add an extra 5–10 mm to this width (if using a non-match paper) before you cut the paper.

2. When hanging the paper, make sure you push it well into the corner using a straight edge or similar.

3. You can then use the leftover part of the paper to start papering again on the next wall, hanging the paper right into the corner

4. Check with a plumb bob and mark your straight line before you hang the length on the next wall.

5. Make sure you cover the overlap and take care to match up any pattern.

PRACTICAL TIP

When papering on ceilings and staircases, make sure you have the right sort of stable platform to work from. It should run the entire length of the room or length of paper.

KEY TERMS

Star cut

– a series of small triangular cuts made from the centre of the paper to form a star shape to negotiate a light fitting, pipe or ceiling rose.

Half star cut

– used if only half the length of paper is covering the feature.

Figure 7.54 Cutting paper around a light fitting

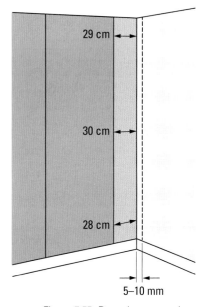

Figure 7.55 Papering around an internal corner

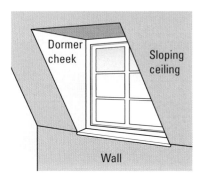

Figure 7.56 Papering a dormer window

KEY TERMS

Dormer cheek

– the vertical side of a dormer.

External angles

When papering around an external angle, once again ensure you have a significant overlap (25–50mm). In the same way as for an internal corner, once you've papered around the corner, mark a plumb line on the next wall before applying your next section of wallpaper over the top of the overlap.

Ceiling rose

Working around ceiling roses is similar to the process for light switches. The paper must first be laid over the feature, then pressed against it so that you are left with an impression on the paper. This section should be cut out of the length, but make sure you leave enough overhang around (25–50mm) at the edges. This overhang is then cut into a star cut, trimmed off, and pressed down flat around the edges. You may need to use a half star cut if the feature is wider than the paper.

Chimney breasts

As mentioned, when working with strong room features such as chimney breasts, it is important to centralise your pattern, but also take into account how many drops you are able to hang across the feature. You may need to adjust your central starting point if the paper falls too close to the edges. Apply the process for external angles as you make your way around the chimney breast, re-marking plumb lines each time you turn a corner and turning the corner by at least 50mm.

Borders

Border papers are often applied horizontally at the top of a room. You may need to go around door fittings or down stairs which requires the paper to be cut at an angle using mitre cuts (see Fig 7.51) to ensure that there are no gaps in the paper.

Reveals

When papering a recessed window you will need to overlap two lengths of paper to make a perfect edge and then splice through the overlap to remove the offcuts.

Sloping ceiling

When papering a sloping ceiling, the ceiling will meet the wall and there might not be a definitive line marking the boundary between the wall and ceiling. Make sure that you follow the pattern from the ceiling to the wall. At the joint, overlap the paper being sure to match the pattern and then cut the joint.

Dormer window

Paper the wall/sloping ceiling either side of the window and above the window. Finally, paper the **dormer cheeks**. You will always lose the pattern around a dormer.

Alcove/niche/arch

Start by preparing the two walls either side of the arch and the wall above the arch to leave an overlap of about 25 mm all around. Use paperhanging shears to make small wedge-shaped cuts in the overlap so that they can be folded into the underside of the arch. Finally find the centre of the arch and measure the depth and width of the arch so that you can cut four strips of paper to fit the underside of the arch. The same method and principles can be applied to papering an alcove or niche. See practical task 1 on pages 243–4 for step-by-step instructions on how to *Apply wallpaper to an arch*.

> **PRACTICAL TIP**
>
> It is tricky to get a pattern matched along the curved join of an arch. It is therefore best to choose a paper with a random pattern or to use a complementary pattern inside the arch.

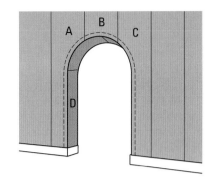

Figure 7.57 Papering an arch

> **PRACTICAL TIP**
>
> Papering around objects is similar to hanging paper on walls and uses many of the same techniques.

Hanging processes for wide-width vinyl

Paper-backed and fabric backed wide-width vinyls are hung in pretty much the same way. The main difference is the way in which the joints are created. For step-by-step instructions refer to practical task 2 *Apply wide-width vinyl wall covering to walls* on pages 244–7.

Cutting from rolls (descending order)

Most wide-width vinyls are made in large batches. As a result of the manufacturing process, there may be subtle differences in colour between rolls at the beginning and end of a batch. Manufacturers number each roll. To minimise the risk of shading, manufacturers suggest that the wall covering is hung in order of roll number to minimise the risk of shading between rolls.

You should organise the rolls that you are intending in order from the highest roll number to the lowest roll number (e.g. 34, 33, 32) and work in this order.

Directional hanging advice

The information supplied in the roll label will give you vital information, such as the direction of hanging (i.e. straight hung or reverse hung), details of the design and **colourway**, the batch number and the roll number.

Reverse hanging

If a wallcovering needs to be reverse hung, you will need to alternate lengths in opposite directions (Fig 7.59).

As the drops are cut, each drop needs to be numbered at the top so that it is clear whether it needs to be straight hung or reverse hung. The drops should be hung around the room in number order.

Adhesives

Adhesives for wide-width vinyls tend to be ready-mixed and should be applied using a roller according to the manufacturer's instructions. Use a paintbrush to cut into the skirting and ceiling and to get into internal angles.

Positioning the first drop

The first drop should be positioned at the starting point in an internal corner with the direction of work away from the corner. Use a plumb line/spirit level to make sure it is straight. The wallcovering should be applied using a plastic spatula, pure bristle brush or felt roller in a vertical direction only to smooth out

> **PRACTICAL TIP**
>
> It is important to prepare the surface correctly to ensure a good finished result. Refer back to Chapter 5.

> **KEY TERMS**
>
> **Colourway**
> – the combination of colours that a pattern is printed in.

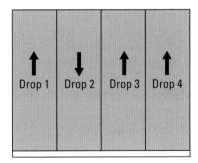

Figure 7.58 Reverse hanging

> **PRACTICAL TIP**
>
> To check that the drops are being reverse hung in the correct order, the top of each drop should be peeled back and the pencil number checked.

PRACTICAL TIP

As each drop is hung it is important to keep checking for shading. However you should also check for shading between rolls prior to hanging to avoid having to take any defective paper down. Check each roll on the pasting table before applying adhesive. If you find any defective rolls these should be hung in less noticeable places.

PRACTICAL TIP

Once you have hung three drops, inspect them for any issues before continuing with the rest of a job. Manufacturers won't accept any claims of defects once more than three drops have been hung.

PRACTICAL TIP

When decorating columns using wide-width wall coverings remember that they should be reverse hung to avoid shading of the joints.

any air bubbles and form a good bond. Leave the wallcovering to settle for 15 minutes before cutting off the wastage at the top and bottom as it may have been stretched during application.

Trimming the selvedge

Wide-width vinyl coverings are not usually pre-trimmed and come with a selvedge (see Fig 7.93 on page 248). This needs to be trimmed by overlapping each drop parallel to the previous drop. Use a metal straight edge or three pins used as markers that are placed top, middle and bottom 50 mm in from the edge. Drop two can then be lined up using the pins.

Jointing methods

The material should be overlapped and trimmed according to the manufacturer's instructions (e.g. splicing/double cutting) so the substrate is not damaged. Some manufacturers may recommend specific tools (e.g. a plastic protective strip that is placed behind paper-backed vinyls). See Figs 7.87 and 7.91 on pages 246 and 247.

* Fabric-backed vinyl: double-cut joints are used and made with a joint cutter (see page 246).

* Paper-backed vinyl: a plastic protective strip is placed behind the joint to protect the substrate and the joint is cut using a sharp knife and straight edge (see page 247).

Use of full width material

Full width material must be used above doors and windows. Offcuts must not be used otherwise the rolls will be out of sequence and there is an increased risk of shading.

Offcuts/part rolls should only be used for standalone features such as columns or pillars.

Free-standing column/pillar

Look back to page 227 to remind yourself of the planning and preparation you need to do before papering a free-standing column. Make sure you have allowed for an even number of drops to cover the column plus a 50 mm overlap. When pasting each drop allow an overlap and ensure that alternate drops are reversed. Smooth out the external corners horizontally using a spatula to remove any air bubbles and allow to settle before cutting joints.

Ceiling of above average span (i.e. 3 m)

Concertina folding should be used to apply lengths of paper of up to 10 m to a ceiling. You should make sure there is a good working platform the length of the ceiling to work from.

Internal/external angles/reveal

* Internal angles: an overlap of 2 mm should be left and cut using a plastic spatula and a sharp knife.

* External angles: these should be dealt with in the same way as the corners for the free-standing column (see above).

Windows/reveals

The process is different from that for standard wall coverings and the order in which the wallcovering should be applied is shown in Fig 7.60 below. For step-by-step instructions refer to practical task 2 *Apply wide-width vinyl wall covering* on pages 244–7.

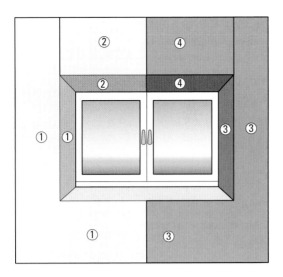

Figure 7.59 The order for hanging wide-width vinyl to a window reveal

Hanging processes for specialised wall coverings

Lincrusta®

There are some key differences between conventional paperhanging techniques and the way in which Lincrusta® is hung:

* It is soaked with water prior to applying Lincrusta® adhesive.

* It should not be taken around external corners due to the thickness of the product. It should be cut at corners and if necessary gaps should be filled with linseed oil putty.

For step-by-step instructions refer to practical task 3 *Hang Lincrusta® paper to walls and around internal and external angles* on pages 248–50.

Hanging Supaglypta®

* Each length is soaked for approximately 15 minutes after pasting and before hanging so that it becomes more pliable and easier to hang.

* Soaking time for different lengths should be kept consistent so that each drop stretches before and during hanging by the same amount. This is important otherwise the paper may expand by different amounts and cause matching problems.

* It is also essential not to smooth it too much when applying it to surfaces as this could cause the pattern to be flattened.

* A very thin gap should be left at the joint which will be filled when the surface is painted, giving a seamless finish.

For step-by-step instructions refer to practical task 4 *Hang Supaglypta® paper on a ceiling* on pages 251–2.

Hanging woven glass fibre

* Drops should be cut with a 50mm excess at the top and bottom.

* Paste should be applied evenly to the wall for the first drop to the same width as the fabric.

* Apply the first drop and smooth down with a plastic spatula.

* Trim neatly at the ceiling and floor using a sharp blade and straight edge.

* Paste the next section of wall and hang the next drop with a butt joint.

* Continue hanging the drops using the same method.

Hanging paper-backed fabric papers

* Paper-backed wall materials (such as felt, hessian, flock and silk) need very little soaking.

* Hang in the same way as standard wallpaper with butt joints and use a felt roller to smooth down.

* Apply to internal and external angles in the same way as standard paper, making sure to wrap the fabric around external angles otherwise the fabric may fray.

* Most manufacturers recommend alternate strips are reversed to avoid shading.

* Avoid getting any paste onto the fabric as it can cause damage.

For step-by-step instructions refer to practical task 6 *Hang specialist flock paper* on pages 234–5.

Hanging metallic paper
Paste the wall:

* Thorough preparation is necessary. If the walls are not smooth, apply a lining paper with PVA adhesive containing fungicide to ensure a smooth surface for this reflective wallcovering.

* PVA adhesive should be applied to the wall with a short pile roller and left to become tacky.

* The dry paper should then be hung using a plumb line and butt joints.

* It should be smoothed down with a felt or rubber roller.

* Any excess paste should be cleaned off with warm clean water as the paper is being hung.

Paste the paper:

* Thorough preparation is necessary. If the walls are not smooth, apply a lining paper with PVA adhesive containing fungicide to ensure a smooth surface for this reflective wallcovering.

* PVA adhesive should be applied to the back of the paper with a short pile roller and left to become tacky.

* The dry paper should then be hung using a plumb line and butt joints.

* It should be smoothed down with a felt or rubber roller.

* Any excess paste should be cleaned off with warm clean water as the paper is being hung.

Hanging hand-printed papers

* Application is the same as for standard papers.

* Avoid getting adhesive onto the surface of the paper to avoid damage.

* Seam rollers should never be used as they will cause polishing at the joints.

Hanging mural papers

* The mural should be laid on a large, clean area to match up the panels.

* Each joint should be marked on the wall with a pencil.

* A clear, non-staining adhesive should be applied to the back of the wallcovering with a short pile roller and the first panel of the wallcovering should be hung using a plumb line.

* The remaining panels should be hung in a similar way with butt joints and the joints smoothed out gently to avoid damaging the image.

Applying liquid wallpaper

* The surface should be prepared and primed as for standard wallpaper.

* The wallpaper base mixture is mixed with water to an even consistency.

* The mixture should be left for 3–4 hours and then applied to a prepared wall using a trowel.

Defects

Table 7.7 below describes the different kinds of wallpapering defects you might come across, their causes and how to avoid them.

Defect	Cause	How to avoid
Creasing (Fig 7.60)	• Uneven surfaces • Too much brushing • Not enough smoothing • Heavy-handed seam rolling	• Ensure surfaces are properly prepared. • Take care when handling, brushing, smoothing and rolling the paper.
Overlapping edges	• Over brushing • Heavy-handed seam rolling	• Use wallpaper brush only enough to remove air bubbles and smooth the paper. • Don't push too hard with the seam roller, or avoid using a seam roller at all • Make sure that you are applying to a plumb line
Blisters or bubbles (Fig 7.21)	• Careless smoothing • Lumps in paste • Uneven pasting • Wrong adhesive • Blistering on lining paper • Incorrect soaking time	• Ensure you brush all the air bubbles out. If some will not come out, gently lift paper and smooth down again. • Remove any lumps from paste before applying to wall. • Ensure paste covers the entire back of the paper. • Always read manufacturer's instructions before choosing adhesive. • Ensure your lining paper has adhered properly before applying decorative paper. • Check that you are using the right soaking time
Tears (Fig 7.61)	• Using thin paste • Blunt cutting tools • Careless paper handling	• Check the consistency and type of your paste. • Ensure your tools are kept clean and sharp so they cut cleanly. • Don't allow your paper folds to drop suddenly.
Polished edges	• Paste seeping through the joints • Using too much paste • Over brushing • Heavy-handed seam rolling • Too much rubbing with a rag or sponge	• Use the right type, consistency and amount of paste. • Any paste that gets on the decorative surface should be wiped clean with a warm, wet sponge. • Don't apply too much pressure to your seam roller • Be gentle when cleaning up your paper
Open joints or joint gapping (Fig 7.62)	• Poor alignment of paper • Uneven surfaces • Thin paste • Careless smoothing • Stretching the paper • Overbrushing • Incorrect soaking time	• Ensure that you slide the paper so it meets the edge. • Prepare your surfaces well and/or use lining paper. • Choose and prepare your paste according to instructions. • Smooth from the centre to the edges, making sure the whole length is brushed. • Do not pull or tug at the paper as it will retract later when dry
Loose edges/ peeling (Fig 7.63)	• Wrong paste used for paper • Not enough adhesion • Old or thin paste • Over porous surface	• Always check the manufacturer's instructions before choosing your paste. • Mix paste as per instructions and ensure it is fresh if using starch adhesive. • Ensure surface is well prepared and/or lined.

Irregular cutting (Fig 7.64)	• Using blunt tools • Careless cutting technique	• Always keep your tools clean and sharp. • Don't rush your cutting, take care to follow your measurements and cut in a straight line.
Paste staining/ surface marking (Fig 7.65)	• Careless brushing • Dirty paste table • Wrong paste used • Paste too thin • Oversoaking of paper	• Align your paper edges to the table when pasting. • Wipe down your table after pasting each length. • Use the correct paste with the correct consistency.
Corners incorrectly negotiated	• Poor planning • Inaccurate marking of plumb line	• Make sure you have thought out your starting point, looking ahead to any obstacles or features. • Allow your plumb line to be steady before marking.
Inaccurate plumbing	• Incorrect use of tools • Not checking plumb lines frequently	• Make sure there's enough chalk on your chalk line. • Wait for the plumb line to settle before marking. • Mark lines for your starting point and when working around corners.
Dry edges (Fig 7.66)	• Not enough paste • Lack of paste • Careless brushing	• Make sure your paste goes right to the edges. • Ensure that all of the paper is covered in paste.
Delamination (see Fig 7.22)	• Oversoaking of paper	• Follow manufacturer's instructions for correct soaking time.
Sheen patches (Fig 7.67)	• Careless application of paste • Getting paste on face of paper	• Align the edges of your paper with the pasting table. • Keep your pasting table clean between lengths.
Poor matching (Fig 7.68)	• Uneven or poorly prepared surface • Wrong type of paste used • Misses or uneven pasting • Oversoaking of paper • Too much brushing leading to stretching	• Substrates should be properly prepared and/or lining paper used. • Choose your paste according to the paper manufacturer's instructions. • Follow the soaking time recommended for the paste and paper. • Ensure that all of the paper is covered in paste.
Mould growth (see Fig 5.64 on page 160)	• Damp walls • Insufficient surface preparation • Condensation • Wrong type of paste • Stale paste	• Sources of damp should be found and repaired. • Old paste must be completely removed before hanging paper. • Use expanded polystyrene to put a barrier between the cold wall surface and the wall covering. • Use a paste that contains a fungicide. • Don't use old or stale paste, especially if starch based.
Inaccurate angle cutting	• Incorrect measuring and calculation of angles so paper is too short	• Measure and work out angles to make sure there is sufficient paper for hanging in reveals.
Shading (Fig 7.69)	• Variations in colour between different rolls of paper of the same pattern/colour due to printing	• Reverse alternate drops when hanging.
Loose fibres (Fig 7.70)	• Overworking the joints in fabric papers • Forming joints on external angles with fabric papers	• Turn the angle and form joint on the adjacent wall surface. • Do not overwork the joints.
Flattening of emboss (Fig 7.71)	• Using a seam roller on embossed papers • Overuse of brushes and rollers on joints	• Avoid using a seam roller on embossed paper. • Use a felt roller to smooth down. • Clean with minimal water to avoid damaging the embossed surface.

Table 7.7 Defects in wall coverings

Figure 7.60 Creasing

Figure 7.61 Tears

Figure 7.62 Open joints

Figure 7.63 Loose edges/
peeling

Figure 7.64 Irregular cutting

Figure 7.65 Paste staining/
surface marking

Figure 7.66 Dry edges

Figure 7.67 Sheen patches

Figure 7.68 Poor matching

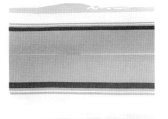

Figure 7.69 Shading

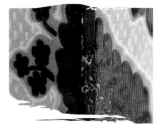

Figure 7.70 Loose fibres

Figure 7.71 Flattening of
emboss

STORING MATERIALS

Refresh your memory about storing materials in Chapter 6. Remember these points when storing wallpaper materials:

* Never store rolls of paper on their ends as it will damage and crease the edges of the paper.

* Don't store rolls of paper in too high a pile or they may get squashed out of shape.

* Keep rolls of paper out of direct sunlight as it can fade the colours and patterns.

* Keep the rolls of paper on a rack and not on the floor as frost and damp can cause the paper to degrade or grow mould.

* Keep rolls of paper in their wrapping to protect from dust and to save the manufacturing details (batch/shade number, international symbols and so on).

* There are some types of wallpaper that have a limited shelf-life, so stock should be rotated to make sure the oldest papers are used first.

* To avoid wastage unused wallcoverings and adhesives should be saved and stored for future use. They may be useful for repairing any future damage.

- Specialist wallcoverings may be left with a narrow strip left on the edges of rolls to protect the edges of the roll from damage during production and transportation. This is known as a selvedge. In the case of flocked wallpaper, the selvedge protects it when it is being stored 'on end'. Flocked wallpaper rolls should be stored on their ends to prevent the delicate pile from getting damaged.

- It is essential to keep tools in a clean and cared for condition to ensure that the quality of your work is of the highest standard. For example:

    blunt tools may cause tearing and irregular cutting

    rusty tools will stain wallcoverings

    a build-up of products on tools may damage the paper.

CASE STUDY

I now understand why Maths and English are so important!

Thomas Regan works for Accrington and Rossendale College as a painter and decorator

'I studied Maths and English up to level 2 when I was at College. I must admit they weren't my favourite subjects but they have helped me significantly in my job role. I am responsible for ordering materials which means I need to work out quantities of materials required for a job and how much the job will cost in total. I also need to order the materials and communicate with other tradespeople and colleagues to ensure that the work runs smoothly and so that everyone is aware of any changes that might occur. Part of my job is to look after an allocated budget for the year, which means I have to source materials from various suppliers and ensure that this is done within budget. When I was at college, I wasn't sure why I had to study Maths and English but now I have realised how beneficial it has been for me and how it helps me on a day-to-day basis at work.'

PRACTICAL TASK

1. APPLY WALLPAPER TO AN ARCH

OBJECTIVE

To apply standard wallpaper with a straight match to an arch.

PPE

Ensure you select PPE appropriate to the job and site conditions where you are working. Refer to the PPE section of Chapter 1.

TOOLS AND EQUIPMENT

Tape measure	Scissors
Plumb bob or spirit level	Sponge
Pencil or chalk line	Cutting knife
Working platform	Cutting straight edge
Paste table	
Bucket of paste	
Stirring stick	
Paperhanging brush	

STEP 1 Apply wallpaper to the walls adjacent to the arch allowing a 50 mm overlap into the arch opening.

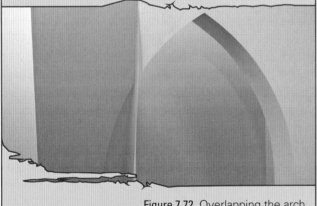

Figure 7.72 Overlapping the arch

STEP 2 For the pieces of paper either side of the arch, make a horizontal cut (at 90°) top and bottom into the overlap.

Figure 7.73 Cutting the overlap

STEP 3 Fold the overlaps and apply them to the inside of the arch.

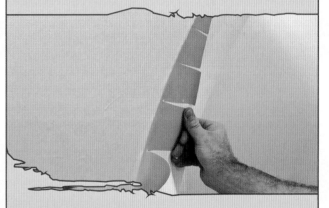

Figure 7.74 Folding and applying the overlap to the inside of the arch

STEP 4 For the paper that is over the top of the arch, make a series of v-shaped cuts into the overlap and fold them over to the underside of the arch.

Figure 7.75 Making wedge cuts into the overlap at the top of the arch

STEP 5 Measure the width and length of the underside of the arch. Cut four pieces of paper – two for the sides and two for the top of the underside of the arch but allow an additional 5–10mm for an overlap.

Figure 7.76 Measuring the underside of the arch

STEP 6 Paste the first piece of paper for the side of the arch and position at the bottom of the curve. Smooth down as normal with a paperhanging brush. Position the second piece of paper for the top inside of the arch in the middle of the underside of the arch. Allow for an overlap and smooth out as normal. Position the other two pieces in the same way on the other side.

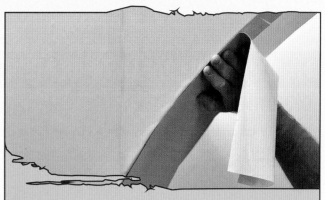

Figure 7.77 Pasting the paper to the underside of the arch

STEP 7 Splice through the overlaps to hide the joints.

Figure 7.78 Splicing through the overlap to hide the joint

PRACTICAL TASK

2. APPLY WIDE-WIDTH VINYL WALL COVERING TO WALLS

OBJECTIVE

To apply wide-width wall covering to walls with internal and external corners.

PPE

Ensure you select PPE appropriate to the job and site conditions where you are working. Refer to the PPE section of Chapter 1.

TOOLS AND EQUIPMENT

Tape measure

Plum bob or spirit level

Pencil or chalk line

Working platform

Wide-width wall covering (paper-backed or vinyl-backed)

Paste table

Ready-mixed paste

Scissors

Pencil

Roller for applying adhesive

75–100mm paintbrush

Spatula/pure bristle brush/felt roller

Sharp knife

Joint cutter

Seam roller

Plastic protective strip

STEP 1 Prepare the surface ready for receiving paper and measure the area to be papered.

REVERSE HANGING

STEP 2 If the wallcovering needs to be reverse hung, unroll the first drop of paper along the pasting table/bench and cut to size.

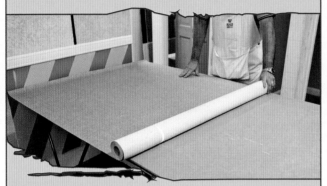

Figure 7.79 Cutting the first length for reverse hanging

STEP 3 To cut the reverse drop, spin the roll round so it is facing the opposite direction and position it at the other end of the table. Measure and cut a length. Repeat steps 2 and 3 for the remaining drops.

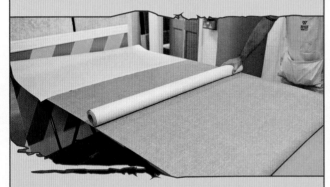

Figure 7.80 Cutting the second length for reverse hanging

STEP 4 As each drop is cut, number the drop on the reverse at the top so that it is clear which pieces need to be straight hung or reverse hung. Hang the pieces in number order.

Figure 7.81 Numbering the drops

HANGING WIDE-WIDTH PAPER TO THE WALL

STEP 1 Decant ready-mixed adhesive to a large paint kettle and apply to the surface using a roller according to the manufacturer's instructions.

STEP 2 Use a paintbrush to cut into the skirting and ceiling and to get into internal angles.

REED TIP

It is more time effective to start applying the adhesive at the bottom of the wall and work your way up. That way the step ladder is ready in place to start with the application of the first drop.

Figure 7.82 Using a paintbrush to cut into the skirting and ceiling with the adhesive

STEP 3 Position the first drop at your starting point which could be in an internal corner. Use a plumb line/spirit level to make sure it is straight.

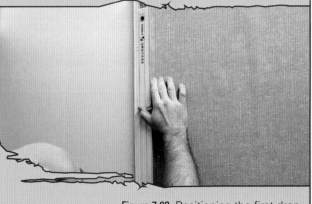

Figure 7.83 Positioning the first drop

STEP 4 Use a plastic spatula, pure bristle brush or felt roller in a vertical direction to smooth out any air bubbles.

Figure 7.84 Using a plastic spatula, pure bristle brush or felt roller to smooth down

STEP 5 Leave the wallcovering to settle for 15 minutes as it may have been stretched. Trim off any excess paper at the ceiling and at the skirting.

Figure 7.85 Trimming excess paper at the skirting

STEP 6 Continue working away from the internal corner positioning the remaining drops. Overlap each drop by 50 mm so that you can cut the joints and trim the top and bottom.

CUTTING JOINTS – FABRIC-BACKED VINYL

STEP 7 Make a slit at the top of the joint using a knife. Insert the joint cutter behind the two drops and keeping the shoe of the joint cutter flat to the wall pull it downwards. As the joint cutter can't go all the way to the skirting, you will need to finish the last bit with a knife or scissors.

Figure 7.86 Using a joint cutter

STEP 8 Peel back and remove the overlap and underlap.

Figure 7.87 Peeling back and removing the overlap

STEP 9 Push the joint together to make sure the butt joint is tight.

Figure 7.88 Forming a tight butt joint

STEP 10 Use a spatula down the length of the joint to smooth it out.

Figure 7.89 Using a spatula to smooth over the joint

CUTTING JOINTS – PAPER-BACKED VINYL

STEP 7 Before hanging the second drop, place a plastic protective strip behind the edge of the first piece to protect the substrate when cutting the joint. Using a sharp knife and straight edge cut the joint.

Figure 7.90 Cutting a joint in paper-backed vinyl using a knife

STEP 8 Peel back and remove the overlap, underlap and plastic strip. Use a spatula to smooth over the joint.

INTERNAL CORNERS

STEP 1 When you get to the first internal corner, overlap the paper around the corner by the width of the spatula (approx 4 mm) and smooth and push into the internal corner so it is tight to the corner.

STEP 2 Cut the joint in the crease of the corner using a sharp knife.

Figure 7.91 Cutting a joint in wide-width vinyl in an internal corner

EXTERNAL CORNERS

STEP 1 At an external corner, mark and cut a right-angled piece from the overlap at the ceiling/coving and at the skirting to allow the paper to be wrapped around the corner. Wrap the paper around the corner and smooth out any air bubbles using spatula.

Figure 7.92 Cutting paper-backed vinyl in an external corner

PRACTICAL TIP

Once you have hung three drops, inspect them for any issues before continuing with the rest of a job. Manufacturers won't accept any claims once more than three drops have been hung.

TRIMMING THE SELVEDGE

STEP 1 Insert three pins – top, middle and bottom – about 50 mm from the edge of the drop. Line the next drop up using the pins and trim the selvedge with a knife and plastic strip or a track and trimmer.

PRACTICAL TASK

3. HANG LINCRUSTA® PAPER TO WALLS

OBJECTIVE

To apply Lincrusta® specialist wallcovering to walls to show two joints.

PPE

Ensure you select PPE appropriate to the job and site conditions where you are working. Refer to the PPE section of Chapter 1.

TOOLS AND EQUIPMENT

Lincrusta® wallcovering

Lining paper or size

Plumb line

Pencil

Sharp knife

Metal straight edge

Bucket of warm water

Sponges

Lincrusta® adhesive

Synthetic bristle paintbrush (50–75mm)

Mohair roller

Rubber roller

Cutting board

Linseed oil putty

STEP 1 Measure the perimeter and plan out drops to take into consideration internal, external corners and obstacles (such as switches and sockets).

STEP 2 Prepare the surface thoroughly by cross lining non-porous surfaces and applying a layer of size to any porous surfaces prior to hanging Lincrusta®. (Look back at Chapter 5 to remind yourself.)

STEP 3 Plan your starting point and mark using a plumb line and pencil.

STEP 4 Measure and cut your lengths, including those for corners, allowing for 50mm at the top and bottom, which will be trimmed later. Trim off any selvedge using a knife and straight edge.

Figure 7.93 Trimming the selvedge on Lincrusta®

STEP 5 For internal corners measure the width from the previous drop up to the corner and cut the paper to fit this width. The offcut will go on the adjacent wall.

> **PRACTICAL TIP**
>
> Refer to the table on the back of the manufacturer's instructions in the roll for the recommended trim measurements for different designs.

STEP 6 Mark the top of each cut length with a pencil to make sure you hang them the correct way up.

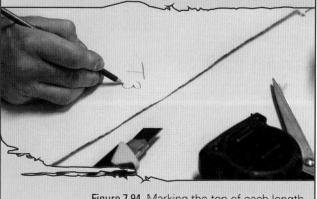

Figure 7.94 Marking the top of each length

STEP 7 Wet each of the lengths with a sponge and warm water and leave to soak for 20–30 minutes to allow the paper to expand. Leave the wetted lengths back to back (i.e. so that the wet sides are facing one another).

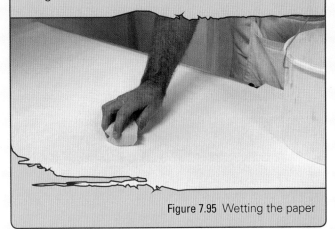

Figure 7.95 Wetting the paper

STEP 8 Once soaked, wipe the back with a dry sponge to remove any water.

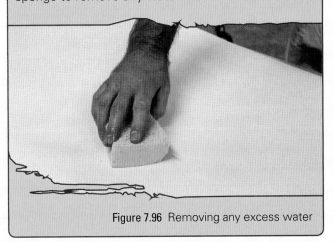

Figure 7.96 Removing any excess water

STEP 9 Apply Lincrusta® adhesive using a 50–75 mm synthetic bristle paintbrush or a mohair roller on larger areas.

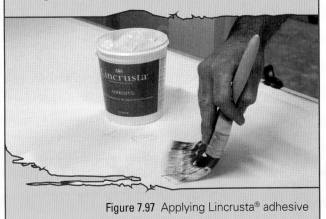

Figure 7.97 Applying Lincrusta® adhesive

STEP 10 Hang the paper and smooth down with a rubber roller to remove all air bubbles. Work in a horizontal direction from the middle to the outer edges.

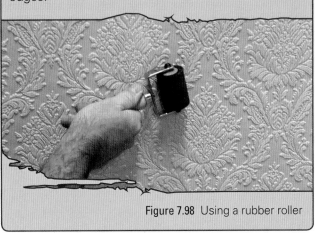

Figure 7.98 Using a rubber roller

STEP 11 Mark the excess at the top and bottom with a pencil line and straight edge. Place a cutting board against the wall behind the wallcovering and cut the marked line with a straight edge and sharp knife.

Figure 7.99 Trimming the excess at the ceiling

INTERNAL ANGLES

Lincrusta® paper should never be taken around corners due to the thickness of the paper.

STEP 1 Position the piece of paper you have cut (in step 4 on page 248) to go into the external corner.

Figure 7.100 Positioning the paper up to an internal angle

STEP 2 Using a plumb line position the offcut on the adjacent wall to form a butt joint.

Figure 7.101 Positioning the offcut on the adjacent wall to form a butt joint

STEP 3 If any surface pattern is standing proud, use a sharp knife at a 45° angle to trim it.

Figure 7.102 Trimming the surface pattern

EXTERNAL ANGLES

STEP 1 Position the pasted paper up to the external angle and mark and cut any excess material using a sharp knife.

Figure 7.103 Positioning the paper up to the external angle

STEP 2 Hang the offcut on the adjacent wall using a plumb line.

Figure 7.104 Positioning the offcut on the adjacent wall

STEP 3 If there are any gaps, apply some linseed oil putty and mould with your fingers.

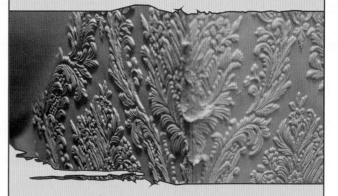

Figure 7.105 Filling the gaps

PRACTICAL TIP

Borders and friezes should be cut to the right length and centred on the main wall. If the wall is very long (i.e. over 2 m in length), the friezes should be cut into 3 m lengths.

PRACTICAL TASK

4. HANG SUPAGLYPTA® PAPER ON A CEILING

OBJECTIVE

To hang Supaglypta® specialist wallcovering on a ceiling over 3 m with a centre piece or ceiling rose.

PPE

Ensure you select PPE appropriate to the job and site conditions where you are working. Refer to the PPE section of Chapter 1.

TOOLS AND EQUIPMENT

Supaglypta® paper   Paste table/board

Bucket of paste

Stirring stick

Paperhanging brush

Scissors

Cleaning-up cloth

Decorator's crutch

Cutting knife

Cutting straight edge

Pencil

Tape measure

Chalk line

Sponge

Working platform

Dust sheets

Plastic rubbish bag for unwanted offcuts

A box for useful offcuts

Watch or clock

PRACTICAL TIP

It is important to make sure a ceiling is strong enough to support the weight of the product. If the material is being applied to a ceiling of above average span, look back to page x to remind yourself of best practice.

STEP 1 Clear you work area of any obstacles. Position steel trestles and working platform in a safe and appropriate position.

STEP 2 Prepare ceiling accordingly (refer to Chapter 5 if necessary). Once preparation is completed, apply a coat of wallpaper size to the ceiling area and allow to dry.

STEP 3 Work out your starting point, taking into consideration the ceiling rose or light fitting and working away from the natural light. Measure one roll length from the middle of the ceiling rose at two points on one side of the ceiling rose. Mark using a pencil.

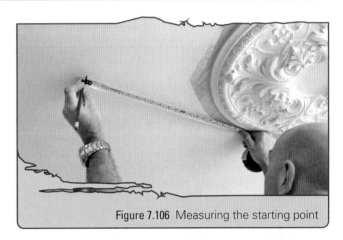

Figure 7.106 Measuring the starting point

STEP 4 Snap a chalk line running through the two points.

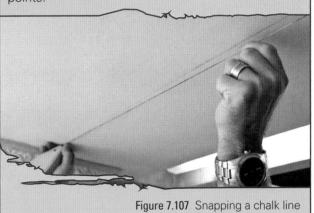

Figure 7.107 Snapping a chalk line

STEP 5 Apply an even coat of wallpaper adhesive to the lengths of paper and leave to soak for 15 minutes.

STEP 6 Hang the first piece adjacent to the wall using the chalk line as a guide on the other side. Use concertina folding and a decorator's crutch. Trim any excess at the coving.

Figure 7.108 Offering up the first piece of paper

STEP 7 Wipe off any excess adhesive from the coving using a damp sponge.

STEP 8 Offer up the second piece to the other side of the chalk line over the ceiling rose – again using a concertina fold and decorator's crutch.

Figure 7.109 Offering up the second piece of paper

STEP 9 Make a star cut to accommodate the centre piece or ceiling rose and wipe any excess adhesive from the centre piece/ceiling rose.

Figure 7.110 Star cut

STEP 10 Continue hanging lengths in the same way until the ceiling is covered.

PRACTICAL TASK

5. HANG SPECIALIST WEFTLESS PAPER

OBJECTIVE

To apply specialist weftless paper to a wall.

PPE

Ensure you select PPE appropriate to the job and site conditions where you are working. Refer to the PPE section of Chapter 1.

TOOLS AND EQUIPMENT

Paste table/board   Weftless paper

Bucket of paste
Stirring stick
Paperhanging brush
Scissors
Cleaning-up cloth
Cutting knife
Cutting straight edge
Pencil

Tape measure
Sponge
Dust sheets
Plastic rubbish bag for unwanted offcuts
A box for useful offcuts
Working platform

STEP 1 Measure, mark and cut lengths matching the pattern.

STEP 2 Apply adhesive to the wall using a roller to a width of 50 mm wider than the width of the paper. Cut in using a paintbrush at the coving/ceiling and skirting.

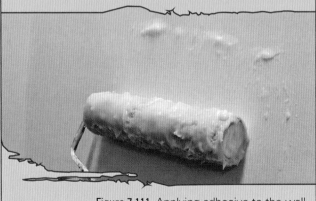

Figure 7.111 Applying adhesive to the wall

STEP 3 Roll up the first drop so that the face is inside the roll.

Figure 7.112 Rolling up weftless paper

STEP 4 At your starting point, offer up the roll with the face towards you to the already pasted surface and apply at the ceiling/coving.

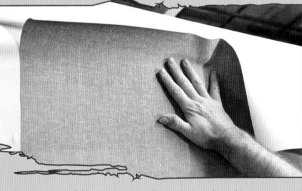

Figure 7.113 Offering up the first piece of weftless paper

STEP 5 Unroll the wall covering. Using a paperhanging brush, smooth out any air bubbles down the middle and from the middle to the outside as you go.

Figure 7.114 Smoothing out weftless paper

> **PRACTICAL TIP**
>
> Take care when smoothing not to go beyond the edge of the paper to avoid getting adhesive on your paperhanging brush.

STEP 6 Trim the excess at the ceiling/coving and skirting with a knife or paperhanging scissors.

STEP 7 Hang the next lengths in the same way.

> **PRACTICAL TIP**
>
> For internal and external corners, the process is the same as that for wide-width vinyls. See page 244.

> **PRACTICAL TIP**
>
> Have warm soapy water to hand to wipe off any adhesive from the face of the paper.

PRACTICAL TASK

6. HANG SPECIALIST FLOCK PAPER

OBJECTIVE

To apply specialist flock paper to a wall.

PPE

Ensure you select PPE appropriate to the job and site conditions where you are working. Refer to the PPE section of Chapter 1.

TOOLS AND EQUIPMENT

Paste table/board

Flock paper

Bucket of paste

Stirring stick

Paperhanging brush

Felt roller

Scissors

Cleaning-up cloth

Cutting knife

Cutting straight edge

Pencil

Tape measure

Sponge

Dust sheets

Plastic rubbish bag for unwanted offcuts

A box for useful offcuts

Working platform

STEP 1 Measure, mark and cut lengths matching the pattern.

> **PRACTICAL TIP**
>
> When working with flock paper it is essential to keep your work area and dust sheets spotlessly clean.

> **PRACTICAL TIP**
>
> The lengths of paper should be loosely rolled up and placed horizontally on the floor to avoid creasing in preparation for hanging.

STEP 2 Apply adhesive to the wall (see Fig 7.119) using a roller to a width of 50 mm wider than the width of the paper. Cut in using a paintbrush at the coving/ceiling and skirting.

STEP 3 Roll up the first drop so that the face is inside the roll.

STEP 4 At your starting point, offer up the roll with the face towards you to the already pasted surface and apply at the ceiling/coving.

Figure 7.115 Offering up the first roll of flock paper

STEP 5 Unroll the wall covering. Using a felt roller in a downward motion, smooth out any air bubbles.

Figure 7.116 Smoothing out flock paper

> **PRACTICAL TIP**
>
> Take extra care when smoothing out flock paper not to go beyond the edge of the paper to avoid getting adhesive on your felt roller. Any adhesive on the surface of the flock will damage the fibres.

STEP 6 Trim the excess at the ceiling/coving and skirting with a knife or paperhanging scissors.

STEP 7 Using a paperhanging brush, brush from bottom to top to brush the nap back up. (You will have flattened the nap with the felt roller.)

Figure 7.117 Brushing the nap back up

STEP 8 Hang the next lengths in the same way.

TEST YOURSELF

1. What is duplex wallpaper?

   a. Pre-pasted paper

   b. Foundation paper for covering defects

   c. Paper with two layers that are glued together

   d. Paper with a single layer

2. What is Lincrusta®?

   a. A cloth-backed textured solid relief wallpaper

   b. A paper-backed textured hollow relief wallpaper

   c. A paper-backed wallcovering made from jute

   d. A paper-backed textured solid relief wallpaper

3. What is the purpose of the selvedge on some types of wallpaper?

   a. To protect the roll from damage during production and transport

   b. It contains information about the direction of hanging

   c. So that the paper can be stored on its end on the floor

   d. All of the above

4. Which of the following types of adhesive should not be used to apply heavyweight textured papers?

   a. Starch paste

   b. PVA

   c. Ready mixed

   d. Cellulose paste

5. Which of the following defects can be caused by incorrect consistency of adhesives?

   a. Blistering

   b. Open joints

   c. Polished edges

   d. Loose edges

6. Which of the following defects can be caused by using the wrong type of adhesive?

   a. Polished edges

   b. Loose edges/peeling

   c. Delamination

   d. Dry edges

7. When hanging wide-width vinyl, why do you need to hang it in roll number order?

   a. To ensure a good pattern match

   b. To make sure the colourway matches

   c. To reduce the risk of shading

   d. All of the above

8. When would you use a joint cutter?

   a. To double-cut joints on fabric-backed vinyl

   b. To double-cut joints on paper-backed vinyl

   c. To trim selvedge on fabric-backed vinyl

   d. To trim wastage on paper-backed vinyl

9. What is the cause of flattening of emboss on embossed paper?

   a. Oversoaking the paper

   b. Using the wrong type of paste

   c. Overbrushing

   d. Blunt cutting tools

10. Why should flocked paper be stored on end?

    a. To prevent damage to the delicate pile

    b. To prevent it from degrading in the sunlight

    c. To maximise the amount of storage space

    d. All of the above.

11. In which direction should you work when using butt joints and why?

    a. Towards the light to reduce shadows

    b. Away from the light to reduce shadows

    c. Away from the light to enhance shadows

    d. In any direction as shadows don't matter

12. What action is required at external corners with Lincrusta® paper?

    a. It should be taken round external corners and any air bubbles should be smoothed out before cutting joints

    b. It should be taken around corners to prevent it fraying

    c. It should be cut to a finish flush at the corner and any gaps filled with linseed oil putty

    d. It should be taken around corners and any cracks filled with linseed oil putty

Chapter 8

Unit CSA L3Occ121
PRODUCE ADVANCED DECORATIVE FINISHES

LEARNING OUTCOMES

LO1/2: Know how to and be able to produce high quality ground coat finishes for advanced painted decorative work

LO3/4: Know how to and be able to produce broken colour effects using water-borne and solvent-borne scumbles

LO5/6: Know how to and be able to prepare plates and apply stencils

LO7/8: Know how to and be able to produce replica graining

LO9/10: Know how to and be able to produce replica marbling

LO11/12: Know how to and be able to apply metal leaf

LO13/14: Know how to form painted lines and bands

LO15/16: Know how to and be able to produce texture designs and smooth finishes using brush, roller and comb and trowel

INTRODUCTION

The aims of this chapter are to:

* help you produce ground coats with a good quality finish
* show you how to produce a range of broken colour effects, replica graining, marbling and textured plain finishes
* show you how to apply stencils and metal leaf.

PRODUCE HIGH QUALITY GROUND COATS

Decorative effects are used to add visual interest to a room. Some finishes are designed to look like different surface materials, such as marble, fabric, wood and other textures. They were often used in the past when these natural materials were very expensive. While these effects were very popular in the twentieth century, due to the unavailability or cost of raw materials, there is still a need for some painters to have these skills so that older buildings can be maintained.

Before applying specialist decorative finishes, it is essential that the surface and ground coat have been well prepared.

PPE

Look back at the previous chapters about the most appropriate PPE for working with paint products. In particular, when creating decorative effects you will need to protect your hands as you will be coming into close contact with paints and glazes. There are several **barrier cream** products available that will help to protect the skin if it comes into contact with contaminants and chemicals in paints, solvents and scumbles.

If you have to do any extra surface preparation, such as abrading, then you will also need to protect your eyes, nose and mouth.

Protecting the work and surrounding area

You should protect your working area from paint drips and spatter. This includes the floors, but also any furniture and fittings. You will probably have already laid all your dust sheets and removed fittings when preparing your surface and applying primers and undercoats. Look back at Chapter 6, pages 174–8 for more information about preparing the work area.

Preparation processes

In Chapter 5, you learnt about wet and dry abrading, spot priming and making good surfaces, ready to receive coatings. It is especially important when creating decorative finishes that your surface has been well prepared. If you don't have a high quality surface and ground coat finish, your decorative effects can be ruined.

Defects in decorative work

Table 8.1 below lists the common defects in decorative work that may result if the quality of the ground coat is not of a high quality.

KEY TERMS

Barrier cream

– an ointment or lotion that creates a physical barrier between the skin and substances which can cause skin infections or dermatitis.

Defect	Cause	How to avoid
Uneven colour	Small holes in the surface that have not been filled properly. This stops the top coat from spreading evenly and shows up as dark spots in the finish.	• Go over your surface carefully, looking for any indentations, cracks or holes • Make sure that all surface imperfections have been filled with the right product • Abrade the filled area so it is flush with the wall
Ropiness	The ground coat has not been applied carefully, leaving misses or brush marks, not laying off, or it has been applied to a surface that is not completely dry.	• Wait for your surface and any layers of paint to dry properly • Apply the correct type of ground coat • Apply all primers, undercoats and ground coats carefully, making sure you lay off each time
Sinking	The decorative coating disappears on porous surfaces, such as filler. This will leave an uneven finish for your decorative coating.	• Any patches that have been filled must be spot primed so that top coatings do not seep in or evaporate
Bittiness	When dust or grit has appeared on or under the surface you have painted. It will make the finish look lumpy and uneven.	• Once you have finished abrading, thoroughly dust down the surface • Keep the whole area clean as dust and debris can be kicked up and attach itself to a wet surface

Table 8.1 Defects in decorative work

Application methods and finish

The application methods you use can affect the quality of your finish. You will need to use the right products, tools and equipment, and apply coatings correctly to avoid the defects covered in Chapter 6, pages 193–6. The decorative effect that you are trying to achieve may not work if you have not prepared your surface or applied your products correctly.

For example, brush marks will need to be removed in order to achieve the effect you want. Using a stipple brush or roller with a very fine sleeve will help make sure your ground coat is even and will remove the brush marks. When it comes to applying the specialist coat, the result will be much better.

Types of ground coats

It is important to choose your ground coat carefully. Not all paints are well suited for specialist coatings. The best types of ground coat to use are eggshell paints (either oil or water based) or something with a slight sheen such as a silk emulsion. Use a colour that is a tone lighter than the lightest part of your finish. A neutral shade is best so it does not compete with the colour of the decorative finish.

Tools and equipment

Many of the tools needed for applying ground coats have been covered in earlier chapters, such as rollers, rubbing blocks, paintbrushes, buckets, dusting brushes, paint stirrers, strainers and kettles. They have the same use in applying ground coats and decorative finishes. Sponges and hair stipplers are covered on page 262 under broken colour effects.

Tack rags

A tack rag or tack cloth is used for cleaning dust, dirt and debris off your surface. It is 'tacky' or slightly sticky so that the dust will stick to it better.

Scumble

– a type of coloured glaze, scumble is a mix of linseed oil, white spirit and pigment (colour) that creates an opaque coating. It is applied on top of a ground coat and used for decorative finishes.

Glaze

– a transparent and thin coating of linseed oil and white spirit painted over the top of paint. A glaze looks glossy or shiny, is clear in colour and can provide a protective coating.

Opaque

– when something cannot be seen through, e.g. a coating that is not transparent. There are different levels of opacity depending on the amount of colour that is present.

Translucent

– another word for transparent or clear, this is something that allows the light to pass through, allowing the surface or object behind to be seen.

Extender

– a chemical additive that when added to paint, glaze or scumble will slow down the drying process.

Glycerine

– a chemical based liquid that has no colour or smell. It has many uses in the food and pharmaceutical industries. It is used as a thinner or extender in water-based glazes or scumbles.

Drier

– a chemical additive that will speed up the drying process when added to a paint, glaze or scumble.

PRODUCE BROKEN COLOUR EFFECTS

Glazes and scumbles

To achieve broken colour effects, you can use either an acrylic (water-based) or an oil-based **scumble**. A scumble is created by adding colour to a clear **glaze**. It can be tinted with water- or oil-based colourant or you can simply use a small amount of coloured paint. For an oil-based glaze, you would mix linseed oil and white spirit to thin it out.

A scumble will usually be **opaque** due to the pigment or colour that has been added. If the coating is **translucent** or transparent, then it will be a glaze. An acrylic scumble will be milky-white when mixed, but then dries clear.

Methods of extending and reducing drying time

Extenders and driers

Sometimes it will take a long time to achieve your effect, or you may be working on a large surface. There is a risk that your scumble or glaze could dry out, so you may need to use an **extender** such as **glycerine** or a proprietary retarding agent and add it to your water-based scumble. You can also extend the drying time of water-based scumble by using the light spray method, where water is sprayed onto the work, or you can use a wet rag before adding scumble to the rag when creating a rag rolling effect.

If you need longer drying time for an oil-based scumble, you can add more linseed oil.

Sometimes you may want your finish to dry more quickly, for instance if it is very cold or if the coatings are old. You can add a **drier** to your oil-based scumbles to speed up the drying time. It's important to note that if you change the viscosity and add too much drier, then it can cause cracking of the surface if it dries too quickly.

Factors affecting drying time

Whether or not you use extenders or driers will depend on the conditions you are working in:

* The room temperature – a cold room may mean you need to speed up your drying time; a warm room may mean you need to slow it down.

* The thickness of your material – your scumble may be too thick or too thin for the job at hand; oil-based scumbles will stay workable for longer.

* How large the space is that you are decorating – a large space may mean you need to extend your drying time to keep the scumble workable; a small job means you may finish before the scumble dries.

* The number of layers you are applying – if you are using several colours, then you need a faster drying time between coats.

* The number of people who are working on the same task – more than one person working on a job will speed up your work time and may mean you do not need to use an extender.

Factors affecting materials used in producing broken colour effects

If you are using an oil-based coating for your decorative finish and it is white, then it might go yellow over time. Using an acrylic coating will prevent yellowing, however, it will not be as long lasting as an oil-based coating. Water-based coatings are damaged more easily by scratches or knocks. For this reason, if producing a decorative finish on woodwork, it would be better to use an oil-based coating. Even if your ground coat is water-based, you can still use an oil-based scumble on top. Note, however, that you cannot use a water-based scumble over the top of an oil-based ground coat as it will not adhere to the surface and will cause cissing.

You should always measure up and estimate the size of your job as accurately as possible so that you can make the right amount of glaze or scumble. It is better to overestimate the amount of coating you will need and have some left over.

If you don't mix sufficient scumble, then you would have to stop the job part way through which will affect the drying time and leave you with an obvious line between the two stages of applying the finish. Also, when mixing the colour in, it is quite difficult to get the right amount, so if you run out there is the risk that your new batch will not be exactly the same colour.

In the same way, you should plan ahead and estimate how long it will take to apply the decorative finish to the area. For instance, you wouldn't start the job with just a half hour until your lunch break! If the job is particularly large then several people will need to work on it to ensure a consistent finish.

REED TIP

It is important to be able to think on your feet and try to find solutions to problems rather than just seeing the problem.

Tools and equipment for producing broken colour effects

Tools and equipment	Description and use
Hair stipplers (Fig 8.1)	A brush made of hog hair, it is used to get rid of brush marks from oil-based paints and glazes. It is also used to create the decorative stipple finish which looks a little like suede.
Sponges (natural and synthetic) (Fig 8.2)	Used for creating a broken colour effect by either applying or removing paint using the sponge. Natural sponges produce better quality decorative finishes, though synthetic ones can also be used effectively.
Mohair roller	These roller covers have a short pile which is used for applying oils to surfaces. The short pile stops the oil from flicking off.
Chamois leather	Often known as a 'shammy', it is s special type of gentle and highly absorbent leather. It can be used for creating a rag-rolled effect because it is a lint-free cloth.
Lint free cloth	Lint free cloths are used for producing a rag-rolled effect. They are made of fabric that will not leave fibres behind on the surface.
Dragging brush (Fig 8.3)	Also known as a flogger, a dragging brush has coarse bristles that can be made of nylon, fibre or horse hair. It is used for creating a grain pattern.
Palettes (Fig 8.4)	A flat surface to place small amounts of paints on when working with stencils. They can be easily held in one hand.
Plastic pots	These are small containers to hold your decorative coatings in. The most commonly used are sign writers' pots.
Masking tape	Protects surfaces by 'masking' them from paint. Care needs to be taken when removing not to damage the effect by peeling away the top coating. Comes in various widths from 6 mm to 300 mm and can be either standard tack or low tack.

Table 8.2 Tools and equipment for broken colour effects

Figure 8.1 Hair stipplers

Figure 8.2 Sponges

Figure 8.3 Dragging brush

Figure 8.4 Paint palette

Broken colour techniques

Rag rolling

Rag rolling is an effect created by using a bunched up piece of rag, paper, chamois leather or lint-free cloth on a glaze or paint that has been applied to the surface. Your surface finish will be affected by the equipment and materials you use, e.g. using paper will give you a sharper effect than using a chamois.

It is also possible to achieve the rag rolling effect using paint rather than a scumble.

Subtractive rag rolling

When the rag touches the surface, it takes some of the scumble away (i.e. subtracts it) and leaves a textured effect that looks like crushed velvet. Subtractive rag rolling is also called 'ragging off'. The scumble or paint is applied to the whole surface to be decorated, finishing off with a stippler to get rid of brush marks. The rag is then crumpled up and moved in different directions, lifting the glaze from parts of the surface. The rag will need to be cleaned off regularly.

Additive rag rolling

Adding scumble or paint to the ground coat, using a rag, is called additive rag rolling or 'ragging on'. The rag (or whatever material you are using) is crumpled up and dipped into the scumble or emulsion paint (although paint on its own won't give you a good finish). It should become soaked through with the coating, then wrung out and rolled into the right shape. When placing the rag onto the surface, it should be used in different directions so that the effect is random rather than accidentally creating a pattern. Take care not to go over the same patch of surface, or the effect will be uneven. The rag should also be reshaped to help achieve a uniform, random effect.

Sponge stippling

Stippling creates a soft, broken colour effect. It can be done with a hog-hair stippler or a sponge.

When using a stippler, a coating is added to the ground coat, then the stippler is dabbed at the surface which removes small dots of paint. The stippler needs to be blotted from time to time to remove excess paint from the bristles. The final texture should be very even and soft, like suede, though it may vary from fine to coarse. You may get a better result by using an oil-based glaze or scumble when using a stippler. For a more even finish, you can use the stippling process on the scumble before using a subtractive sponge effect.

Using a sponge will create a more striking broken colour effect. The finish will also depend on whether a natural or synthetic sponge is used. A natural sponge will provide a more interesting and irregular effect.

Figure 8.5 A subtractive rag rolling effect

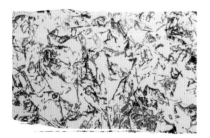

Figure 8.6 An additive rag rolling effect

If you *add* paint, scumble or glaze to the ground coat using the sponge, this is called additive sponge stippling. If you remove or *subtract a* coating from the surface, this is called subtractive sponge stippling.

This mottled look can be enhanced by using more than one colour, once each layer of coating is dry. You can use different types of sponge for the different layers. Using more than one colour gives a marbled effect.

Bagging

Bagging is a useful technique for hiding any small surface imperfections. Similar to rag rolling or sponging, the effect is created by dabbing the wet scumble with a crumpled plastic bag or cellophane. As with other techniques, it can be subtractive or additive. Note, however, that as you are using a non-absorbent material, it doesn't actually remove the scumble from the surface as a sponge or rag would, but moves it around the surface to create the effect.

Dragging

Dragging creates a series of textured fine lines on the surface. It is achieved by pulling a dragging brush down through the scumble, which leaves the colour of the ground coat just showing through. Dragging should only be done on a very well-prepared and smooth wall otherwise the imperfections will show up.

Glaze and wipe

Glaze and wipe is a technique used on **relief surfaces**, such as embossed wallpapers and mouldings. It brings out the design features by creating a greater contrast between them. Usually, an oil-based scumble will be used because it has a longer working time.

Once the ground coat is dry, the scumble is applied with a brush or roller. A hog's hair stippler is then used to create a very smooth surface with even colour. Finally, a lint-free cloth is wiped over the scumble before it is dry. This takes the scumble off the higher level of the surface and leaves a darker colour on the indented part of the surface.

Application faults and problems

Working on decorative finishes requires a lot of care. If you don't take the time to get the surface prepared, ground coat applied, coating thickness right, prepare enough coating, plan your work, and use correct technique, the finished result will not look good. Listed below are some problems that tend to occur when applying decorative finishes.

Loss of wet edge

As mentioned earlier, it is important to plan your work so that you don't have to stop partway through the job. If you do, then you will lose your wet edge – the coating will start to dry at the point you stopped. This stopping and restarting point will show up in your finish. Make sure you have the time and materials to keep working the surface until you reach a sensible stopping point, e.g. the corner of a room.

Banding/tracking

When rag rolling, bagging and sponge stippling, it is important that you overlap your work when adding or subtracting your scumble. If you don't, then you will end up with an irregular pattern in the finish. Make sure that you go over one-third of the already worked area when starting a new section.

Figure 8.7 Bagging effect

Figure 8.8 Dragging effect

KEY TERMS

Relief surfaces

– a surface with raised parts that stick out from the background, such as textured or embossed wallpapers. A surface that stands out more is called 'high relief'; a surface that stands out less is called 'low relief'.

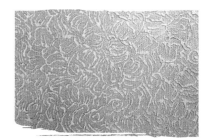
Figure 8.9 Glaze and wipe effect

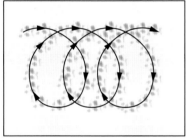

Figure 8.10 Avoiding banding or tracking

When sponging, you can work in a circular motion to help avoid this same fault.

When stippling, make sure you work in a random pattern and not in defined sections or rows, otherwise you will see bands of colour in the finish.

Slip/skid marks

When applying coating to your surface, it is important not to press too hard or the roller might slide across the surface. This will leave marks and give you an imperfect starting surface to work from.

The same applies when using a stippler: if you press too hard or use the brush at an angle, then the bristles can splay and slip. This will leave you with lines rather than dots.

Problems with incorrect removal of masking

Before applying decorative finishes, you may have prepared some surfaces with masking or decorator's tape. If you take the tape off too late or carelessly, then you can not only damage the decorative effect itself, but can even remove the ground coat. Both of these problems would need to be fixed and repainted. These problems can also be reduced by making sure you use the right masking product, for example a low-tack tape, so that any possible damage is minimised.

When removing the tape, roll it back onto itself, rather than away from the surface. If you try to remove it too quickly, you may take layers of coating off.

You must dispose of your masking tape responsibly. Usually you would dispose of it along with other contaminated materials at the end of a job by taking it to the local council waste site or recycling service, especially if you have been working with solvent-based products.

Cleaning and storing your tools and equipment

Refresh your memory about cleaning and storing your painting tools and equipment by looking again at Chapter 6, pages 202–6. If you keep your tools clean, dry and store them appropriately they will last you longer and give you a better finish.

In addition, it is important to note that rags soaked in solvent should not be stored bunched up anywhere, including in your pockets, because of the risk of spontaneous combustion. Make sure you open up the rag and allow it to dry before throwing it away. Remember when you are working with solvents and solvent-based coatings that any contaminated equipment should either be washed out or thrown out. When cleaning your tools, bear in mind the environmental guidelines and avoid pouring any scumble, glaze, paint or solvent into the drain.

If you are using a natural sponge, take good care of it – they are quite expensive.

When using a dragging brush or flogger, make sure you keep the bristles dry by wiping them on a piece of lint-free cloth as you are working. This will keep the bristles in shape.

Rollers and brushes should be cleaned and stored in the normal way, depending on the type of coatings you have used.

If using solvent-based scumbles, a chamois or lint-free cloth should be spread out for the solvents to evaporate before disposing of them. Note, however, that a chamois will no longer be soft if it has been used with a solvent-based product, so it's better to use these cloths with water-based scumbles only.

PREPARE PLATES AND APPLY STENCILS

Stencil work is a way of decorating a room with designs and colours that may form a border, a single image, or a pattern that covers an entire surface. Because stencils can be made of almost any design, they can be a unique expression of a customer's taste and style. Stencils are an easy way of repeating a pattern consistently across a surface.

More complex designs can be created with a multi-plate stencil where each plate is used for a different colour. For step-by-step instructions on how to create and cut out a stencil, see practical task 6 *Apply a multi-plate stencil* on page 299.

Figure 8.11 A multi-plate stencil

Tools and equipment for stencil work

Table 8.3 lists the tools and equipment needed for stencil work. In addition, a palette is needed – see page 261.

Tools and equipment	Description and use
Plate materials, acetate and films (Fig 8.12)	Acetate: Clear and flexible sheets that can be used to make a stencil. One of the benefits of acetate film is that it can be photocopied onto. it is also highly durable and waterproof.
	Frisk film is a type of low-tack stencil material that sticks well to your wall or surface, which avoids paint bleeding underneath, leaving you with a very sharp edge for your design.
	Mylar: Polythene sheet that is inflexible. The benefits are that it is durable and transparent. It is the best material to use for multi-plate stencilling.
	Paper or card can be used for creating stencil plates, but it must be treated first with linseed oil or knotting solution so that it is strong, waterproof and therefore reusable. Cheap and simple to produce.
	Proprietary stencil card or paper (made of oiled manila) is commonly used for creating stencils because it is specifically designed for this purpose.
	Metal sheet: Light metals such as brass, thin aluminium or zinc sheet are used in ready-made stencils.
Pencil	Use a pencil to make light marks on the wall to help with planning where your decorative finishes will begin and end, as well as the top and bottom of the stencil to keep it in line. You must remember to clean off these marks before stencilling.
Ruler	You will need a ruler to mark the central point of the area you are planning to stencil, and to make general measurements in your stencil area.
Tape measure	You will need a tape measure to pinpoint the positions for your stencil work, especially the starting, finishing and central points in the room.
Chalk and line	As when working with wallpaper, when stencilling you may need to pinpoint and mark the centre of the room as your starting point, or indeed any relevant point in the room. Chalk lines are a quick way of doing this.
Craft knife (Fig 8.13)	Sharp knives used for cutting paper, card or acetate stencil plates. Should be used with cutting mats to prevent damage to surfaces. For safety, cuts should be made away from the body.
Hot knife (Fig 8.14)	An electric pen/knife that gets hot. It is ideal for cutting stencils out of Mylar. It is used like a pencil to trace over the stencil design and can be moved in different directions making it good for cutting detailed stencils with curves or jagged edges. This makes it faster and easier to cut with.
Cutting mat or glass plate (Fig 8.15)	Used for laying your stencil design and card onto before cutting so you do not damage the surface below. Cutting mats are made of composite materials that give a smooth heat-resistant surface. They are sometimes referred to as 'self healing' mats as any nicks left by the knife blade are not permanent. Glass plates make an ideal cutting surface as they are solid and don't allow the knife to penetrate. This results in a clean cut and reduces the wear and tear on the blade so that the edge of the blade remains sharp.

Table 8.3 Tools and equipment for stencil work

Figure 8.12 Plate materials

Figure 8.13 Craft knife

Figure 8.14 Hot knife

Figure 8.15 Stencil cutting mats

Figure 8.16 A positive stencil design

Figure 8.17 A negative stencil design

Figure 8.18 Treating stencil paper

Positive and negative stencils

Positive stencils are where a design is cut out of the stencil so that the gaps are filled in which leaves the design on the wall (see Fig 8.16).

Negative stencils are where it is the *background* of a design that is cut out. When the stencil is painted over, it is the wall colour behind that forms the design (see Fig 8.17).

Methods of transferring stencil designs

* Trace – the design is traced onto tracing paper, then the paper is attached onto stencil card covered in chalk. The design is redrawn over the traced lines which removes the chalk from the stencil card.

* Pounce – the design is traced onto tracing paper and cut out, then placed on top of the stencil card. The paper stencil is dabbed with a pounce pad or quilt which is filled with chalk powder. This is a technique used by signwriters for gilding. French chalk is used due to its fineness.

* Photocopy – the stencil design is laid onto the photocopier plate and the stencil paper is placed into the paper tray.

* Illuminated projection – an overhead projector is used to cast the stencil design image onto a piece of paper attached to the wall, which is then traced.

Treating paper

Once you have cut out your stencil, it is important to treat the paper so that it doesn't expand or warp from contact with the paint. Depending on the material you've used, you will need to coat it in linseed oil or shellac (knotting solution) to create a waterproof surface. The whole stencil plate should be treated, both front and back, as paint can soak in from underneath.

Enlarging and reducing

There are several ways of enlarging or reducing the size of your stencil design.

* Grid – this is a freehand method of redrawing a design. A grid is drawn on the surface of the original design, and then a new grid of the correct size is drawn onto new card. The design is then redrawn by using the grid squares as a reference point between the two sizes. Accurate measurement is essential to ensure that the design is transferred correctly and that the material has been scaled up and down correctly.

* Illuminated projection – by using an overhead projector, the image can be enlarged or reduced and projected directly on the surface to be painted.

* Photocopy – you can easily enlarge or reduce a design on the photocopier, and then cut out the new size.

Cutting considerations

There are several factors to bear in mind when creating an accurate stencil:

* Cleanliness – by keeping your stencil clean and in good condition, you will be able to reuse it for longer.

* Hand position – keep your supporting hand (i.e. the hand holding the stencil down) behind and out of the way of your cutting hand and keep moving the hand to keep it away from the blade; cutting away from your hand will avoid injuries, but you can also wear a mesh glove for extra protection.

* Direction of cutting – when cutting into any corners or points in the design, cut away from the corner. Cut away from ties and yourself.

* Blade sharpness – if your blade is blunt, then you can tear the stencil card or paper, or have to go over the same lines.

* Repair of broken **ties** – as these connecting strips hold a stencil plate together, they need to be repaired using masking tape if they are broken to prevent the stencil plate from collapsing and because any broken ties will affect the finished design.

* Size and sequence of pattern – start with cutting out the small areas and vertical lines of your stencil to gain confidence in cutting. These areas are the easiest to repair if you make a mistake.

* Free movement of stencil plate – turn the stencil and not your hand so you are always cutting at a comfortable angle; this will help you to cut more smoothly.

* Margin widths – make sure there is enough of a border around the whole design.

* Base materials – make sure you have securely attached your stencil paper or card to a cutting mat or glass mat. You can use a low-tack tape or adhesive spray and this will stop it slipping while you cut. Never cut on unsuitable surfaces, such as cork or wood, as your blade could snag in the surface and break. This might cause an injury. When cutting on glass be sure to use toughened safety glass within a frame (which will hold the pieces together if it breaks). Ordinary glass should not be used as it can break easily.

Planning considerations

When preparing any stencil work, it is important to think about the space you are working in. If you begin applying the design without planning first, you could end up with patterns that do not meet up around the room, or match up as you go along. You could end up with an unbalanced look if there are gaps or irregular spaces between each motif. When you are planning your stencil work, keep in mind the following:

* Room dimensions – are the walls different sizes? How much paint will you need to complete the decorative finish for this room? Where is the central focal point of the room?

* Access requirements – will you be working up high? Down low? What access equipment and PPE will you need?

* Location of doors and windows – are the doorways or windows going to interfere with the design? Will your stencil become halved? Will it affect your starting point?

* Corners – will the design meet up well at the corners of the room? Will this affect your starting point?

KEY TERMS

Ties

– connecting strips that hold the design together. Without them the stencil would collapse.

PRACTICAL TIP

Never put a knife in your pocket uncovered. Always be sure to wrap it first.

* Number of repeats/connections – what is your starting point and finishing point? Are there any overlapping points in the stencil? How many times will the stencil be applied vertically and/or horizontally?

* Stencil size – how big is your stencil? If it is large, will you need to apply only part of one of the repeats?

* Spacing – how far apart should each placement of the stencil be? Do you need to increase or reduce this space to ensure the design meets well at the corners? Is your stencil going to cover the entire wall, or just one part of it?

* Order of application – which colour should you apply first? Dark colours should be applied before light colours and when blending, always start with the strongest colour.

Marking out

Marking out your space is an essential step before you start stencilling. Each position for your stencil should be carefully marked with pencil or chalk so that the pattern will be regular and unbroken. The stencil must also be level all the way across the surface.

Use a chalk line to mark the area where your stencilling is to start. This will often be the central point of the wall or the focal point of the room. See page 230 to remind yourself how to use a chalk line.

Depending on the design you will be stencilling, you may need to mark either horizontal or vertical lines, for example:

* if it is to be used as a border around the middle of the wall, then you will need to carefully mark a horizontal line that is level all the way around the room

* if the design repeats down the wall, then you will need vertical lines marked along the wall at each point the stencil should be placed

Stencils may come with their own **registration marks**, or you may need to cut small Vs at the centre points of your stencils. These marks are the key to getting an accurate match up of your design.

Methods for securing stencil plates

Not all stencils will need to be fixed to the wall while you work. A smaller stencil may be simply held up against the wall. If you are using larger stencils, then you will need to attach them to the wall to avoid slipping or wrinkles in the work.

You can use a spray adhesive which does not damage paintwork and which will stay sticky for a while, meaning you can move it around the room until it dries. This is a quick and easy way to fix stencils, but if the stencil is large, it may not be strong enough to hold it. Beware that securing stencils with spray may leave a residue on the surface.

If using masking tape, do not use a high tack tape as this may remove paint from the surface. Use a low tack tape that will not remove any paint from your surface and will be strong enough to hold up larger stencil plates.

Applying paint to stencils

See practical task 6 on page 299 for step-by-step instructions on how to *Apply a multi-plate stencil.*

Producing graduated effects

You may want to produce an effect in which colour is graded from light to dark across the stencil. This technique should be practised as it is a hard one to master.

Application faults

See Table 8.4 below for faults that can occur when applying stencils, their causes and how to avoid them.

Application fault	Causes	How to avoid
Creep (Fig 8.19)	When too much paint has been applied over the stencil, the extra paint can bleed under the holes and blur or ruin the design.	Make sure you blot your brush or roller before applying paint to the stencil. Roll or dab the excess paint onto a paper towel or your palette first.
Smudging (Fig 8.20)	Paint rubs off the stencil when it is being removed from the wall.	When removing and placing your stencils, do it carefully so that any wet paint does not touch the surface. You should also check that the paint on the stencil has dried or been wiped off before placing it on the wall again.
Paint lifting (Fig 8.21)	Paint can lift off the surface when the masking tape holding the stencil on is pulled off.	Make sure you have used a low tack tape. Don't rip it off the wall, but remove it slowly and gently. This will also avoid tearing your stencil. You may consider using adhesive spray instead.
Uneven colour (Fig 8.22)	When paint hasn't been applied correctly and evenly.	Keep an eye on your work and make sure all of the holes in your stencil have been filled and that you have used the same amount of paint for the whole stencil.
Bittiness (Fig 8.23)	Where there is dirt or debris on the surface.	Your wall should be properly prepared before starting. Make sure the surface and the area around you is clean before starting.
Undue texture	When too much or too little paint has been applied through the stencil, e.g. if the brush or roller has been overloaded.	Blot your brush and roller each time you load it. Stencilling is usually a light effect, so less is more – you can load your brush or roller again if needed.
Uneven weight of colour over repeats	When using more than one colour or a multi-layered stencil, the colours may have been applied more lightly or heavily between repeats of the design.	Try to use the same amount of paint for each stencil section. Start off lightly and add more paint only if you need to. If uneven weight of colour has occurred, you would need to completely redo the job.
Buckled/ curled stencil plate	If your stencil has become old and worn or not cleaned and stored correctly, it can become misshapen and cause the design to be wrong or irregular.	Clean, dry and store your stencils to keep them flat for the next time you use them. You may need some extra adhesive to hold them flat against the wall, but do not use too much as it can leave a residue. It may be better to cut a new stencil.

Table 8.4 Application faults

Figure 8.19 Creep

Figure 8.20 Smudging

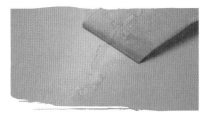

Figure 8.21 Paint lifting

Figure 8.22 Uneven colour

Figure 8.23 Bittiness

Cleaning stencil tools and equipment

Cleaning painting tools and equipment has been covered in Chapter 6 (see pages 202–6).

PRODUCE REPLICA WOOD GRAINING AND MARBLING EFFECTS

Certain types of timber and marble can be very heavy, large and unwieldy to work with, very expensive, and difficult to source. For these reasons, it is often much easier to replicate the look of the natural materials by using decorative painting techniques. Sometimes these effects may have already been used in the same space or building, in which case it would be better to match the existing work than to change materials.

Wood grain

As a tree grows it produces layers of wood cells that align themselves with the direction of the trunk. This is what gives wood its grain.

The grain pattern in wood appears differently depending on how it has been cut:

* Flat grain – where a timber log has been sawn (**plain sawing**) from one end to the other parallel to the grain direction. It is the most common method of sawing timber and shows the annual rings and reveals the best grain patterns.

* Quarter grain is cut parallel to the grain but through the radius of the growth rings (**quarter sawing**). It is more expensive and less common. It has a subtle figure (grain pattern) but gives a beautiful ray pattern in oak.

* End grain – where timber has been cut at a 90° angle to the grain – you will be able to see the rings.

Oak and mahogany grain

Oak and mahogany grain effects are designed to look like natural wood grain.

Oak graining

Oak coloured scumble is applied to the ground coat with a brush and then a dragging brush to create a **straight grain** effect, then a comb is dragged through the wet scumble to imitate the interesting **figure work** that appears in the wood. A flogger is used to recreate the pore marks, and mottlers and softeners are used to finish the effect (see page 276). Once the scumble has dried, a varnish is applied.

Mahogany graining

An even layer of mahogany scumble is applied to the ground coat with a brush and flogged to remove the brush marks. This leaves pore marks. A cutter is then used to create a series of elongated curves that imitate the figure work found in mahogany heartwood. The effect should be softened – starting from the centre. A mottler is used to remove small amounts of the scumble to create highlighted areas usually found in mahogany.

See practical tasks 1 and 2 on pages 290–3 for step-by-step instructions on how to *Produce figure work mahogany and oak wood grain effects*.

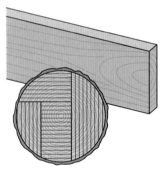

Figure 8.24 Plain sawn wood

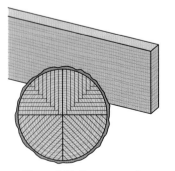

Figure 8.25 Quarter grain wood

Figure 8.26 Straight grain

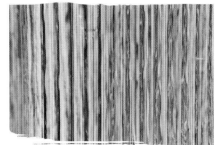

Figure 8.27 Light oak graining

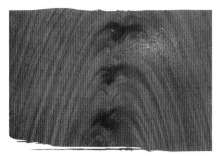

Figure 8.28 Mahogany graining

Characteristics of marbling

Marble is a rock that comes from limestone that is formed in layers. The limestone within the earth's surface is exposed to high pressures and temperatures and becomes marble. The distinctive coloured patterns found in marble are formed by this process. Marbling has been used for centuries for decorative purposes, such as for walls, floors and decorative features.

* Laminated – marble has a layered structure which comes from the way in which the limestone (that it comes from) has been formed.

* Brecciated – marble contains angular fragments which results in contrasting streaks of texture that appear randomly within the marble. Brecciated marble forms when nearby minerals or materials (such as shells or coral) become embedded in the limestone, which over time form bands within the marble.

* Crinoidal marble – originates from the sea floor and contains fossils and fine veins.

* Stalagmitic marble – also known as onyx marble, it comes from stalagmites and stalactites – icicle-shaped structures that are formed as water that has passed through cracks in the rock meets the air at the top and bottom of a cave. It is a pearly white colour and is used as a decorative stone.

* Verde antique – often mistakenly called serpentine marble, is not actually a marble. It is a decorative stone that is dark green in colour with white veins. It is used in construction.

* Variegated – means containing irregular patches or streaks and can be used to describe the patterns found in marble resulting from the veins.

Types of marble

There are many different types of marble that are unique to different regions all over the world.

Carrara marble

Carrara marble comes from Italy and is white or blue/grey in colour. The more pure the marble, the whiter it is. There can also be some discoloration in Carrara which gives it some green, blue or yellow tinges.

Vert de Mer marble

Vert de Mer marble is a black and green marble effect with distinctive white veins and a lot of detail.

Sienna marble

Sienna marble also comes from Italy. It is yellow in colour with red, blue and white veins.

DID YOU KNOW?

The difference between scumble and varnish
Scumble is the coloured base for creating decorative effects, whereas varnish is a coloured coat that adds shine and protects the finished surface. It can enhance the colour of the effect.

Figure 8.29 Brecciated marble

Figure 8.30 Crinoidal marble

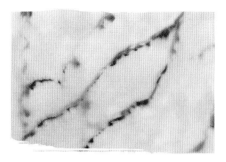

Figure 8.31 Carrara marble effect

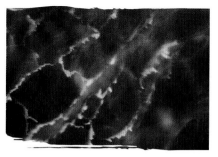

Figure 8.32 Vert de Mer marble effect

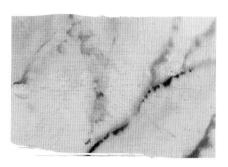

Figure 8.33 Sienna marble effect

Other types of marble effects

Figure 8.34 Black and Gold (or Portoro) marble effect

Figure 8.35 Rouge Royal marble effect

Figure 8.36 St Annes marble effect

Figure 8.37 Breche Violet marble effect

Scumbles for wood graining and marbling

Scumbles for both wood graining and marbling (see above) can be either oil or water-based materials.

Oil-based scumbles

The following are ingredients in oil-based scumbles:

* Oil-based glaze– a transparent coating applied to the ground coat to provide a two-tone effect. Colour is added to suit the effect you are producing.

* Oil colourant – comes in a range of colours suited to copy the colour of various wood types and are added to an oil-based glaze. They come in standard colours but can be adjusted further to better match existing colours if needed.

* Oil graining colour/medium – when creating a marble effect, it is necessary to make the scumble flow so it blends with the other colours to produce a soft effect. This is achieved by mixing white spirit, linseed oil and a liquid drier along with the colour stain.

* Proprietary scumble – it is possible to purchase pre-tinted oil-based scumbles in a range of standard wood colours, specifically designed for wood graining.

* Solvent-borne varnish – a varnish will seal and protect the scumble glaze and reduce yellowing and comes in finishes from matt to gloss. It can be either colourless or tinted with the appropriate colour.

* White spirit – used as a thinner or cleaning solvent for oil-based glaze or scumble.

* Linseed oil – used to extend the drying time of the scumble.

* Driers – these may be added to speed up the oxidation (drying) process, especially when working in cold conditions.

Water-graining mediums

The following are ingredients in water-based scumbles for wood and marble effects:

* Glaze – clear emulsion glaze can be tinted with colorants and has a longer wet edge time.

* Acrylic colorant – this can come pre-mixed in a concentrated liquid form to be added to the glaze.

* Dry pigments – powdered colours can be used to add to a glaze, but they must first be soaked in water overnight until they turn into a paste, then thinned with water and a binder (such as Fuller's earth, stale beer or vinegar).

* Water-graining colour/medium – you can also use stale beer for a medium mixed with water/acrylics. The beer allows the medium to spread and helps with binding the pigment to the surface.

* Crayons – these can be used in oil-based marble effects to create some of the veins, especially when underlining strong veins in Sienna marble.

* Binders – binders hold the glaze and colourants together and make them adhere better to the surface to prevent cissing (see below). Binders for acrylic paint include Fuller's earth, whiting, and stale beer.

* Acrylic varnish – as for solvent-borne varnish, but should be used over water-based scumbles.

* Glycerine – can be added to extend the drying time of the scumble.

* Proprietary retarding agents – these are extenders specifically designed for slowing down the drying time of the acrylic scumble.

When you are using water-based scumbles, there is the risk of cissing, where the paint separates while wet because it has no key to attach to the surface below, e.g. when using over the top of an oil-based paint. Using stale beer mixed with water helps with the spreading and binding of the scumble. Fuller's earth (a clay-like material) and detergent, which is wiped onto the surface after abrading, are also used as binders for water-based graining materials. Both of these will stop cissing.

Figure 8.38 Buff ground colour

Whiting is another ingredient that can act as a binder. It is a chalky powder used to make glazing putty, and it can also be added to thicken paint or make it more opaque. Most white, water-based house paints use whiting as the pigment.

Colours

It is important to choose the right colours for ground coats so that the final effect is as realistic as possible. The right ground coat will enhance the colour of the wood and marble you are trying to create.

The colour of certain types of wood is very distinctive, e.g. mahogany is a reddish-brown and oak ranges from a light honey colour to a darker, warm golden brown. The ground coat for a timber effect is usually a neutral colour such as 'buff', a light yellow-brown.

Figure 8.39 Mahogany ground colour

Wood grain effect	BS 4800 colour	RAL colour
Mahogany	04 D 44 Misty red/tawny	RAL 8016 Mahogany brown
Light oak	Tint of 08 C 35	RAL 1001 Beige
Medium oak	08 C 35 Fudge/Butterscotch/Bamboo	RAL 1011 Brown beige

Table 8.5 Commonly used colours for wood grain effect

Figure 8.40 Artist's oil

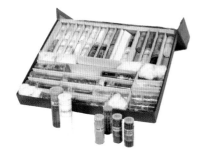

Figure 8.41 Powder pigment

There is a selection of commonly used colours for creating some timber effects and the veins in marbling. These tend to be earthy colours such as raw sienna, ochre, umber or burnt umber.

Carrara marble in its purest form is white with grey veins, therefore the appropriate ground colour to use is white. Vert de Mer marble is black and green with white veins, therefore the appropriate ground colour to begin with is black. Sienna marble is yellow with different coloured veins so an appropriate base colour is yellow or ochre.

Marbling effect	BS 4800 colour	RAL colour
Carrara	00 E 55 White	RAL 9010
Sienna	06 C 37 Sienna	RAL 9010
Verte De Mer	00 E 53 Black	RAL 9005

Table 8.6 Commonly used colours for marble effect

Colorants

Colorants used for scumbles include:

* Artist's oil – these come in a very wide range of colours and are typically manufactured for the benefit of artistic painters who use oils as their medium. They use high levels of pigment and provide a stronger tint.
* Acrylics – for use in water-based scumbles.
* Poster colours – these are water-soluble paints with a glue-size binder. It tends to be a cheap material, comes as a liquid or powder and is often used as children's paint.
* Powder pigment – these are pure colours mixed with water which can then be added to your scumble. You could also mix them with linseed oil for an oil-based scumble.
* Universal stainers – these are a liquid colourant that can be added to an oil-based glaze.

Pigment colours for different types of marble effects

Marble type	Pigment colours
Cararra	Payne's grey, titanium white, raw umber, yellow ochre
Vert de Mer	Prussian blue, yellow ochre, titanium white
Sienna	Burnt sienna, yellow ochre, ultramarine, Indian red, raw umber, lamp black, titanium white
Black and Gold	Lamp black, yellow ochre, titanium white
Rouge Royal	Lamp black, titanium white, raw umber, red ochre, yellow ochre
St Annes	Payne's grey, titanium white
Breche Violet	Titanium white, lamp black

Table 8.7 Pigment colours for different types of marble effects

Gilp

Gilp is a mixture of one part linseed oil to two parts turpentine or white spirit and 10 per cent liquid oil (terebine) driers. The gilp is applied over the ground coat when creating marble effects. Its purpose is to prevent the oil-based paints from drying out to give you a longer working time.

TOOLS AND EQUIPMENT FOR GRAINING AND MARBLING

Some of the tools for creating replica wood graining have already been covered on page 261: flogger/dragging brush, sponges, lint-free cloth, palettes and plastic pots.

Tools and equipment	Description and use
Rubbing-in brushes (Fig 8.42)	A rubbing-in brush is a normal paintbrush, that is old and worn. They are called rubbing-in brushes because you rub the scumble into the surface to help it spread more evenly.
Mixing brushes	You will need various paintbrushes on hand to mix your paints, glazes and scumbles.
Pencil overgrainer	Long thin brushes used to create fine detail in the graining work and give a secondary grain effect.
Varnish brushes	Varnishing brushes are used specifically for applying varnish, not paint. These brushes tend to be tightly packed with 100% natural bristle filling. A slightly oval-shaped head of the brush will help when getting into slightly uneven timber surfaces. There are brushes specially designed for applying stains and varnishes, e.g. the Hamilton namel var brush.
Lining fitch (Fig 8.43)	A fitch is a type of fine brush used for detailed work. They come in various sizes. A lining fitch has a flat, diagonal brush end, designed for painting bands.
One stroke brushes	A short flat brush used for edges and shaping. Due to its shape it makes a unique chisel edged flat brush mark. One-stroke brushes are used to add figure work to both oak and mahogany by adding colour.
Mottlers	Mottlers are soft-bristled brushes used to remove areas of coloured glaze to create highlighted areas found in certain types of wood.
Cutters (Fig 8.44)	Made from hog's hair or camel hair, cutters are used to paint in heartwood grain and create fine highlights, such as in mahogany. They come in a range of sizes from depending on the effect required.
Softeners (Fig 8.45)	Softeners are soft brushes made of hog or badger hair. They are used to make the edges of graining or marbling effects more gradual, or softer. Badger softeners should be used for water-based materials; hog's hair softeners should be used with oil-based materials. You may also hear them referred to as 'blenders'.
Combs (metal, rubber, card) (Fig 8.46)	Combs are used for creating a wood grain effect. They come with teeth of different sizes to suit different effects.
Check/tick roller (Fig 8.47)	This is a tool used for recreating the look of old, weathered wood. It is made of loose metal discs with jagged edges. Some rollers are fitted with a mottle, which coats the discs in colour.
Sable pencils and writers (Fig 8.48)	These are fine brushes used by artists and signwriters. Sable hair is a traditional material taken from the winter coat of the sable, known for being extremely soft. The brushes are either shaped into a fine point or with a flat 'chisel' edge, and are used for creating veining when producing a marble effect
Feathers (Fig 8.49)	Large, stiff feathers, such as goose wing feathers, are used when creating a marbling effect. Feathers can also be used for certain grain markings when producing a wood grain effect.
Dipper	A small plastic or metal container that clips to the side of the palette for holding oil, turpentine or gilp.
Veining horn	A plastic tool with a round end and a square end. It is used to produce oak grain patterns. It can also be used during marbling when colour is wiped out on the marbling to expose the ground coat.
Crayons	Crayons can be used in oil-based marble effects to create some of the veins, especially when underlining strong veins in Sienna marble.
Palette knife (Fig 8.50)	For mixing colours used in marbling on the palette.

Table 8.8 Tools and equipment for wood graining and marbling

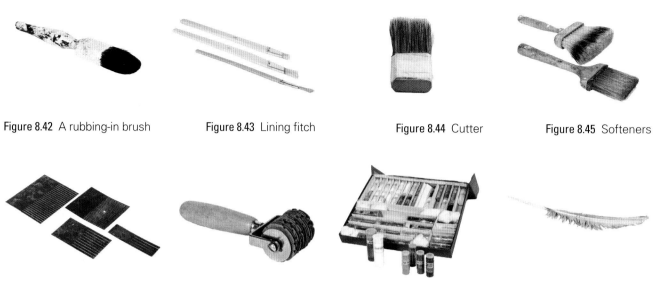

Figure 8.42 A rubbing-in brush

Figure 8.43 Lining fitch

Figure 8.44 Cutter

Figure 8.45 Softeners

Figure 8.46 Graining combs

Figure 8.47 Check roller

Figure 8.48 Sable pencils and writers

Figure 8.49 Feathers

Figure 8.50 Palette knife

APPLYING GRAINING AND MARBLING

Processes for creating wood graining and marbling effects

Process	Description
Oil-in or rubbing-in	Using a rubbing-in brush once you've applied scumble to the ground coat. Because the brush is soft and worn, it reduces friction and you will end up with a more even, softer effect. When creating a wood grain effect, work with the brush in one direction only.
Flogging	Once the scumble has been applied, use a flogging brush to gently tap the surface in a straight line with the brush at a shallow angle (45° or less). Complete one strip at a time and move slowly in one direction only, from bottom to top. Flogging will imitate the look of wood pores, and is used in wood graining.
Combing	Using a steel or rubber comb, drag the comb through the scumble to reproduce the look of straight grain patterns in wood. for a more detailed timber, hold the comb at a 30° degree angle.
Softening	After the scumble has been rubbed in, flogged and combed, using a badger or hog hair softener, gently brush over the scumble in various directions. This will fade any sharp edges from the graining patterns in wood grain. When marbling, the softener is used similarly to blend out the strong lines created by veining.
Mottling (Fig 8.51)	Using a mottle at a 45° angle, wipe away narrow strips of the glaze in a side-to-side motion to add light and dark areas to the wood effect. Once you have done this apply the mottler over the area using 's' shape movements to blend the effect so that isn't uniform. A mottler can also be used to enhance the grain over heartwood.
Overgraining (Fig 8.52)	The overgrainer is dragged through the scumble at 90° to the surface to create fine detail in the graining work and give a secondary grain effect.
Wiping out (Fig 8.53)	Draw a veining horn wrapped in an absorbent lint-free cloth through the glaze to make short twisting lines or **rays**. Also used in quartered oak figure work.
Painting in heartwood (Fig 8.54)	Oak grain: See 'Wiping out' above. Mahogany: Use a cutter, create long curved shapes that imitate the **heartwood** grain pattern in mahogany.

Table 8.9 Processes for creating wood graining and marbling effects

Figure 8.51 Mottling

Figure 8.52 Overgraining

KEY TERMS

Rays
– flecks that cross the grain pattern. Found in some hardwoods, especially oak.

Heartwood
– heartwood is a layer of dead cells that support the tree. It produces a rich-coloured wood.

Figure 8.53 Wiping out

Figure 8.54 Painting in heartwood

Direction of natural grain

When applying grain to a variety of structural components there are a few general rules that you must follow. Before applying grain, think what it would look like if real wood had been used and apply the grain effect in that direction. For example, on a panelled door the grain on the rails would run horizontally and the grain on the stiles would run vertically. Table 8.10 below outlines the sequence that you should follow when applying wood grain to common structural features:

Structure	Order of applying wood grain effect
Panelled doors	Start by applying wood grain effect to edges and panels. Then apply to mouldings, muntins and stiles. Follow the natural direction of the wood grain (see above).
Windows	Start at the top and work downwards. Finish with the window sill. Apply top and bottom of the frame horizontally and the sill horizontally and the sides of the frame vertically.
Dado rails	Apply in sections and apply horizontally to the whole of the section.
Architraves and skirting	Apply horizontally to skirting and then apply vertically to architraves.

Table 8.10 The sequence for applying wood grain to a range of structural components

Importance of cleanliness and sharpness when graining

When applying varnish the success of the finished result depends on the cleanliness of your brushes and the working environment:

- Previously used varnish should be strained.
- Use old bristle brushes as new brushes may shed small hairs.
- Do not use synthetic brushes as these may shed bristles.
- Use separate brushes for varnishing only.
- Use a tack rag to remove any dust on the surface.
- When varnishing, place clean paper under the surface being varnished to prevent the brush from picking up dust.
- Keep the work area covered and vacuum to avoid dust.

Sharpness of the grain is also important in the finished effect:

- If the solvent is too thick and is evaporating too quickly, thin it down.
- Make sure that your wiping out is clean and clear of smudges and marks.

CLEANING AND MAINTAINING TOOLS FOR GRAINING AND MARBLING

Refresh your memory about cleaning and storing your painting tools and equipment in Chapter 6 pages 202–6, and on page 264 in this chapter. Keeping your tools clean, dry and stored appropriately will mean they last you longer and give you a better finish.

METAL LEAF

The process of applying thin sheets of metal leaf for decorative purposes is known as gilding. Metal leaf can be applied to both external surfaces, such as decorative iron work and signwriting, and internal surfaces, such as plaster cornices, ceilings and wall. It can also be applied to timber, glass, porcelain and a variety of other surfaces.

History of gilding

The Ancient Egyptians were the first to use gilding over 3,000 years ago. The Egyptians are known to have used gold leaf on Pharaoh's tombs, coffins and other objects. The Ancient Greeks and Romans applied gilt to ceilings of important buildings such as temples and palaces. The art of gilding has also been practised for centuries across Europe – for example by the Italians, the French and the Dutch. In Britain, gilding has been used since the seventeenth century for interior decoration. It continues to be popular today to give a luxurious effect in both modern and traditional interiors.

Figure 8.55 The art of gilding has been practised for centuries and gives a luxurious finish

Production of gold leaf and other types of metal leaf

The process of producing gold leaf and other types of metal leaf is similar. Outlined below is the process for gold leaf.

There are two main steps in the production of gold leaf:

* rolling
* beating.

Rolling

* The metal and its **alloys** are melted in a furnace.
* The liquid gold is poured into a mould and left to cool.
* It is then passed through a roller many times to produce a very thin sheet. The gap between the rollers is reduced each time until the required thickness is reached (about one thousandth of an inch in thickness for gold).

Beating

* The rolled metal sheet is cut into smaller squares (about 2.5 cm in size for gold) and placed delicately using wooden pincers inside a 'skin' made from a polyester film, such as Mylar.
* The film 'skin' is then wrapped in several layers of animal skin to hold the packet together and it is beaten into a very fine sheet.
* The very fine sheets are then cut into squares to be supplied as books.

Metal leaf formats

Books

Books come in two forms:

* Loose leaf – each leaf sits loosely within the book on paper. It needs to be picked up using a gilder's brush/tip and placed on a gilder's cushion for cutting or transferred directly to the sized surface. Loose leaf is used in water gilding (see page 283).
* Transfer leaf – leaf that is lightly pressed to a fine paper backing sheet. It can be removed from the paper by hand. The transfer leaf is placed face down onto the sized surface and the backing paper is rubbed to transfer the metal onto the surface. This type of leaf is easier to use on flatter surfaces and easier to use than loose leaf. Transfer leaf is used in oil gilding (see page 283).

Ribbon leaf

Metal leaf can also come as a long ribbon on tissue paper/waxed paper. Ribbon leaf can come in rolls up to 75 m in length and vary from 6 mm in width to 100 mm.

Powder

Metallic powders are also available and can either be applied to a sized surface or mixed with varnish and applied like paint. Common metallic powders include bronze powder and aluminium powder.

KEY TERMS

Alloy

– a blend of two or more metals (e.g. gold, silver and copper in the case of gold leaf).

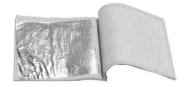

Figure 8.56 Loose leaf

Figure 8.57 Transfer leaf

Figure 8.58 Metallic powder

Types of metal leaf

Gold leaf (Fig 8.59)	The purity (**carat**) of the gold and colour of the gold leaf depends on the amount of the alloys (silver or copper) added. For example, white gold which is only used on interior surfaces may be an alloy of gold and nickel. Gold leaf is available in a range of carats from 6 carat to 18 carat and 24 carat – which is pure gold and rarely used in decorating. The most common gold leaf used is 23 carat gold which is mixed with a little silver to make it easy to work with. 23 carat gold can be used on most exterior surfaces.

Size and format:	**Advantages and disadvantages:**
• Books contain 25 leaves • 80 mm² • Available as loose leaf and transfer leaf	• Gold leaf cannot be applied by hand because of the fineness of the sheets. It needs to be applied using a **gilder's tip**. • It gives a high-quality finish and doesn't tarnish.

Copper and aluminium leaf (Fig 8.60)	Copper and aluminium leaf are cheap and easy to produce. Aluminium is often used instead of silver on exterior surfaces. As aluminium is more readily available, the squares are larger in size and are much thicker, so they can be applied by hand.

Size and format:	**Advantages and disadvantages:**
• Books contain 25 leaves • 140 mm² • Available as loose leaf and transfer leaf	• Aluminium doesn't **tarnish**. • Can be applied by hand. • Cheaper than gold. • Copper tarnishes so a protective layer needs to be applied.

Dutch metal (Fig 8.61)	This is another name for brass leaf and is an alloy of 85% copper and 15% zinc. It can be applied by hand and is sometimes used as a substitute for gold.

Size and format:	**Advantages and disadvantages:**
• Books contain 25 leaves or packets of 500 or 1,000 leaves • 140 mm² • Available as loose leaf and transfer leaf	• As the copper tarnishes, it needs to be sealed with a protective layer. • It can be used as a substitute for gold. • It is difficult to use with water gilding.

Platinum	Platinum leaf gives a darker finish than other white metals but has a warm red tone.

Size and format:	**Advantages and disadvantages:**
• Books contain 25 leaves • 80 mm² • Available as loose leaf and transfer leaf	• It does not tarnish. • It is quite hard to work with and is quite hard to get hold of.

Silver (Fig 8.62)	Silver leaf gives a mirror-like finish when water gilded.

Size and format:	**Advantages and disadvantages:**
• Books contain 25 leaves • 80 mm² • Available as loose leaf and transfer leaf	• It is thicker than gold alloys and can be applied by hand. • It tarnishes easily and should either be sealed or used on interior surfaces. • As it tarnishes it should not be used with oil gilding as it will turn brown.

Table 8.11 Types of metal leaf

DID YOU KNOW?

Higher carat gold, because it doesn't tarnish, doesn't need a protective coating and can be used on both interior and exterior surfaces.

Figure 8.59 Gold leaf

Figure 8.60 Aluminium leaf

Figure 8.61 Dutch metal leaf

Figure 8.62 Silver leaf

Tools and equipment

Tools and equipment	Description
Gilder's cushion (Fig 8.63)	A flat surface covered with a layer of chamois leather or goat's skin used for cutting and preparing metal leaf. One end has a guard made from parchment to stop the metal leaf from being blown away. Underneath are straps for the thumb and fingers to hold the board.
Gilder's tip (Fig 8.64)	Thin, flat brushes usually made from squirrel, camel or badger hair. The hairs are held in place by two thin pieces of cardboard. It is used to pick up metal leaf and transfer it to the surface. In three sizes: short hair: 40 mm, medium-length hair: 50 mm and long hair: 60 mm.
Camel hair mop (Fig 8.65)	Made from squirrel hair or camel hair, it is used for applying mordant. A dry camel hair mop is also used for removing any excess leaf from the surface.
Gilder's knife (Fig 8.66)	A long, smooth blade with no sharp edge, it used to cut the metal leaf in to pieces. The handle is balanced so that when the knife is set down the blade will not touch the surface.
Pounce bag (Fig 8.67)	A piece of calico cloth filled with French chalk and tied into a small ball. It is used for dusting tacky surfaces before gilding to prevent gold from sticking to certain areas on the surface.
Cotton wool	Used for preparing surfaces used when burnishing and water gilding i.e. used to smooth and wipe down surfaces. Also used for pressing metal leaf into place.
Glass container	Glue size is decanted into a glass container.

Table 8.12 Tools and equipment for gilding

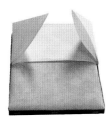

Figure 8.63 Gilder's cushion

Figure 8.64 Gilder's tip

Figure 8.65 Camel hair mop

Figure 8.66 Gilder's knife

Figure 8.67 Pounce bag

Mordant

Mordant or size is the adhesive that holds the metal leaf in place. There are different types of size:

* Oil goldsize: gives a good shine to the end result. It needs to be thinned using varnish. It is available in various drying times or **tack times** e.g. 3 hours, 12 hours or 24 hours which vary depending on atmospheric and surface conditions. The longer the drying time the shinier the finish. This size lasts better than acrylic size. Oil-based size can be tinted with oil colours to produce a base tint.

* Japan goldsize: a traditional oil-based size that has a shorter drying time. The addition of varnish will extend the drying time. It is suitable for smaller areas. Drying times are from half an hour to 8 hours.

* Acrylic goldsize: a fast-drying water-based size with a quick tack time of 15–20 minutes. It can be used for water gilding and with metal powders.

* Gelatine size: water-based size that contains gelatine and is used in water gilding. Gelatine is made from boiling animal hooves, bone and skin.

* Isinglass: a substance that comes from the dried bladders of fish. It is used in water-based size to apply loose leaf metal leaf to glass.

* Glue size: rabbit skin glue size is one of the ingredients for gesso solution and gilding water. It comes in granules.

* Gilding water: a liquid made from water, methylated spirit and rabbit skin size used in water gilding. Also known as gilder's mordant.

* **Gesso**: a fine chalk powder that is an essential part of gesso solution.

Tack time testing

Knowing the tack time of the size is important in order to ensure that the metal leaf adheres properly. For example, metal leaf that is applied to size that is too wet will be drowned and any shine will be lost. The tackiness of the size can be tested using your knuckle. If your knuckle does not stick but you feel a slight pull as your knuckle is pulled away the size has reached the correct tackiness.

Preparation of the surface

The surface that is being gilded should be clean and dry. Any cracks or holes should be filled and abraded so that the surface is smooth and level. For further information on surface preparation, see Chapter 5.

Applying barrier coat

When oil gilding or water gilding, before applying the size you should prepare the surface with one of the following barrier coats to ensure the metal leaf adheres only to the intended areas:

* Egg glair is size made by adding egg white to warm water and mixing thoroughly. Once the gold leaf has been applied the surface should be washed down with clean water to remove the egg glair.

* French chalk can be used instead of egg glair. The process of applying it is similar to that for egg glair.

Applying metal leaf

Metal leaf is applied in one of two ways:

* water gilding
* oil gilding.

Water gilding

This process needs more preparation than oil gilding but it produces a higher quality finish. It is only used internally, for example in the internal décor of stately buildings, as water (in the atmosphere) can dissolve the size. Water gilding can be used to apply all types of metal leaf. The process involves applying numerous coats (up to 12 coats) of gesso to the surface followed by a number of coats (up to 8 coats) of **bole** to provide a very smooth finish on which to apply the metal leaf. The leaf is then applied using **Gilder's water** which activates the glue. If a matt finish is required, it is left as it is or if a shiny finish is required it is polished (**burnished**).

Oil gilding

Oil gilding can be applied to most building surfaces, both internally and externally. It is often used on exterior signs and architecture. An oil-based or acrylic size is applied to a primed surface or a surface that has been prepared with layers of gesso. Metal leaf is applied to the surface when the size has reached the appropriate tackiness. After gilding, loose fragments are brushed from the surface and the surface is buffed with cotton wool. This method does not allow the leaf to be burnished and therefore does not provide the same high-shine finish as water gilding.

See practical task 4 on pages 295–7 for step-by-step instructions on how to *Apply gold loose leaf and transfer leaf*.

Backing up

A layer of coloured paint, clear oil-based varnish, shellac or size should be applied using a pure bristle brush or a sable hair brush to protect and seal the metal leaf. If a coloured paint is used this can affect the tone slightly. For example, red will add warmth to gold, yellow will add a mellow tone to gold and black will make it appear bolder.

Gilding defects

Gilding requires considerable skill to ensure that the end result is free from defects and is long lasting. Some common defects are listed in Table 8.13 below:

KEY TERMS

Bole

– a base colour for gilding. Traditionally it was made from clay. Today paint may be used instead. It adds colour and warmth to the gold and silver leaf.

Gilder's water

– a liquid made from water, methylated spirit and rabbit skin size that is used in water gilding.

Burnished

– polished using a hard smooth surface.

Defect	Cause	How to avoid
Misses/patchy application/ poor adhesion (Fig 8.68)	Patches where metal leaf has failed to adhere.	Make sure size is the right tackiness and avoid letting it become too hard as it won't adhere properly. Add colour to size so that you can see clearly where it has been applied. Repair by 'faulting' (see page 296).
Wrinkled edges (Fig 8.69)	If oil size is too thick or if it contains too much oil, the size remains soft under the leaf causing it to wrinkle. Can also be caused by drafts.	Apply size in a thin even layer.
Uneven edges to the metal leaf	Caused by metal leaf not adhering to the size due to lack of size on the surface or due to the size drying too quickly.	Reapply the size to the area.
Scratches (Fig 8.70)	Poor application using the wrong tools or application techniques, e.g. wrong brushes.	Use the correct tools and brushes (e.g. camel hair/badger hair or hog's hair softeners).
Finger prints (Fig 8.71)	Finger prints can be caused by handling the metal leaf. Salts from the skin leave a mark.	Wear cotton gloves when handling metal leaf. Avoid touching silver and copper leaf until it has been sealed with varnish.
Attached tissue (Fig 8.72)	Pieces of the attached tissue that is designed to protect the metal leaf have been left on the applied leaf.	Make sure that all tissue is removed before applying the leaf.
Lack of lustre (Fig 8.73)	Applying the leaf before the size is dry enough. The leaf 'sinks' into the size.	Allow the size to reach the correct tackiness before applying.

Table 8.13 Gilding defects

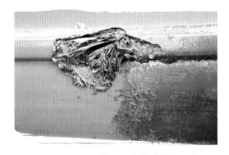

Figure 8.68 Misses/poor adhesion or uneven edges

Figure 8.69 Wrinkled edges

Figure 8.70 Scratches

Figure 8.71 Finger prints

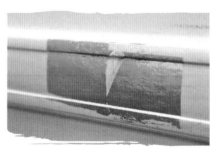

Figure 8.72 Attached tissue

Figure 8.73 Lack of lustre

Storing and cleaning brushes

Specialist brushes, such as the camel hair mop and softener, must be kept in good condition by washing in turpentine or methylated spirit and a little tallow and lard. The mop needs to be oiled so that the solvents that it was cleaned with don't dry out the natural oils. They should be hung up to dry and stored until needed. The gilder's knife should be kept sharp and free of corrosion.

FORM PAINTED LINES AND BANDS

A painted line is where two sections of colour meet, often where you would expect to see a picture rail or a dado rail. You might come across them in older buildings, or in places like a hospital where there would not be an actual moulding on the wall, but where using two colours will give the impression of one. This process is essentially the same as cutting in with a brush before you fill in with a roller.

A painted band is a different colour from your ground coat and is simply designed to add interest to a room or to mimic the effect of a moulding such as a dado or picture rail.

The eye is naturally drawn to imperfections in a paint job, and so it is essential that painted lines and bands are perfectly straight, with no colour creeping over to the other side of the line, and with an even thickness all the way around the room.

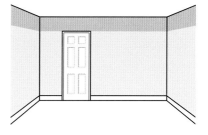

Figure 8.74 A painted line

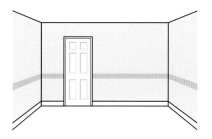

Figure 8.75 A painted band

Marking lines

To mark your lines and bands, use a spirit level and mark around the room with a pencil. You can also use a chalk line with one end taped to the wall.

Chamfered and square-edge straight edges

Painted bands can be created with the help of a straight edge with a chamfered edge. Straight edges will often come with a square edge on one side and a chamfered one on the other.

To paint a fine band, you can hold the chamfered edge up to the wall and using a lining fitch, run the fitch along the edge to create a thin, straight line.

Tools, equipment and materials

Most of the tools, equipment and materials you need have been covered earlier in the chapter. A few extra things to note:

* Water-based paints are the preferred material for painting lines and bands. They dry much more quickly, allowing you to apply a neat second coat after your **dry coat**.

* Chalk can be used for marking your wall. The chalk line can be attached at one end with masking tape while the other end is held and then plucked back.

PRACTICAL TIP

Today, if a painted line or band is required, then it is more common to simply paint over masking tape to achieve the effect. By applying a 'dry coat' of paint to the masking tape, you will avoid any paint seeping in behind the tape, and will end up with a perfectly straight line.

KEY TERMS

Dry coat

– applying a very light, thin coat of paint to a masking tape edge which then dries quickly before the second coat is applied.

Figure 8.76 Chamfered straight edge

- A sash tool can be used when applying painted lines and bands. It is a round brush with a tapered end, allowing for more accurate application of paint
- A mohair pad is particularly good for going around edges. It is a flat rectangular pad, allowing for less splatter than a roller.
- Chamfered straight edge: You may use a straight edge when painting lines and bands on a surface. These can come with a square edge, or one that slopes or tapers off.

Figure 8.77 Sash tool

Cleaning and maintaining your tools and equipment

The sash tool, lining fitch and mohair pad should be washed out straight after the work is finished, and then hung up to dry naturally. The brushes can be coated in a very thin grease or oil, such as vaseline. This will stop them from drying out and will keep the bristles in shape.

The straight edge should also be cleaned off thoroughly – if there is any paint left to dry on it, then it will longer be straight.

Figure 8.78 A mohair pad

PRODUCE TEXTURED DESIGNS AND SMOOTH FINISHES

Textured finishes

Textured surfaces are used on both internal and external surfaces, walls and ceilings. They can be used to provide a decorative finish but are also good at masking surfaces that are not completely sound, e.g. fine cracks or pitted plaster. The coatings used for textured finishes are hardwearing, adhere well to surfaces, can be overpainted, and have a degree of fire and mould resistance.

Surface preparation

In order to use textured coatings on surfaces, you must consider the following:

- Is it clean? The surface should be properly cleaned, e.g. degreased and any glue residue removed.
- Has it been keyed? The surface must be rough enough for the coating to adhere to the surface, e.g. a gloss paint should be lightly sanded.
- Is it porous? The surface needs some porosity for the coating to adhere, and to help the coating dry. It should not be too porous, otherwise the coating will dry too quickly, so a sealant may be needed.
- Is it non-porous? If the surface is non-porous, it may need a layer of PVA to help the coating stick to the surface.
- Is it sealed? The surface should be sealed with an appropriate Artex sealer or unibond to help with adhesion and working time of the textured coating.
- Is there **distemper** present? Distemper is an old type of paint that has a chalky finish. It must be removed or sealed before new coatings are applied, otherwise they will not adhere.

KEY TERMS

Distemper

– a type of paint which is used on lime plaster, often found in old buildings. It is made of powdered chalk and glue-size and applied as a gel. If you touch the surface, you will see a white powder come off on your hands.

- Is it a loose, friable surface? External walls may be friable, meaning it is flaky, peeling or chalky. Because a textured coating would not adhere to this, it should be raked out and brushed down to remove any loose material, then primed with a sealant.

- Is there new plasterboard? You may need to use PVA because the surface is not porous enough, as well as jointing filler and tape.

Texture designs

- Stipple – using a rubber stipple brush, the coating is bounced to leave a rough and bumpy finish. It is often used on ceilings to hide small defects such as indentations.

- Swirl – a swirl effect is also used on ceilings, after stippling. The stipple brush is turned through the coating in clockwise and anti-clockwise circles, avoiding rows of circles.

- Bark – often used on walls, a textured bark roller is rolled down the surface in one direction, leaving a raised pattern that looks similar to bark on a tree.

- Broken leather – a thicker coating is applied, then the stippler is covered in a plastic bag and the coating is twisted using random strokes in various directions.

- Circle/fan – using a comb, use your wrist to create semi-circular shapes. The second swirl is positioned so it overlaps the first one. With the second row ensure that the top of the fan shape meets the bottom of the fan shape from the first row

- Basket weave – using a comb at 90° drag the comb in both horizontal and vertical directions to achieve a woven appear. You will need to plan out the pattern prior to starting.

For step-by-step instructions on how to *Apply a half circle or fan textured effect to a ceiling* see practical task 5 on pages 297–9.

PRACTICAL TIP

Note: asbestos used to be present in textured coatings. While this is no longer the case, if you are working on an older surface that already has a textured coating, then certain safety guidelines must be followed. If you suspect that a textured coating contains asbestos, see your supervisor before doing any work on it.

DID YOU KNOW?

For more information on asbestos and what to do if you encounter it, go to: *www.hse.gov.uk/ asbestos/essentials/ index.htm*

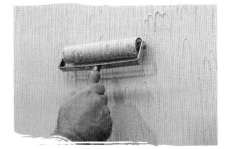

Figure 8.79 Stipple effect

Figure 8.80 Bark effect

Figure 8.81 Swirl effect

Figure 8.82 Broken leather effect

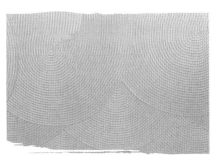

Figure 8.83 Circle/fan effect

Polished plaster

Polished plaster is a smooth decorative finish. There are different finishes from **Venetian plaster**, with a highly polished finish, to **Marmorino** and other textured polishes.

Polished plasters are applied in thin layers using a spatula or trowel and then polished to leave a smooth stone-like surface.

Tools and equipment

Tools and equipment	Description and use
Paddle/bumper (Fig 8.84)	Also known as a mixing tool, it is a round metallic or plastic plate with a series of holes that can be attached to a handle, such as a broom handle. It is used for mixing textured paints. The paint passes through the holes when the bumper is plunged up and down.
Bark roller (Fig 8.85)	A roller made of foam with a textured or patterned surface. When rolled on wet texture paint it leaves a timber bark effect.
Lacer (Fig 8.86)	A lacer or lacing tool is a plastic triangle used for scraping off the sharp tips that can occur with textured finishes. The lacer is used for removing these spikes or high-points as the paint hardens, but before the paint has dried.
Rubber stipple brush (Fig 8.87)	The filling is made of thin and flexible rubber 'bristles', set into a wide, flat base. Unlike most brushes, the handle is rounded and attached at two points to give you a solid grip.
Texture comb	A plastic comb with a flexible blade containing a series of grooves. As the comb is dragged through the paint it creates rows of patterned lines on the surface.

Table 8.14 Tools and equipment for textured finishes

Figure 8.84 Paddle or bumper

Figure 8.85 A bark roller

Figure 8.86 A lacer

Figure 8.87 A rubber stipple brush

Materials for creating textured finishes

Texture materials

A thick paint-like coating is used to create textured effects. They are usually water-based and are available either in a powder form that you mix yourself or as a ready-mixed preparation. The advantage of using a powder is that you can vary the thickness of the mixture, which will affect your patterned finishes. Ready-mixed material will give you a consistent finish and save you preparation time; however, it is more expensive to buy in this form.

Textured paint materials are often given the generic name of 'Artex', which is a well-known brand specialising in textured coatings. There are similar products for exterior use known as 'high-build' which can be used to cover up fine cracks and uneven surfaces.

When working with a texture-effect paint, it should be applied in sections of 1 m × 0.5–1 m. This is to maintain the correct consistency (i.e. so it keeps a wet edge and does not dry too quickly before you can apply the patterned effect). If you need to slow down the drying time of your mixture, keep the temperature of the room down and reduce ventilation by closing and covering windows. Once the paint has been applied and worked on the whole surface, increasing ventilation and temperature will help speed up the final drying time.

Masking and protection

Working with textured effect paints can be extremely messy. As such, you will want to apply masking tape and papers to any adjoining surfaces such as skirting boards. The floor surface should also be covered with drop sheets.

Finishing processes

Lacing

Once a textured effect has been finished and has dried, it can leave very sharp edges. When applying stipple and bark effects to walls in particular, it is important that these sharp edges are not left on the surface as they can be a hazard.

To remove these sharp edges, once the effect has been finished and has partly dried, a lacing tool is dragged along the surface. The lacing tool will even up the finish and take off any spikes. It is best to do this before the coating has fully dried, otherwise the final effect would be damaged when removing the sharp edges.

Applying a margin

Textured effects require the use of a tool such as a stipple brush or roller to make the finished pattern. As you have already learnt when applying paint with a roller, it is not possible to reach the very edges of your surface without using a small tool such as a brush.

Once a section of textured paint has been applied and worked with a tool, you will need to use a small brush (13–25 mm) to put a band around the edges. This margin will leave you with a neater finish.

Margins should be applied at the edge of the ceiling, in corners, along skirting boards and around obstacles such as light fittings and switches.

Figure 8.88 Lacing

Figure 8.89 Applying a margin

Health and safety

Refer to pages 104–5 of Chapter 4 for information on health and safety including the Work at Height Regulations 2005, ACoP, COSHH and the importance of following the manufacturer's instructions.

Bear in mind that you will come into contact with solvents and solvent-based products. Remember that they are a fire hazard, as well as the risks to your respiratory and skin health.

In addition, it is important to note that rags soaked in solvent should not be stored bunched up anywhere, including in your pockets, because of the risk of spontaneous combustion. Lay rags out in a well-ventilated area so that the solvents can evaporate.

CASE STUDY

Sustainability – helping the environment and budgets!

Thomas Regan works for Accrington and Rossendale College as a painter and decorator

'Sustainability is a big buzz word in our industry. I try to work in a sustainable way as far as I can. For example, I try to calculate just enough paint or wallpaper to complete a job so that there is as little waste as possible when a job is completed. This is also a cost-effective approach and saves money from my budget. I will try to re-use materials, such as plastic sheeting, sandpaper, solvent for washing out as much as possible. Where I can I use low VOC paints as these are much kinder to the environment. I believe that demonstrating sustainability shows good practice and professionalism, helps the environment and also the budget!'

1. PRODUCE A FIGURE WORK MAHOGANY WOOD GRAIN EFFECT

OBJECTIVE

To produce a water-based mahogany heartwood effect on a panelled door.

PPE

Ensure you select PPE appropriate to the job and site where you are working. Refer to the PPE section of Chapter 1.

TOOLS AND EQUIPMENT

Hog hair softener

Sable writer (chisel edge)

Fine sized artist brush

25 mm worn paintbrush

2 × 50 mm rubbing-in brushes

Flogging brush

Synthetic sponge

Paint kettles

Quantity of clean lint free rags

Terracotta red oil-based eggshell paint

Vandyke brown oil tube colour

Oil-based mahogany scumble

Raw linseed oil

White spirit

Burnt umber oil tube colour

STEP 1 Prepare and coat the door with a ground coat using a foam roller – terracotta red oil-based eggshell paint. Remember there should not be any blemishes on the surface of the finished panel.

Figure 8.90 Finishing off a quality ground coat

STEP 2 Prepare the scumble as follows:

- Place a small amount of water into a clean paint kettle (enough to coat the panel you are working with).
- Mix in the required amount of burnt sienna.
- A small amount of acrylic scumble glaze can be added to the mix to extend the working time (open time).

Figure 8.91 Using rubbing-in brush to apply scumble

STEP 3 Mask off the panel to complete figure work graining and apply an even coat of the mahogany scumble to the panel using a partly worn 50 mm paintbrush.

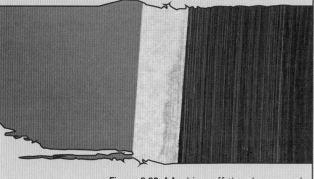

Figure 8.92 Masking off the door panel

STEP 4 Using a cutter, open out the feathered shape found in this type of timber.

Starting at the bottom, apply a triangular shape using one or two passes (depending on the size of the panel).

Figure 8.93 Forming the feathered shape

STEP 5 Carefully soften any hard edges of the effect using a hog hair or badger softener.

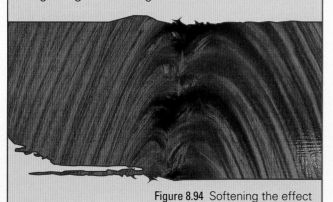

Figure 8.94 Softening the effect

STEP 6 Use a mottler to mottle the side grain.

Figure 8.95 Mottling the side grain

STEP 7 Soften the complete panel using a badger or hog hair softener.

STEP 8 Complete the panel by applying one or two coats of eggshell varnish for protection and final appearance.

Figure 8.96 Applying coat of varnish

2. PRODUCE A FIGURE WORK OAK WOOD GRAIN EFFECT

OBJECTIVE

To produce figure work oak graining on a panelled door using solvent-based materials.

PPE

Ensure you select PPE appropriate to the job and site where you are working. Refer to the PPE section of Chapter 1.

TOOLS AND EQUIPMENT

Buff ground coat – yellow ochre (06 C 33) eggshell

Foam roller

Burnt umber oil tube colour

Raw sienna oil tube colour

25 mm worn paintbrush

Rubbing-in brush

Paint kettles

Flogger

Two metal combs – 25 mm and 100 mm

Veining horn (or a wipeout tool)

Lint free cloth

Badger or hogs hair softener

White spirit

Wax or solvent-based varnish

STEP 1 Prepare and coat the panel with a buff ground coat – yellow ochre (06 C 33) eggshell using a foam roller. Remember there should not be any blemishes on the surface of the finished panel.

STEP 2 Mix the scumble of burnt umber and raw sienna – equal parts – and white spirit.

Figure 8.97 Applying the scumble

STEP 3 Mask off the panel to apply figure work graining and apply an even coat of the mahogany scumble to the panel using a rubbing-in brush.

STEP 4 Prime a flogger with the scumble and flog the entire surface bottom to top.

Figure 8.98 Flogging

STEP 5 Comb the entire panel vertically from top to bottom at a 30° angle with the 100 mm comb.

Figure 8.99 Combing 1

STEP 6 Using the smaller 25mm comb, comb diagonally from top to bottom

STEP 7 Using a veining horn (or a pencil eraser or wipeout tool), wipe out the medullar rays.

Figure 8.100 Wiping out 1

STEP 8 Dab a lint-free cloth in between the rays where you have wiped out.

Figure 8.101 Wiping out 2

STEP 9 Soften left to right and up and down with a softener.

Figure 8.102 Softening

STEP 10 Apply a protective layer of wax or solvent-based varnish to protect.

PRACTICAL TASK

3. PRODUCE A SIENNA MARBLING EFFECT

OBJECTIVE

To produce a Sienna marble effect on a panel using water-based materials.

PPE

Ensure you select PPE appropriate to the job and site where you are working. Refer to the PPE section of Chapter 1.

TOOLS AND EQUIPMENT

Rubbing-in brush

Flat fitch (size 6)

Lint-free rags

Sable pencil brushes (size 0 and 2)

Hog hair softener

Palette

White eggshell (water based?)

Raw sienna tube colour

Burnt sienna tube colour

Burnt umber tube colour

Yellow ochre tube colour

Black tube colour

Fuller's earth

Linseed oil

Turpentine/white spirit

Liquid oil driers

Clear or white glaze

Cotton bud

STEP 1 Prepare the surface with a buff ground colour O8 B 15. Lightly abrade the ground coat once dry to prevent cissing or wipe over the surface with Fuller's earth.

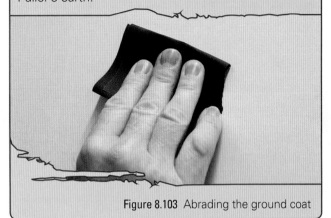

Figure 8.103 Abrading the ground coat

STEP 2 Mix together the gilp – one part linseed oil, two parts turps/white spirit and 10% liquid oil driers.

STEP 3 Mix together some yellow ochre tube colour with gilp and apply evenly to the surface in wide diagonal strips using a fitch.

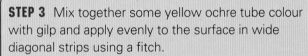

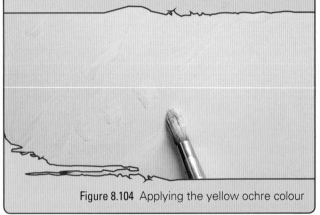

Figure 8.104 Applying the yellow ochre colour

STEP 4 Soften in all directions with a hog hair softener so there are no hard lines of colour or brush marks.

Figure 8.105 Softening

STEP 5 Mix the burnt umber with the gilp and apply in diagonal strips using a fitch.

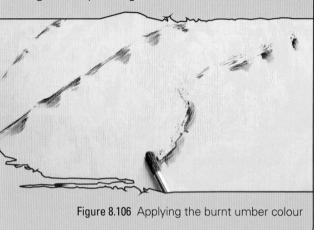

Figure 8.106 Applying the burnt umber colour

STEP 6 Soften in all directions using a hog's hair softener.

STEP 7 With a cotton bud wipe out oval-shaped stones of varying sizes to reveal the ground colour.

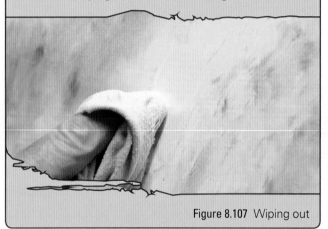

Figure 8.107 Wiping out

STEP 8 Mix some raw sienna with burnt sienna and add veins of varying widths around the stones using a writing pencil. When applying the veins, vary the pressure on the brush to get different widths.

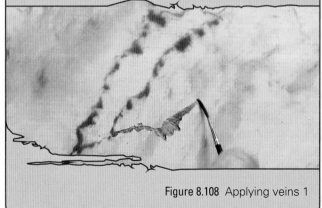

Figure 8.108 Applying veins 1

STEP 9   Continue to add veins to add depth.

STEP 10   Add white ghost veins in the opposite direction.

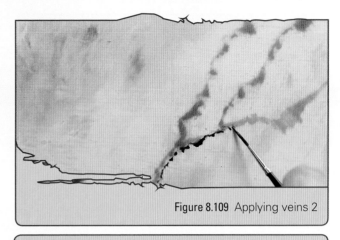

Figure 8.109 Applying veins 2

STEP 11   Once dry, apply a layer of varnish.

PRACTICAL TASK

4.   APPLY GOLD LOOSE LEAF AND TRANSFER LEAF TO A SURFACE

OBJECTIVE

To apply loose metal leaf or transfer leaf to a door moulding using oil gilding.

PPE

Ensure you select PPE appropriate to the job and site where you are working. Refer to the PPE section of Chapter 1.

TOOLS AND EQUIPMENT

Yellow ochre base coat O8 C 35

Natural bristle brush (size will depend on area being covered)

Ready mixed gold size/mordant/gilp

Pounce bag

Gold leaf – loose leaf and/or transfer leaf

Ox hair filling-in brush

Gilder's pad

Gilder's knife

Gilder's tip

Gilder's mop

Sable brush

Coloured paint or clear oil-based varnish

LOOSE LEAF

STEP 1   Prepare the surface so it is non porous and smooth and apply a barrier coat.

STEP 2   Apply a yellow ochre base coat of O8 C 35 to the surface.

STEP 3   Apply the ready-mixed gold size/mordant/gilp to the surface using a 12–18mm ox-hair filling-in brush or a sable fitch. Leave until it becomes tacky according to manufacturer's instructions.

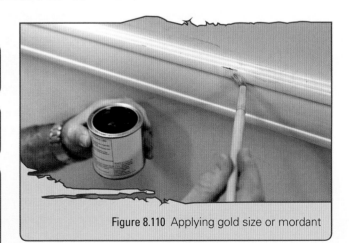

Figure 8.110 Applying gold size or mordant

STEP 4 Test for tackiness using the back of the fingers.

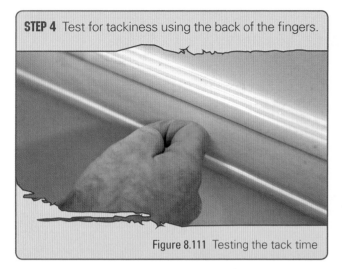

Figure 8.111 Testing the tack time

STEP 5 Place the metal leaf onto the gilder's pad and cut into pieces of the required size using a gilder's knife.

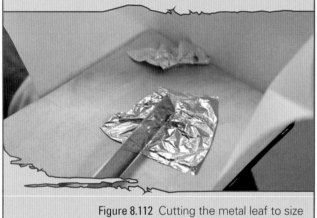

Figure 8.112 Cutting the metal leaf to size

STEP 6 Wipe the gilder's tip across your forehead to apply some grease.

STEP 7 Use a gilder's tip to pick up and apply the loose leaf gold to the surface.

Figure 8.113 Applying the gold leaf to the surface 1.

STEP 8 Overlap each piece.

Figure 8.114 Applying the gold leaf to the surface 2

STEP 9 Allow the surface to dry for a couple of hours and, using a dry gilder's mop, brush over the surface in gentle circular motions to remove the excess gold leaf.

Figure 8.115 Removing the gold leaf

STEP 10 Use a small sable brush and small pieces of gold leaf to fill in any holes or bare patches.

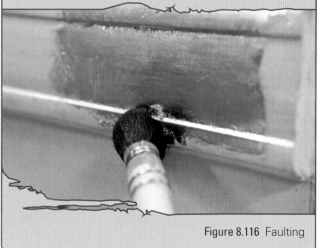

Figure 8.116 Faulting

TRANSFER LEAF

Steps 1 to 5 are the same as for loose leaf.

STEP 6 Apply the face of the transfer leaf to the surface.

Figure 8.117 Applying the leaf 1

STEP 7 Gently rub the back of the leaf with the gilder's mop to transfer the leaf to the surface.

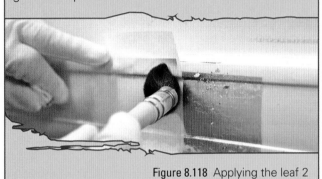

Figure 8.118 Applying the leaf 2

STEP 11 Protect and seal the metal leaf. Apply a layer of coloured paint or clear oil-based varnish.

STEP 8 Keep applying the leaf until the whole area is covered.

STEP 9 Allow the surface to dry for a couple of hours and using a dry gilder's mop brush over the surface in gentle circular motions to remove the excess gold leaf (see Fig 8.121 above).

STEP 10 Use a small sable brush and small pieces of gold leaf to fill in any holes or bare patches (see Fig 8.122 above).

STEP 11 Protect and seal the metal leaf. Apply a layer of coloured paint or clear oil-based varnish.

PRACTICAL TASK

5. APPLY A HALF CIRCLE OR FAN TEXTURED EFFECT TO A CEILING

OBJECTIVE

To apply texturing material to a ceiling with a half circle or fan textured comb effect.

PPE

Ensure you select PPE appropriate to the job and site where you are working. Refer to the PPE section of Chapter 1.

TOOLS AND EQUIPMENT

Builder's bucket

Cotton dust sheets

Access equipment – working platform and trestles

Bumper

Texturing comb

Medium pile roller

Long pile roller

Roller extension handle

Paintbrush (25 mm)

Paint kettle

Sealant

Powdered texture materials

STEP 1

- Cover the floor area immediately below the ceiling area you will be coating.
- Erect and position the working platform.
- Inspect and prepare the ceiling as needed.
- Remember that the ceiling must be free of any high points before the application of the sealing material.

STEP 2 Apply a coat of the sealing material to the ceiling following manufacturer's instructions (typically one or two coats) by cutting in around the edge of the ceiling area using a paintbrush then coating the remaining area with a medium pile roller.

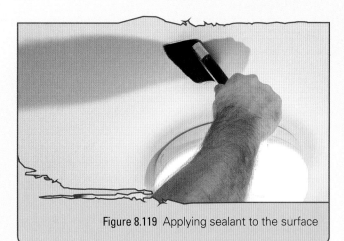

Figure 8.119 Applying sealant to the surface

STEP 3 Half fill a builder's bucket with clean water. Slowly add the texture material and mix the powder and water immediately using a stripping knife. This will stop lumps forming in the mixture.

Mix enough texture material to cover the intended area. Allow the mixture to stand for approximately 10 to 15 minutes.

Mix the material to a stiff consistency.

Figure 8.120 Mixing powder materials

STEP 4 Before applying the material to the ceiling, add some cold water to the material to loosen it.

Figure 8.121 Loosening the mix, ready for application

STEP 5 Apply a coating of the texture material to the application tools (brush, roller, comb) and apply to the ceiling.

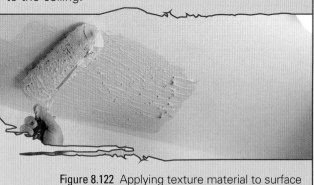

Figure 8.122 Applying texture material to surface

STEP 6 Starting at one side of the ceiling, hold a primed texture comb at a right angle to the ceiling and push the texture comb into the texture material on the ceiling. Rotate the comb in a semi-circle to create a fan design. Lift the comb once you have completed the first fan. Place it down next to the completed fan and use the same technique to apply the patter n to the rest of that row.

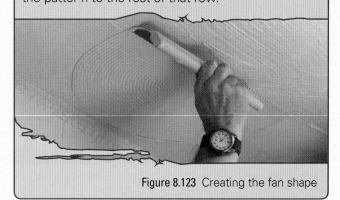

Figure 8.123 Creating the fan shape

STEP 7 Continue using the same technique to apply a fan pattern to the next row. The pattern for the next row must be offset so that the ends of the semi-circles in the previous row are overlapped by the middle of the semi-circles in row two. This pattern needs to be completed for the whole ceiling area.

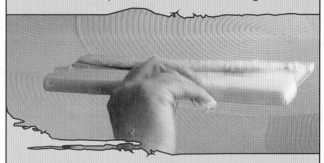

Figure 8.124 Creating the second row of the fan pattern

STEP 8 Using a 25 mm paintbrush, apply a smooth band around the edge of the ceiling and around any ceiling light fittings.

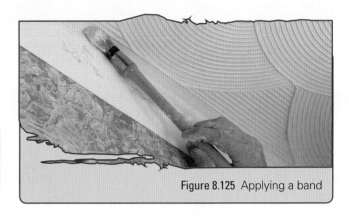

Figure 8.125 Applying a band

PRACTICAL TASK

6. APPLY A MULTI-PLATE STENCIL

OBJECTIVE

To set out and apply a multiplate stencil to a wall.

PPE

Ensure you select PPE appropriate to the job and site where you are working. Refer to the PPE section of Chapter 1.

TOOLS AND EQUIPMENT

Masking tape	Tape measure
Pencil	Spirit level
Chalk	Clean-up rags
Cutting knife	Working platform
Cutting mat	Small palette
Filling knife	Stencil brushes
Tracing paper	Emulsion paint (different
Stencil card	from ground colour)
Paper towel	

STEP 1 Prepare, enlarge and cut out the stencils, as done at Levels 1 and 2.

STEP 2 Measure the position of the stencil accurately using a spirit level and mark using a chalk line. Secure to the wall with masking tape.

STEP 3 Select and mix the tints/shades for the monochromatic colour scheme – i.e. different shades of one colour. Place the different colours onto the palette.

STEP 4 Using a stencil brush, apply the first colour to the first stencil plate. Add another colour to give depth if needed.

Figure 8.126 Applying the first colour

STEP 5 Once the first application is dry, apply the second stencil plate to the correct area and line up with the registration marks on the first stencil plate.

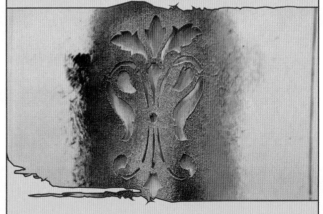

Figure 8.127 Applying the second colour

STEP 6 Apply the next desired colour to the second stencil plate using the stencil brush.

TEST YOURSELF

1. In which of the following situations might you use an extender?

   a. The room temperature is too cold

   b. If you are applying more layers

   c. The room temperature is too warm

   d. If you are working in a small space

2. Which of the following statements about stencil cutting mats is true?

   a. Cutting mats made of composite materials are not heat resistant

   b. Toughened glass plates reduce wear and tear on knife blades

   c. Cutting mats made of composite materials are easily damaged by knife blades

   d. Cork is the best material for a cutting mat

3. Which of the following pigment colours are used to create Sienna marble?

   a. Prussian blue, white, yellow ochre

   b. Burnt sienna, yellow ochre, Indian red, ultramarine, Prussian blue, white

   c. Black, white, raw umber, red ochre, yellow ochre

   d. Burnt sienna, yellow ochre, ultramarine, Indian red, raw umber, black, white

4. What is gilp used for?

   a. To extend the drying time when using oil-based paints to create wood graining effects

   b. To speed up the drying time when using oil-based paints to create marbling effects

   c. To extend the drying time when using water-based paints to create marbling effects

   d. To extend the drying time when using oil-based paints to create marbling effects

5. Which of the following BS 4800 codes is the ground colour for mahogany wood grain effect?

   a. 08 C 35

   b. 04 D 46

   c. 04 C 44

   d. 04 D 44

6. Which of the following statements about applying wood graining effects is true?

   a. When applying wood grain to a window sill you should apply the grain horizontally

   b. When applying wood grain to a window you should work from the bottom upwards

   c. When applying wood grain to a panelled door you should start with the mouldings, muntins and stiles

   d. When applying wood grain to a window sill you should apply the grain vertically

7. Which of the following types of metal leaf cannot be applied using oil gilding?

   a. Aluminium

   b. Silver

   c. Gold

   d. Platinum

8. Which of the following gilding defects is likely to be caused by applying metal leaf to size that is too wet?

   a. Lack of lustre

   b. Misses

   c. Scratches

   d. Patchy application

9. What is a gilder's tip?

   a. A long smooth blade for cutting metal leaf into place

   b. A squirrel hair or camel hair brush used for applying mordant

   c. A thin squirrel hair or camel hair brush used to pick up and transfer metal leaf

   d. Advice on how to carry out gilding

10. Which of the following textured finishes is not applied with a brush or roller?

    a. Stippling

    b. Swirls

    c. Fans

    d. Bark effect

Chapter 9

Unit CSA L3Occ123
APPLY WATER-BORNE PAINT SYSTEMS USING AIRLESS EQUIPMENT

LEARNING OUTCOMES

LO1/2: Know how to and be able to prepare the work area for applying paint systems using airless spray equipment

LO2/3: Know how to and be able to select components and produce a working airless spray units

LO3/4: Know how to and be able to prepare and apply water-borne coatings by airless spray

LO7/8: Know how to and be able to rectify faults in spray equipment and defects in applied coatings

LO9/10: Know how to and be able to clean, maintain and store airless spray equipment and materials

INTRODUCTION

The aims of this chapter are to:

* show you how to set up, use, clean and maintain airless spray equipment

* help you understand paint defects associated with airless spraying.

PREPARING THE WORK AREA

Airless spray systems provide a faster and more effective way to apply coatings to large buildings/surfaces than by paint and roller and other spray paint systems, such as HVLP. Airless spray painting is particularly efficient when covering large surface areas without any obstacles, e.g. doors, windows etc. Although airless spraying provides less '**overspray**' or 'bounce back' than the HVLP method, there is still some overspray meaning that it is very important to protect your work area thoroughly.

Your work area will need to be prepared as for painting with brush and roller, but take note: any surface that is not to be painted must be covered. Look at Chapter 6, pages 174–8, to remind yourself of domestic and commercial factors to be considered, applying, using and removing masking tape, protective sheeting, its uses and how to store it and the objects, people and activities you are protecting.

Also, you must cover windows, skirting boards, architraves, door frames and other items in the room that are not being sprayed, for example pipes and cables.

When applying masking tape and masking papers, you will need to allow for some overlap so there are no gaps that the overspray can penetrate. You may also want to use a masking shield or spray shield for big jobs when it would take too long to apply masking tape, such as around exterior corners or ceiling lines. Masking shields are made of aluminium and have a wooden handle so you can move them around the area as needed.

If you are spraying exterior surfaces, the area must be enclosed in a tent of taped-up polythene sheeting to protect surrounding buildings, cars and any passing members of the public. Because it is an enclosed space, the tent will need to have a local exhaust ventilation (LEV) or fume extraction and natural ventilation system. An LEV should be set up by a competent, qualified person. Any person working in the tent will have to wear respiratory protective equipment (RPE).

SETTING UP AND USING AN AIRLESS SPRAY UNIT

Airless spray system versus HVLP spray system

In an airless spray system, paint is pumped into a spray gun at high pressure and forced out of a very small tip. Unlike the HVLP method, no air or air compressor is involved which is why it is called 'airless'. Some airless spray systems are powered by mains electricity (i.e. cabled) while others are cordless battery-operated systems. The cordless systems require less equipment and so are easier to carry from one job to another.

As airless spraying 'fires' material under high pressure onto the surface, rather than 'blowing' it onto the surface like HVLP spraying, it produces less overspray and is much faster. These benefits make it a better system to use for large jobs. However, HVLP does give better control when spraying small objects and provides a finer finish than airless spraying.

Components of an airless spray system

The main parts of an airless spray unit are shown in Figure 9.1 and explained in the table below:

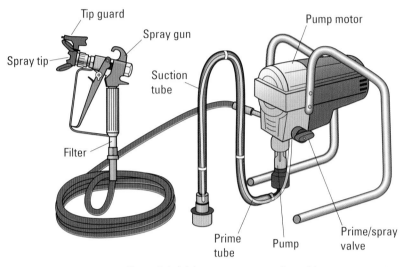

Figure 9.1 Main components of an airless spray system

Figure 9.2 Main components of a spray gun

KEY TERMS

Piston pump

– the rod or 'piston', driven by the motor, moves up and down creating a vacuum, which draws fluid into the fluid line and gun.

Diaphragm pump

– during the upstroke of the motor a vacuum draws paint up the hose into the paint chamber. On the down stroke of the motor the paint is discharged to the fluid line/hose and gun.

Component	Function
Main components	
Fluid pump	The pump delivers a stream of high pressure fluid through the hose to the spray gun. Pumps can be either **piston pumps** or **diaphragm pumps**.
Pump filters	These protect the pump from harmful debris and particles, which helps to increase the lifespan of the pump and maintain consistent spray pressure. The filters are located at the point where the hoses are connected to the pump. These wear out with use and need to be replaced.
Gravity feed hopper and filter	This feeds the gun using gravity. It is usually positioned at the top of the gun. It is used instead of a suction feed tube and a container of paint when spraying small areas to save having to prime the suction tube, which uses up a large amount of paint.
Suction feed tube and filter	This connects the fluid pump to the paint container. It contains a filter that filters out any debris in the paint.
Fluid line	The fluid line is the hose for carrying the paint under high pressure from the pump to the spray gun. As the hose is under high pressure, it should be laid out as straight as possible and should never be bent sharply. It should also be stored straight and kept clean to prevent clogging. Available in different lengths and diameters. Different lengths and diameters affect the pressure to the gun so you need to make sure the length of hose is suitable for the pump you are using.
Whip-end	A shorter hose that is narrower in diameter. It can be attached to the fluid hose to give more flexibility to the length of the hose. It will make it easier to move around corners and angles and makes movement of the gun easier, which reduces tiredness in the hands.

Component	Function
Gun	The gun is made up of a gun body, a tip, a tip safety guard (or tip holder) and a gun filter. Paint passes through the gun and is delivered through the spray tip.
Gun in-line filter	The gun in-line filter filters out large particles from the spray material and prevents the tip from getting clogged up.
Trigger	The trigger releases the pressure in the system, which releases the paint through the tip at high pressure to atomise it.
Trigger locking device	A safety lock located on the trigger that prevents spraying by accident. When you are not spraying, make sure the locking device is closed.
Trigger guard	It reduces the risk of the trigger being pressed if the gun is dropped or bumped.
Fluid tips	Fluid tips are available in different sizes to control the amount of spray and spray pattern. Tips are made from tungsten carbide – a highly resistant material that is not easily abraded. Fluid tips usually have a three-digit number on them. The first number is the fan width (i.e. the width of spray). The second two numbers give the width of the tip itself in thousandths of an inch. Different tip widths are recommended for different types of paint coatings.
Tip safety guard	A safety device that prevents the risk of accidentally placing your fingers or any other part of your body on the spray tip. Also holds the tip in place.
Ancillary components	
Swivel head fluid tip	These are also known as 'reversible tips' or 'turner tips'. They can be reversed or rotated 180° to allow the fluid line to 'clean itself' by blowing pressure out of the tip.
Extension pole	To allow you to reach ceilings, stairways or high-up areas, for example in an industrial warehouse, without the need for access equipment. Poles can be extended up to 3 m in height and can be angled.
Pole gun	A gun with an extended barrel for use when spraying high or recessed areas. They are available in two different lengths – 3 and 6 feet.
Rollers	These can be used in areas where spray painting is not possible (i.e. you cannot risk overspraying). They are connected to the pump in the same way as the spray gun and the roller is fed paint by the pump to allow continuous rolling.

Table 9.1 Main components of an airless spray system

Assembly sequence for component parts of an airless spray unit

For step-by-step instructions on how to *Set up airless spray paint equipment for use* see practical task 1 on pages 315–16.

Adjustments to ensure correct application

Once you have sprayed a test area you may need to make some adjustments to ensure the correct **atomisation** of the paint. The **flow rate** controls the atomisation of the paint. The flow rate is determined by the size of the spray tip and the pressure from the pump.

Spray pattern is determined by the spray tip size and atomisation of the paint.

Adjustments to the atomisation of the paint can be made either by adjusting the pressure in the airless system or by adjusting the **viscosity** of the paint.

KEY TERMS

Atomisation
– the process in which a substance is changed into very fine particles, creating a mist.

Flow rate
– the speed and ease with which a coating will pass through the hole in the spray tip.

Viscosity
– the thickness of a liquid.

TIPS FOR PREVENTING PREMATURE TIP WEAR

Use a low amount of pressure.
Strain the paint to remove anything that might clog up the tip.
Use filters and clean them after use.
Remove fluid tips before cleaning the pump.
Clean spray tips with a soft bristle brush.

Figure 9.3 Swivel head fluid tip

Pressure

The lowest possible pressure should be used to completely atomise the paint. The pressure can be adjusted to achieve the best atomisation. The right pressure has been reached when the spray pattern is elliptical or oval in shape. If more pressure is needed, use a tip with a smaller hole (or orifice) rather than increasing the pressure.

* Using too high a pressure wastes paint by causing overspray and it can also cause premature tip wear.

* Too low a pressure can result in an uneven spray pattern (i.e. it has tails). If you are getting 'tails' you will need to increase the pressure to increase the paint flow. If you are still getting an uneven spray pattern at maximum pressure, try using a tip with a smaller hole.

Viscosity

* If the paint is too viscous it will not atomise finely – resulting in paint defects. The viscosity can be lowered by adding solvents or by heating the paint. There is more information about viscosity on page 307.

Tip size and fan size

* Adjustments can also be made to the tip size and **fan size** depending on the width of the pattern and flow rate required.

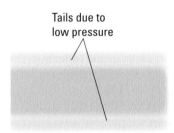

Tails due to low pressure

Figure 9.4 Tails – uneven bands at the top and bottom of the sprayed area

KEY TERMS

Fan size

– the amount of material coming out of the tip (i.e. the angle of spray).

Hazards associated with airless spray systems

Look back at Chapter 4 to remind yourself about the Approved Code of Practice (ACoP). You should also be familiar with Control of Substances Hazardous to Health, outlined in Chapter 1.

A thorough risk assessment should be carried out before starting work to prevent breaches in health and safety. Working with airless spray systems not only presents the hazards involved with paint being atomised into the atmosphere, such as inhalation, ingestion and irritation, but also the hazards of working with equipment that is under high pressure, including the potential for fire and combustion.

Fluid injection hazard

Airless spray systems generate extremely high pressure. Spray from the gun or leaks from equipment that are under high pressure can penetrate through the skin like a needle and enter into your body. This can cause serious injuries and could even result in the amputation of limbs.

REED TIP

Fluid tips and filters are expensive for an employer to replace. Good practice will help prevent excessive premature wear.

KEY TERMS

Earthed

– connected to the ground.

Prevention

* Never point the spray gun at a person or at any part of your own body and never put your hand or fingers over the spray tip.

* Make sure that the spray tip guard is always in place when spraying.

* A loaded spay gun when not being used should be positioned so that the tip is pointing towards the ground.

* Always de-pressurise the system using the correct procedures before cleaning, removing the spray tip or servicing the equipment.

* Never try to stop leaks with your hands or body and do not try to stop leaks with a cloth in your hand.

* Ensure the trigger guard is always in place when not spraying.

* Check safety devices before use and if any device is not working it must be adjusted, repaired or replaced.

* Fluid line hoses should be maintained so that they do not get damaged as fluid can leak at high pressure from damaged hoses. Hoses should never be bent or kinked (which may develop into ruptures) and should be kept in as straight a line as possible and they should be left in a safe place to avoid being damaged by other machinery (e.g. if equipment is wheeled over them).

Fire or combustion

The high speed flow of fluid through the pump and fluid line creates static electricity. If the spray equipment isn't **earthed** properly, sparks may be produced. Sparks can ignite fumes in the solvents within the paint, dust particles and other flammable substance and cause a fire or explosion.

Prevention

* Make sure all equipment, objects and people are earthed.

* Make sure that all electrical equipment is well maintained.

* Ensure that the area is properly ventilated.

* Remove all sources of ignitions, e.g. cigarettes.

Electric shock

When working with electrical equipment there is always the risk of electric shock.

Prevention

* Ensure all equipment for airless spraying is maintained and checked regularly for signs of damage.

* Never operate electrical equipment when it is wet or in wet areas.

Inhalation, irritation and ingestion

Fumes from coatings or other materials may cause irritation if they come into contact with the skin, nose, mouth or eyes.

Prevention

* Wear the correct personal protective equipment (PPE) and respiratory protective equipment (RPE), i.e. respirators.

* Follow manufacturer's instructions and appropriate health and safety legislation – Health and Safety at Work Act and CoSHH.

PPE and RPE

To avoid exposure to overspray, you should wear overalls and gloves to protect your skin, and safety glasses/goggles to protect your eyes.

In addition, you must protect yourself from the fumes caused by the paint sprayer by using respiratory protective equipment (RPE). A simple dust mask is unlikely to be suitable unless you are working on a very small area that is well-ventilated. Proper RPE is costly to buy and to maintain, as well as being awkward or uncomfortable to wear for long periods of time, but it is essential that you wear it when using airless spray equipment.

There are three main types of respirators you can use, depending on the level of fumes and length of time you will be exposed.

* A filter respirator has disposable cartridges that filter the air. They are best used in well-ventilated areas, not confined spaces. Because the cartridges have to be replaced, filter respirators can be expensive.

* An air-fed respirator is more suitable if you are working with more toxic coatings, e.g. solvent-based paints. It is a full head set with a regulator that protects the nose and mouth as well as the eyes. It is joined to an air filtration unit which uses compressed air and is normally connected to an electrical supply.

* A powered respirator also sends filtered air to a full head set but has a battery-powered motor. The motor and filter can either be part of the head set itself or can be separate and connected by a flexible air hose. These provide excellent protection against toxic gases, vapours and dust but are very expensive to buy.

Figure 9.5 A filter respirator

Figure 9.6 An air-fed respirator

Figure 9.7 A powered respirator

Viscosity

As mentioned above the viscosity can affect the atomisation of the paint by affecting the flow rate. The more viscous a liquid is the less flow it has. Getting the right viscosity is therefore important to ensure the correct paint flow through the system, and therefore the correct atomisation. Otherwise, if the paint is too viscous you may 'clog up' the system resulting in blockages or a poor quality finish. If the paint is not viscous enough, it will evaporate and dry before the surface has been coated. Refer to the manufacturer's instructions for information on viscosity and flow rate.

Temperature, humidity and ventilation

Temperature, humidity and ventilation all have an effect on viscosity and the drying process. In Chapter 6, pages 189–90, we looked at the drying process in more detail. In general, the warmer, drier and more ventilated an area is, the faster paint will dry and the less viscous a paint will be. Cool, dark and damp conditions will slow down the drying process and the paint will be more viscous.

Checking and adjusting viscosity
Ford cup
The viscosity of paint can be measured using a viscometer, also known as a Ford cup. The paint is poured into the cup and drains through a hole in the bottom. The viscosity can be measured by timing how long it takes to drain through.

Ratio stick
You can also use a ratio stick to measure how much solvent or thinner is to be added. The stick has a series of numbers printed at regular, even intervals.

Figure 9.8 A viscometer

Figure 9.9 A ratio stick

The stick is rested vertically on the bottom of a flat container. First pour in the coating to the first ratio number given (e.g. for a ratio of 4:1, fill to the line that says 4), then keeping the stick in the same position, add the thinner (e.g. for a ratio of 4:1, add the thinner until it reaches the line that says 5). The highest number will be the total of the two ratio parts combined (e.g. for a ratio of 4:1, the total is 5; for a ratio of 4:3, the total is 7).

Straining paint

Straining paint into a clean container prior to spraying prevents debris clogging up the system, in particular the needle tip, causing spluttering and uneven spray patterns and premature tip wear due to the abrasive particles in the paint.

Spray gun application techniques

Good spray gun application technique is essential to get a good finish.

Distance from the surface

Refer to the manufacturer's instructions for guidance on the ideal distance between the surface and the gun to achieve the best results. If the gun is too close to the surface, you will end up with a thick, uneven coat with defects such as runs and sags. If the gun is too far from the surface, then coverage will be fine and uneven with overspray or a 'paint fog'.

Parallel movement

The spray gun must be kept at 90° to the wall. Tilting the gun up or down or '**arcing**' will mean that the distance is constantly changing between the gun and the surface. It will cause an uneven finish with the correct film thickness in the middle of each stroke, but not at the edges. See Fig 9.13 to see the effect of arcing.

Speed of movement

If the movement of the gun is too quick, insufficient paint will be applied the surface, leaving it patchy. If the movement is too slow, too much paint will be applied to the surface, resulting in runs and sags. The speed at which you move your spray gun will also depend on the thickness of the material. Sometimes it is trial and error before you find the right speed for the surface and material you're using. There is no fixed or correct speed.

Overlapping strokes

Each new stroke of spray paint should overlap the previous stroke by 30–50 per cent. This will avoid bands where the edges of the spray fan have not quite met.

Triggering

Triggering controls the whole airless paint system and is important to ensure an even coverage. The trigger should be pulled at the start of each stroke, and released at the end. However, it is important to make sure the gun is moving before using the trigger, otherwise you will cause overspray and paint runs. You should also trigger first before the edge of the surface you are coating, otherwise you will have an edge that is not perfectly covered or has an excess of paint.

Surface obstructions and obstacles

You won't always be spraying a completely flat surface. Some surfaces will have internal and external corners or obstacles such as pipework.

At times you may need to use a paintbrush to help you achieve an even finish and avoid applying too much paint to small areas such as pipes. This 'stripe work'

KEY TERMS

Arcing

– a semi-circular paint band caused by bending your wrist rather than keeping the gun in a straight line. It results in an uneven finish.

Overlapping

– each stroke should be overlapped by 30–50% to avoid banding.

Triggering

– the action of pulling the trigger to spray the paint. You should not trigger until the gun is in motion.

or a 'stripe coat' will stop runs and sags appearing in your work. In particular, you will need to use a stripe coat or 'banding' technique on external corners – this can be done using a paintbrush or spray gun. Because the air pressure will cause the paint to bounce off the corners or pipes, if you spray them first and later spray *up* to them, you will get a more consistent surface finish.

Wet film thickness and dry film thickness

The thickness of coatings is measured throughout the spraying process to make sure the correct thickness of paint has been applied.

Once each layer has been applied, the wet film thickness (WFT) is measured using a gauge while the coating is still wet to make sure it is the correct thickness. This thickness is measured using microns – one micron is one millionth of a metre. By measuring the WFT throughout the process, and adjusting the coatings accordingly, the dry film thickness (DFT) is more likely to be correct.

Once all coats are dry, the DFT is also measured, and if it is not even then additional coatings can be applied.

To measure the thickness, the WFT gauge is pressed into the wet paint. You will see that the two outer edges of the gauge will have paint on them as well as some of the other teeth on the gauge. If, for example, the gauge has paint on the teeth marked 4 and 5, but not 6, then the WFT is somewhere between 5 and 6 microns.

Shut down equipment temporarily to make adjustments

If you need to shut down temporarily to make adjustments to your coating thickness then you must follow the correct procedures to ensure that you fully release the pressure from the system to avoid injury:

* Leave the suction tube and drainage (prime) tube in the paint container.

* Cover the can and hoses with plastic wrap.

* Release the pressure from the system (see below) and place the gun, still attached to the fluid line/hose, completely covered in a bucket of water.

* Place a plastic bag over the suction tube to prevent leakage.

Releasing pressure from the system

* Put the trigger lock on.

* Shut down the power supply to the pump.

* Release the trigger lock.

* Place the gun against the side of a metal bucket and release the trigger to relieve the pressure.

* Put the trigger lock back on.

* Turn the prime/spray valve to 'prime' (to open the fluid pump drain valve) and drain any fluid into a bucket.

* Leave the fluid pump drain valve open until you are ready to spray again.

There is more information on the correct shut down procedure later in this chapter on page 318.

For more information on removing protective materials refer back to Chapter 6.

REED TIP

Work experience, whether in a work placement, as casual paid work or as voluntary work, will help develop key employability skills that employers are looking for, such as problem solving, teamwork, leadership, time management and adaptability.

PRACTICAL TIP

To decrease the WFT:
* increase the speed of spraying
* choose a smaller tip size
* choose a tip with a wider fan
* check the spray gun is the correct distance away from the surface.

To increase the WFT:
* slow down the speed of spraying
* choose a larger tip size
* choose a tip with a narrower fan
* check the spray gun is close enough to the surface.

PRACTICAL TIP

To avoid the material clogging up the tip, every time you stop spraying – even if it is only for a few minutes – lock the gun and place it into a bucket of thinner to prevent the paint from drying out.

RECTIFYING EQUIPMENT FAULTS AND DEFECTS IN APPLIED COATINGS

Sometimes faults may occur with your equipment when applying coatings using airless spray paint equipment and you need to be able to correct these. Where possible you should know how to prevent these faults from occurring.

Fault	Description	Correction	Prevention
Electrical failure	No power to the machine.	Shut down machine. Must be repaired by a qualified electrician.	Ensure machine is well maintained and get it serviced regularly.
Loose or damaged or worn fluid tip or needle (Fig 9.10)	Tip wear, caused by abrasive material in paint solids, will affect the spray pattern. The size of the hole in the tip (the orifice) will increase and the fan width decrease. More paint is used over less area resulting in paint wastage and distorted/rounded paint patterns.	Tighten fluid tip or replace if damaged or worn.	Spray at the lowest possible pressure to atomise the material. Strain the paint before spraying. Clean filters every time the equipment is used. Use a soft bristle brush to clean fluid tips.
Dirty spray tip/air cap (Fig 9.11)	Spray tip becomes blocked so paint spray is defective.	Remove and clean air cap with thinner or solvent. Reverse the tip position, if using a reversible tip, and remove the clog using the machine.	Strain paint prior to use. Clean out equipment thoroughly after use.
Dry needle packing	Needle packing needs lubricating. Can cause leakage of fluid from the gun and runs, sags and defective spray patterns.	Remove and clean using paint thinner.	Lubricate needle packing regularly.
Incorrect fluid tip set-up	When the wrong tip size and fan width is used with a particular pressure or viscosity of material. Can result in increased thickness causing sags and runs, incorrect fluid flow (i.e. the amount of coating applied) and defective spray patterns, such as fingers and tails. Can also cause leakage of fluid.	Remove fluid tip and select a different tip size/fan width.	Make sure you select the correct tip prior to spraying.
Fluttering (Fig 9.12)	Defective paint spray due to air trapped in the system.	Check for loose connection on the fluid line or pump, tighten and then re-prime the pump.	Check connections prior to use.
Defective spray patterns (Fig 9.13)	Uneven paint application caused by: • incorrect pressure settings • incorrect holding of the spray gun • tip wear • incorrect tip size • poor spraying technique • blocked filter.	Increase the pressure or use a smaller needle tip if the maximum pressure cannot be increased. Replace tip. Use correct spraying techniques – hold the gun at the correct angle and overlap previous strokes accurately. Remove filter and clean it.	Spray a test area prior to spraying and adjust the pressure as necessary. Check tips for wear regularly. Check and clean filters after use.

Fault	Description	Correction	Prevention
Fluid leakage	Caused by loose or worn needle packing, loose wet cup or worn piston rod.	Tighten the wet cup. Tighten or replace the needle packing. Replace the piston rod.	Regular maintenance of the equipment.
Kinked hoses (Fig 9.14)	Kinked hose can lead to a build-up of paint and get clogged up. Potentially this can be very dangerous as they could develop into ruptures with the risk of fluid injection.	Straighten out any kinks in hoses.	Keep in as straight a line as possible and when storing leave in a safe place so that they are not moved.
Spluttering	Uneven, intermittent release of paint from the gun can be caused by dirty components or incorrect pressure.	Check the pressure settings are correct and clean any components that may be dirty or blocked.	Strain paint prior to use. Clean out equipment thoroughly after use.

Table 9.2 Equipment faults and spray defects

Good tip

Worn tip

Figure 9.10 Good tip and worn tip

Figure 9.11 A dirty air cap

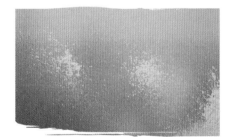

Figure 9.12 Defective spray pattern caused by fluttering

Figure 9.13 Defective spray pattern caused by poor spraying technique

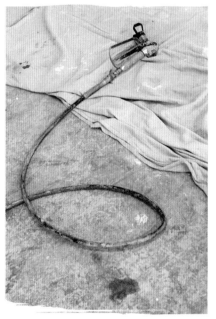

Figure 9.14 Kinked hose

Material faults

There are two main problems you may come across with coatings: contamination and incorrect viscosity.

* It is essential that paints are strained before spraying. Pieces of dirt or old dried paint from a tin can cause the gun to splutter, clog up the needle tip, or give you irregular spray patterns.

* If paint has not been thinned to the correct viscosity, the needle and tip can become blocked. If paint has been thinned too much, the paint will dry up as soon as it makes contact with the air.

Coating defect	Causes	How to fix
Runs and sags (Fig 9.15)	Too much paint has been applied to the surface and runs down the surface, leaving thick lines and beads of paint.	The excess paint must be removed and the surface re-prepared before applying coating again, taking care not to over-paint. Adjust your speed and distance from the surface accordingly.
Dry spray (Fig 9.16)	When the paint has been thinned too much, the spray evaporates as it makes contact with the air and does not all reach the surface. It can also be caused by having your gun too far away from the surface, leaving only a thin coating with the background still showing through.	Follow the manufacturer's instructions when thinning your paint to the right viscosity. Reposition your spray gun so that enough coating is reaching the surface.
Banding (Fig 9.17)	Stripes or bands can be formed where the spray patterns do not meet. This is caused by poor application technique.	Make sure you overlap each pass or stroke by 50%.
Overspray (Fig 9.18)	Overspray creates both mess and wasted materials. There can be several causes, including: • too much air pressure • poor technique (the gun being too far away from the surface, arcing) • incorrect viscosity • windy conditions.	• Reduce your air pressure accordingly. • Check your spray distance (15–25 cm/ 6–10 in) and that you are moving parallel to the surface. • Follow manufacturer's instructions when thinning – adjust if necessary. • Avoid spraying in excessively windy conditions.
Orange peel (Fig 9.19)	This dimpled finish can be caused by: • incorrect air pressure (too much or too little) • incorrect viscosity or wrong thinner used • gun held too close to surface.	• Adjust air pressure. • Follow manufacturer's instructions when thinning to ensure coating is not too thick and that your solvent isn't drying too quickly. • Check your spray distance.

Table 9.3 Defects in applied coatings

Figure 9.15 Runs

Figure 9.16 Dry spray

Figure 9.17 Banding

Figure 9.18 Overspray

Figure 9.19 Orange peel

Spray terminology

Table 9.4 lists some spray terms that relate to the operation of your equipment.

Term	Meaning
Litres per minute – (L/min)	The amount of paint that your equipment can deliver; the higher the number the greater the coverage you will get from a spray gun
p.s.i. – (pounds per square inch)	A measurement of pressure. Airless spray systems are given a pressure rating in psi which is the maximum amount of pressure the system can work at, ranging from about 2000 psi to 4000 psi
Triggering	The action of pulling the trigger to spray the paint; triggering should start when the gun is already in motion but before the area you want to spray (i.e. on the masking paper or for external corners on a masking board)

Term	Meaning
Arcing	Creating a semi-circle (arc) from bending your wrist or elbow, which stops you from keeping the gun parallel to the surface and getting an even coating
Overlapping	Making the spray pattern go over part of the previous stroke; this avoids banding
Spray distance	The gap between the surface and the nozzle of your spray gun; it must be correct and consistent
Gun setup	When all of the components have been put together

Table 9.4 Spray terms and their meanings

CLEANING, MAINTAINING AND STORING AIRLESS SPRAY EQUIPMENT

Shutting down, cleaning and flushing the system

An airless spray system must be cleaned down and flushed through after every use to avoid the paint drying inside the equipment, which may collect in the equipment making it difficult to clean.

See practical task 3 *Shut down, clean and lubricate airless spray equipment* on pages 318–19.

Storage

After cleaning the airless paint spraying system, make sure that all water is drained out of the system and equipment. Any water left in the system will cause corrosion and could freeze in cold conditions causing damage. You may need to pass a **storage fluid** through the system to remove any leftover water in the system.

Store the equipment in a dry, clean indoor area.

Maintenance

Regular maintenance should include:

* inspecting filters
* inspecting wet cup for leakage and tightening the packings if needed
* replacing throat seal oil if needed
* inspecting trigger safety lock
* inspecting high pressure hoses for any signs of damage
* checking for tip wear
* checking electric leads and plugs.

Lubrication

The needle packing should be lubricated regularly with throat seal oil to reduce friction and dissolve any debris. Other components that will need to be lubricated regularly include the threads on the hose and the ball valve in the pump (if it is a hopper fed pump). Regular lubrication reduces abrasion and wear and helps to prolong the lifespan of the system.

PRACTICAL TIP

Be sure to follow the manufacturer's instructions for the particular model of machine you are using to make sure you shut the system down safely.

KEY TERMS

Storage fluid
– a fluid designed to get rid of any leftover water in the system.

PRACTICAL TIP

Store the system with the pump rod in the down location.

Place and fasten a plastic bag over the end of the suction tube to prevent fluid leakage.

Good maintenance will extend the lifetime of the equipment and help an employer keep large costly repair bills down.

Disposal of waste and materials

You will create a lot of waste when working with airless spray equipment, mainly due to all the masking materials you need to use, as well as contaminated rags, left over coatings, lubricants and gun metal cleaner. Remember that these materials will be covered with paints, and potentially some oils and solvents. You must follow the guidance earlier in this book (Chapter 5 on page 176 and Chapter 6 on pages 204–5 to dispose of contaminated waste safely). Remember that there are legal requirements involved when disposing of debris and waste, and guidelines set out in the Health and Safety at Work Act, Environmental Protection Agency (EPA), Control of Substances Hazardous to Health (COSHH) and Health and Safety Executive (HSE).

PRACTICAL TASK

1. SET UP AIRLESS SPRAY PAINT EQUIPMENT FOR USE

OBJECTIVE

To prepare an airless spray system ready for spray application.

PPE

Ensure you select the PPE appropriate to the job and site where you are working. Refer to the PPE section of Chapter 1.

TOOLS AND EQUIPMENT

Manufacturer's instructions	Electric cable (230 V or 110 V)

110 V transformer (if required)

Fluid pump

Fluid line

Suction feed tube and bucket or hopper

Whip end (if needed)

Gun and/or pole gun

Fluid tip

Fluid tip safety guard

Waste bucket

Any other equipment, e.g. extension pole and roller

Throat seal oil

Test board

STEP 2 Connect the suction tube to the fluid pump and a bucket of strained paint.

Figure 9.21 Connecting the suction tube

STEP 1 Connect the **prime tube** to the **fluid pump** and place the other end in a waste bucket.

Figure 9.20 Connecting the prime tube to the fluid pump

STEP 3 Attach the fluid line to the pump.

STEP 4 Connect the gun to the fluid line.

Figure 9.22 Connecting the gun to the fluid line

STEP 5 Fill the **packing** nut/**wet cup** on the gun with **throat seal oil**. Refer to the manufacturer's instructions for the correct amount to be used.

Figure 9.23 Filling with throat seal oil

STEP 6 Plug the pump into the electrical socket.

KEY TERMS

Prime tube
– a tube used when priming the system to allow waster to be removed into a waste bucket.

Packing
– creates a seal within the fluid pump to prevent leakage.

Wet cup
– where leakage collects as the throat packings wear.

Throat seal oil
– a lubricant for the throat packing.

STEP 7 Check the trigger locking device on the gun is in place and attach the tip and tip safety guard.

Figure 9.24 Attaching the tip and tip safety guard

STEP 8 Flush the system through with solvent following the manufacturer's instructions. This is also known as priming the system. Turn the 'spray/prime valve' control knob to 'prime' and switch on the pump. Turn up the pressure knob until the pump starts up and paint starts flowing from the prime tube into the bucket.

Figure 9.25 Priming the system

STEP 9 Once primed, remove the trigger guard and tip and hold the gun over the waste bucket. Switch the 'spray/prime valve' control knob to the 'spray' setting and pull the trigger. When paint begins to flow in a steady stream, let go of the trigger and lock it. The hose has now been filled.

Figure 9.26 Filling the hose

STEP 10 After priming and filling the hose with paint you need to relieve the pressure before installing the tip and tip guard. Turn off the power and turn the pressure control knob to 'prime'.

STEP 11 Point the gun against the side of the waste bucket and pull the trigger to release the pressure. Lock the trigger.

STEP 12 You are now ready to start spraying a test area. Turn the pressure control knob to the required spray pressure and take off the trigger locking device and spray a test area.

PRACTICAL TIP

To reduce the risk of fluid injection (see pages 305–6) *never* hold your hand or any other part of you body in front of the spray tip when checking for a cleared tip or when cleaning the tip. Always point the gun into a waste container.

PRACTICAL TASK

2. APPLY WATER-BORNE PAINT USING AIRLESS SPRAY EQUIPMENT

OBJECTIVE

To apply water-borne paint to surfaces using airless spray equipment.

PPE

Ensure you select the PPE appropriate to the job and site where you are working. Refer to the PPE section of Chapter 1. Also remember that for spray painting you need to wear the correct RPE (see page 307).

TOOLS AND EQUIPMENT

See practical 1 on page 315.

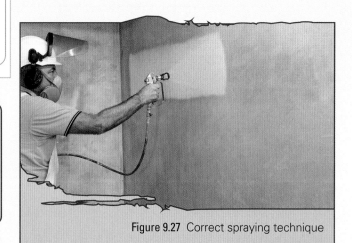

Figure 9.27 Correct spraying technique

STEP 1 With the spray gun already set up (see practical 1), turn the spray/prime valve' control knob to the 'spray' setting. Turn the pressure control knob to the required spray pressure and take off the trigger locking device. Direct the gun towards the surface to be painted.

STEP 2 Holding the gun at 90° to the wall at the distance recommended by the manufacturer, start moving the gun at the right speed before releasing the trigger. Spray a horizontal band.

STEP 3 If needed, adjust the pressure using the pressure control knob.

STEP 4 Using the same technique, apply a second horizontal band and overlap the previous one by 50 per cent.

Figure 9.28 Overlapping the bands

STEP 5 Continue spraying in the same way.

STEP 6 When you have finished spraying, release your finger from the trigger and put the trigger safety guard on.

PRACTICAL TASK

3. SHUT DOWN, CLEAN AND LUBRICATE AIRLESS SPRAY EQUIPMENT

OBJECTIVE

To shut down, clean and lubricate airless spray equipment by flushing the system through with water and repeating the process using spray gun cleaner (solvent).

TOOLS AND EQUIPMENT

See practical 1 on page 315.

PPE

Ensure you select the PPE appropriate to the job and site where you are working. Refer to the PPE section of Chapter 1. Also remember that for spray painting you need to wear the correct RPE (see page 307).

STEP 1 Ensure the trigger safety guard is on and switch off and unplug the equipment at the mains.

STEP 2 Place the suction tube in a bucket containing water and place the prime tube in a metal waste bucket.

Figure 9.29 Shutting down 1: placing the suction feed tube and prime tube

STEP 3 Relieve the pressure in the hose by turning the prime/spray valve to 'Prime', the pressure control knob to minimum pressure and by releasing the trigger against the side of the metal waste bucket.

Figure 9.30 Shutting down 2: releasing the pressure in the gun

STEP 4 Turn the power on and slowly increase the pressure control knob until the pump starts.

STEP 5 Flush through the bucket of water until only clear water leaves the gun.

Figure 9.31 Shutting down 3: flushing until clear water comes through

STEP 6 Get rid of the remaining paint in the hose by switching the power off, turning the prime/ spray valve control knob to 'spray' and releasing the trigger into the paint bucket to get rid of the remaining paint in the hose.

STEP 7 Turn the power switch on. Release the trigger until the water comes through the gun into the waste bucket.

STEP 8 Repeat steps 2–7 using spray gun cleaner fluid.

STEP 9 Check the filters for debris and clean with water or solvent and a soft brush.

Figure 9.32 Cleaning the filters

STEP 10 Dispose of all waste appropriately in line with legislation (see pages 96–8).

STEP 11 Lubricate all working parts, including the tip, trigger nose treads and ball valve if using a hopper.

Figure 9.33 Lubricating the working parts

PRACTICAL TIP

Always check the manufacturer's instructions as there may be specific guidance on how to clean and store each component of the equipment you are using. There may also be a recommended schedule of maintenance that you should carry out in addition to cleaning after each use. Following this will help to prolong the life of your equipment (which is often quite expensive).

TEST YOURSELF

1. What is the main advantage of airless spraying compared with HVLP?

   a. It gives better control when spraying small objects

   b. It is faster so you can cover large areas in less time

   c. It gives a finer finish than HVLP

   d. All of the above

2. What is the effect of using too low a pressure to atomise the paint?

   a. Overspray

   b. Premature tip wear

   c. Uneven spray pattern

   d. Wet film thickness is too thick

3. Why should you never put your hand in front of an airless spray gun when in operation?

   a. To avoid pressure building up in the system and causing sparks, which could cause a fire

   b. The system is under high pressure and spray from the gun could pierce through your skin

   c. To avoid cuts or bruises to your hands

   d. To avoid getting an electric shock from the equipment if it hasn't been earthed properly

4. Which of the following are all components of an airless spray system?

   a. Compressor unit, spray gun/tip, pressure pot

   b. Compressor unit, suction feed tube, fluid line, spray gun/tip

   c. Fluid pump, suction feed tube, fluid line, spray gun/tip

   d. Fluid pump, suction feed tube, spray gun/tip, pressure pot

5. Which of the following is a cause of premature tip wear?

   a. Using too low a pressure

   b. Using paint with a low viscosity

   c. Using too high a pressure

   d. Using too large a tip size

6. What percentage overlap should you leave to get an even finish?

   a. 35%

   b. 40%

   c. 50%

   d. 55%

7. Which of the following statements about the effects of temperature, humidity and ventilation on paint viscosity is correct?

   a. Temperature, humidity and ventilation don't affect paint viscosity

   b. Paint is less viscous in cold, damp and poorly ventilated areas

   c. Paint is more viscous in warm, dry, well ventilated areas

   d. Paint is more viscous in cold, damp and poorly ventilated areas

8. What is the cause of 'tails'?

   a. Arcing the spray gun during application

   b. Trapped air causing spluttering of the gun.

   c. Using the wrong tip size and fan width.

   d. The gun has been held too close to the surface

9. A spray tip is labelled 317. What is the width of the tip in thousandths of an inch?

   a. 31

   b. 7

   c. 17

   d. 317

10. Which is the correct method for releasing the pressure in an airless system?

   a. Turn off the system and spray any remaining paint in the system onto the surface.

   b. Flush the system through with solvent to drain any fluid into a bucket

   c. Place the gun against the side of the metal bucket and release the trigger

   d. Remove the fluid line from the gun

INDEX